Design of Modern Communication Networks

Design of Modern Communication Networks
Methods and Applications

Christofer Larsson

AMSTERDAM • BOSTON • HEIDELBERG • LONDON
NEW YORK • OXFORD • PARIS • SAN DIEGO
SAN FRANCISCO • SINGAPORE • SYDNEY • TOKYO

Academic Press is an imprint of Elsevier

Academic Press is an imprint of Elsevier
The Boulevard, Langford Lane, Kidlington, Oxford OX5 1GB, UK
Radarweg 29, PO Box 211, 1000 AE Amsterdam, The Netherlands
225 Wyman Street, Waltham, MA 02451, USA
525 B Street, Suite 1900, San Diego, CA 92101-4495, USA

First edition 2014

Notice

No responsibility is assumed by the publisher for any injury and/or damage to persons or property as a matter of products liability, negligence or otherwise, or from any use or operation of any methods, products, instructions or ideas contained in the material herein. Because of rapid advances in the medical sciences, in particular, independent verification of diagnoses and drug dosages should be made

British Library Cataloguing in Publication Data
A catalogue record for this book is available from the British Library

Library of Congress Cataloging-in-Publication Data
A catalog record for this book is available from the Library of Congress

ISBN: 978-0-12-407238-1

For information on all Academic Press
publications visit our website at books.elsevier.com

Working together
to grow libraries in
developing countries

www.elsevier.com • www.bookaid.org

To Mária

Contents

Preface

Design of communication network is a complex task, as many of the problems involved are inherently difficult to solve. This leads to the necessity of using a collection of cleverly designed algorithms and methods rather than a single approach. The book is written on at least two levels. Firstly, specific problems related to network design form the main structure of the book. On a second level, various types of algorithms are used, discussed and modified in the course of the text. The structure gives views: one problem based, one method based.

The main intention is that the text will serve as a handbook in engineering with a focus on algorithms and their illustration by examples. On the other hand, rather than being an encyclopedia of algorithms and methods, the text should be possible to read as a discourse in techniques of network design. Even if mainly intended as a handbook, it may also be useful in academia as side literature in telecommunication education programs.

In practice, network design has been more of a handicraft than a science. This is no doubt due to the fact that almost all problems related to network design are $\mathcal{NP}$-hard, or in some other class of problems which are difficult to solve. However, these problems can often be solved approximately. When presenting approximation schemes in this book, the author has taken into account both the precision of the result and the complexity of the algorithm. A simpler, intuitive algorithm has sometimes been favored over more complex algorithms, sometimes even if the precision is lower.

In this text, we will use a mixture of analytical analysis, heuristics, approximations, randomization and common sense to solve difficult problems related to network design. Thus, we may not be able to answer the question "What is the optimal solution to this problem?", but we may be able to answer the question "What is the probability that there is a better solution to this problem?", and if the probability is low enough, we should be close to the answer of the first question. Also, if a proposed network is available to us we can assert its superiority or inferiority to solutions obtained using the methods presented in this book.

It is the hope of the author that the reader will experience the same fascinating journey it has been writing this book. Indeed, at the heart of network design lies the combination of methods, mixed with a fair dose of common sense and experience, needed to solve often very intricate problems. It is an exquisite example of applied mathematics: finding methods that might work!

Most proofs of mathematical theorems have been omitted in the text, as the focus is on application level rather than a theoretical level. These proofs can be found in the cited references. The intention is instead to provide "empirical proofs" of the methods by solving problems using different methods and thereby arriving at the same or similar results. A number of such comparative examples are provided. Proofs have been included in some cases when they provide details that are instructive for constructing an algorithm.

With network design we mean the initial planning or long-term modifications of networks on a rather large scale. The text does not discuss control or routing mechanisms which instantaneously have to react on failure or overload situations. Such aspect may be referred to as operational rather than design related.

The aim of the text is to be as technology independent as possible. That is the reason why there are very little description of actual network technologies such as STM, ATM, IP, and so on, their protocols and functionality. There is a lot of literature available on these topics, and the algorithms presented in this book should be possible to translate into technology specific terms rather easily.

As mathematical prerequisites, the reader is probably familiar with some combinatorics, optimization, fundamental probability theory, queueing theory and analysis.

The author would like to thank V.B. Iversen and F.P. Kelly for their comments and encouragement. He is also grateful to the many researchers and scientists that have made their interesting papers freely available on the Internet.

Christofer Larsson,
Bratislava.

Introduction

The telecommunication industry has seen a tremendous development in recent decades, with the advent of technologies such as cellular technologies, ATM, and in the IP domain, from DiffServ and MPLS to New Generation Networks (also known as All-IP), with increasing capabilities of providing ever more sophisticated services. Changed legislation has opened up markets for new carriers competing with the incumbent ones. The result is that we have networks of different sizes, implemented with different technologies. The efficient design of communication networks is therefore of great importance, whether it is design of new networks or optimization of existing architectures. A good network design ensures that the network has the necessary capabilities of providing services of a certain quality and that it is resilient to faults, at as low cost as possible.

With network design in this text, we mean the collective procedure of planning network architecture (or topology), assigning flows to links in the network, assigning resources (capacity) to network elements, and verifying performance criteria.

The problem is that these are in general very difficult tasks to perform even for small networks, and the difficulty is increasing rapidly with the network size. As a matter of fact, most tasks can't be solved in reasonable time. We will therefore often have to resort to different approaches, such as approximations, randomized algorithms, and simulation. Even if none of these approaches give an exact answer, we can by combining the results from different approximate solutions gain insight into how close to an optimal solution we may be. It is similar to observing a far-away object from different angles in order to determine its actual position.

The reader is assumed to be familiar with the mathematics (in particular, probability theory, basic queueing theory, optimization, and linear algebra) equivalent to the mathematics courses in a Bachelor's degree in computer science or electrical engineering. Most algorithms are suitable for software implementation using some programming language. Higher-level languages such as Matlab (or Octave), or S (or R) are often useful for smaller problems or to test algorithms. Readers familiar with object-oriented programming in general and C++ in particular may find the graph representation and other structures in Shaffer (1997) of use.

1.1   The purpose of this book

The term network design has different connotations. In this text, the term is used for the collective process of constructing a network topology, assigning flows and capacities to network elements, and evaluating related performance metrics. With few exceptions, the objective is that of minimizing the resulting network cost.

Often, due to the lack of a unifying approach and the difficulty of solving these problems, heuristic methods are used in practice. Some heuristics lead to suboptimal network designs (an example is given in Section 1.3).

The purpose of this book is manifold. It is intended to present a framework of methods and algorithms for realistic network design, with focus on practical applications. Many mathematical proofs are omitted in the text, but references are given where these proofs can be found. Examples, illustrations, and pseudocode provide an intuitive picture of how algorithms can be used in practice and indicate how they can be programmed on a computer. It presents a universal 'toolbox' for the network designer that can be used in solving a broad range of network design tasks.

The presentation is, as far as possible, technology independent. It should therefore be general enough to be applicable to different technologies and various implementations.

The foundation of the theory of telecommunication engineering is collection of topics from mainly statistics, probability theory, and optimization theory and has therefore the nature of a cross-scientific monograph. The author has applied two main principles in his selection of topics: the methods should as far as possible be independent of the underlying technology, and presented topics should have a practical relevance to communication network engineering, with numerous examples to illustrate their practical use.

It is also the hope of the author that the different approaches used will give the reader an intuitive understanding of solving difficult problems related to network design and the behavior of complex networks.

1.2   The design process

A network is a certain infrastructure, consisting of nodes and interconnecting links, intended for transportation of some commodities in a way as to meet a distributed demand-supply pattern between the nodes. This network has a topology (or architecture), which describes how the nodes are connected by links. The supply and demand pattern, together with topological restrictions, induce commodity flows in the network which in the case of communication network is known as *traffic*.

The question discussed in this text is how to design an optimal network. The first question that arises is what is meant by 'optimal.' Usually, a network should be designed so that it meets various technical constraints, supports a certain amount of flow (or traffic) subject to various performance criteria such as Grade of Service or Quality of Service and a certain degree of resilience for failure situations and, fulfilling these conditions, has the lowest costs of all possible such candidate network designs meeting these criteria.

Some technical constraints on the network are often relatively easy to formulate in exact terms. Examples of such constraints are how many connections a switch or multiplexor can accommodate, or how many nodes a traffic stream can be allowed to traverse. These constraints are considered 'hard limits' in the sense that if not fulfilled, the network cannot operate properly. It will be assumed that such constraints always can be met for a network candidate of interest in the design process.

Estimating and modeling traffic and formulating Grade of Service (GoS) or Quality of Service (QoS) criteria are difficult problems in their own right. Not only do different types of traffic require different Grade of Service or Quality of Service criteria, but these are generally of a subjective nature. The fundamental question is how the users perceive the service rendered, and what level of quality can be generally accepted by the users. Therefore, these metrics are closely linked to the cost structure of the service.

A somewhat different design criterion is network resilience. The goal is to have the operation of the network disrupted as little as possible, should, for example, a transmission link fail.

Both the service quality for a given traffic volume and the degree of network resilience are proportional to various costs. The costs can also be of various kinds: capital or operational, such as equipment and installation costs, maintenance costs, or rental or lease fees. Therefore, the cost models, which in this text are modeling the cost of transmission over long-distance links, are of vital importance.

To complicate matters further, various *traffic engineering* or *traffic control* policies may be applied to the network, which may have a large impact on its operation, both with respect to efficiency and stability. Traffic engineering such as routing schemes,

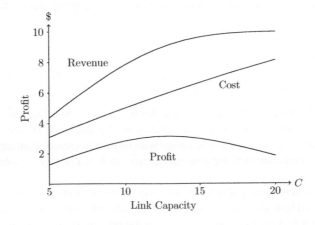

FIGURE 1.1

A simplified optimization problem for a link. Revenue is generated by carried traffic, cost is incurred by the link capacity, and the profit is the difference. The profit has its maximum for $C = 13$.

traffic shaping, or scheduling aim at improving the network operation (and thus service quality and/or stability) by dynamically changing the traffic flow according to the state of the network. Such actions have to be accounted for in the design as well.

In other words, the optimization problem can then be stated as find the feasible network that fulfills the service quality and resilience requirements for a certain amount of traffic, that has the lowest cost.

Mathematically, this can be stated as an optimization problem, such as: Given an amount of offered traffic and a target performance, what is the minimum QoS resources needed?

In mathematical terms, this is an optimization problem, where the level of service quality and network resilience are the side condition, and the quantity of network resources is to be minimized, subject to the cost function. Some examples of cost functions are shown in Figure 1.1.

The optimization problem is nonlinear, both in the side conditions and (usually) in the objective function. In a few cases, network problems can be solved analytically, but for most real cases, various idealizations and approximation have to be made, and we do not have any guarantee of optimality.

Traditionally, traffic engineering has been based on Erlang's pioneering work from 1917. The classical traffic engineering methods have been used successfully for almost a century. With the advent of new technologies, hybrid networks, and complex services with vastly different traffic characteristics, the classical approach fails to model resource and performance requirements. In particular, since different services typically have different performance metrics, modern communication network engineering calls for new methods. Some of the challenges of modern network design are due to the following observations.

Characterization of traffic. Traffic in modern communication networks typically consists of a mixture of traffic types, such as voice and Internet services. These traffic types have wildly different characteristics. How can an aggregated traffic stream be modeled?

Large networks. As networks become larger, and various transport technologies become available, the sheer size of the optimization space of even a medium-sized network becomes enormous. Pen-and-paper designs, although still useful for some benchmarking, are likely to be highly suboptimal.

Dynamic environment. Communication networks are increasingly dynamic: new services are frequently introduced which change the behavior of the network. A network design therefore needs to be robust with respect to dynamic changes. Various network control mechanisms are also interacting, such as priority classes or optimal routing.

Figure 1.2 schematically illustrates the design process. It shows four sub-problems of the design and how all of them affect the final network cost. The subproblems with solid lines are affecting the cost directly, and the ones with dashed lines indirectly. To begin with, a network *topology* is needed. That is, the site locations, where the network nodes are situated and a structure of links interconnecting these nodes. Next,

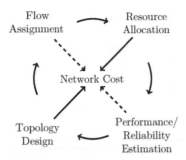

FIGURE 1.2
The interdependence of subproblems in network design.

we assume that we have given flow demands between the nodes, and we perform a flow assignment onto the topology. When the flow has been assigned to the links, we allocate resources to the links. Finally, in the cycle, we need to verify that required performance requirements are met. We can now calculate the cost of the network, or modify topology, flow, and resource assignment further.

In essence, this can be viewed as an investment problem where the offered traffic represents the (potential) revenue and the quantity of system resources represents the required investment. The performance affects the service quality, which is linked to revenue in that the better performance, the more successful calls will be handled by the network. However, this parameter is usually set to a value in the region 90–95% (possibly slightly different parameters for different subnetworks and different services are typically used) for core networks and is not used as an optimization parameter.

1.3 A first example

We consider the design of a small network, consisting of six nodes, depicted in Figure 1.3. The nodes are enumerated $1, 2, \ldots, 6$ and links are labeled (i, j) for any two nodes i and j. The distances between the nodes are given by

$$
D = \begin{pmatrix}
0 & 23 & 20 & 32 & 45 & 51 \\
23 & 0 & 10 & 23 & 23 & 37 \\
20 & 10 & 0 & 15 & 29 & 32 \\
32 & 23 & 15 & 0 & 32 & 20 \\
45 & 23 & 29 & 32 & 0 & 32 \\
51 & 37 & 32 & 20 & 32 & 0
\end{pmatrix},
$$

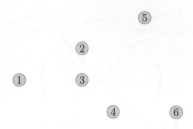

FIGURE 1.3

Six node locations should be connected in a cost optimal way.

where the rows and columns correspond to the node numbers. The flow demand between the nodes is given by

$$
A = \begin{pmatrix}
0 & 2 & 3 & 1 & 2 & 1 \\
2 & 0 & 1 & 2 & 3 & 3 \\
3 & 1 & 0 & 2 & 2 & 1 \\
1 & 2 & 2 & 0 & 3 & 1 \\
2 & 3 & 2 & 3 & 0 & 2 \\
1 & 3 & 1 & 1 & 2 & 0
\end{pmatrix}.
$$

We also need to know the cost structure of the links. For simplicity, we assume that we have at our disposal unit capacity links with a cost proportional to the link distances. There is also a set-up cost c_0 for each link. Letting d_{ij} be the distance of the link—when it exists—connecting nodes i and j, and f_{ij} the aggregated flows on these links, we define the cost

$$
c = c_0 + \sum_{(i,j)} f_{ij} d_{ij}. \tag{1.1}
$$

Finally, we require a method to route flows. A simple and common method is *shortest path routing*, which we use here. The flow between any two nodes is then routed along the shortest total distance connecting these two nodes of all possible routes.

To make a pen-and-paper design of the topology, suppose we come up with the following heuristic: start from any node. Connect this node to the closest node not yet connected. Starting from node (1), say, we connect to the closest node, node (3). The closest node from the node pair (1)–(3) is node (2), which we connect to next, followed by node (4), (6), and (5).

We would also like a node to be connected to at least two other nodes. Nodes (5) and (6) both have single connections, so we connect them together. Now only node (1) remains with a single connection, so we connect it to the next closest node to which it is not connected, node (2). We then have the topology shown in Figure 1.4.

Next, we need to map the flows onto the links. The link distances and the aggregated flows, using shortest path routing, are shown in Table 1.1.

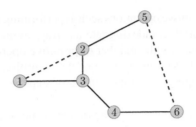

FIGURE 1.4

The resulting topological design.

Table 1.1	Links, their distances, and aggregated flows	
Link (i, j)	**Distance (d_{ij})**	**Total flow (f_{ij})**
(1,2)	23	8
(1,3)	20	10
(2,3)	10	22
(2,5)	23	20
(3,4)	15	26
(4,6)	20	12
(5,6)	32	4

Based on the distances and the flows, we can calculate the cost of this network using Equation (1.1). Although this solution is not that bad, there are many better, cheaper ways to design this small network. We will return to this simple example in subsequent chapters.

1.4 Algorithms for hard problems

This book is about algorithms for network design problems, almost all of which are computationally hard, so it is important to have some way to evaluate their performance and thus be able to compare different algorithms for the same problem. We will therefore have a brief look at algorithm analysis and computational complexity classes.

In algorithm analysis, we imagine that we have a fictituous computer (often called the Random Access Machine [RAM] model) with a central processing unit (CPU) and a bank of memory cells. Each cell stores a word, which can be a number, a string of characters, or a memory address. We assume further that the CPU performs every primitive operation in a constant number of steps, which we assume do not depend on the size of the input.

We use high-level primitive operations such as performing an arithmetic operation, assigning a value to a variable, indexing into an array, comparing two numbers, and calling a method. By counting the number of primitive operations in the algorithm, we obtain an estimate of the running time of the algorithm. (The actual execution time is obtained by multiplying the total number of steps by the CPU execution time for primitive operations.)

It is common to analyze the worst case running time of an algorithm because it is usually much easier than finding the average running time, and also because the worst case running time is a more useful measure, since it provides an upper bound for all possible inputs. This is particularly important for the classes of problems that we consider in this book.

Let the input size, n, of a problem be the number of bits used to encode an input instance. We also assume that characters and each numeric type use a constant number of bits. Note that the size of n is an integer.

Let the input size of a problem be n, where n can be the number of bits of some representation of the problem. We will avoid rigorous technical arguments and will be satisfied with an algorithm construction if we have done our best, or if we have taken reasonable precautions to make it as efficient as possible. What we mean by that—in very imprecise terms—is that we try to find as short and efficient a representation as possible for the problem instances. We also assume that any evaluation of a solution to the problem is performed efficiently.

Define the worst case running time of an algorithm A to be the time A runs with an input instance of size n, where the worst case is taken over all possible inputs having an encoding with n bits.

We will express the worst case algorithm running times using the big-O notation. Suppose we have a function $f(n)$, representing the number of steps of an algorithm. Then we say that $f(n)$ is $O(g(n))$ of another function $g(n)$, if there exists a real constant $c > 0$ and an integer constant $n_0 \geq 1$, such that $f(n) \leq c \cdot g(n)$ for all $n \geq n_0$. This should be interpreted as $f(n)$ is bounded above by a constant c times $g(n)$ for large enough n, and the constants need not be specified when stating the upper bound.

Complex algorithms are typically broken down into subroutines. Therefore, the following fact is useful in the analysis of composite algorithms. The class of polynomials is closed under addition, multiplication, and composition. That is, if $p(n)$ and $q(n)$ are polynomials, then so are $p(n)+q(n)$, $p(n) \cdot q(n)$, and $p(q(n))$. Thus, we can add, multiply, or compose polynomial time algorithms to construct new polynomial time algorithms.

Any sum of functions is dominated by the sharpest increasing one for large n. Therefore, we have, for example, $\log(n) + \log(\log(n)) = O(\log(n))$, $2^{10} = O(1)$, and $n + n^2 = O(n^2)$.

An algorithm is called efficient if it runs (at most) in time $O(n^k)$ of the input size n and some $k > 0$, that is, if it is bounded above by a polynomial.

Table 1.2 lists some common names and running times used to describe algorithms.

Table 1.2　Common complexity classes of algorithms	
Constant	$O(1)$
Logarithmic	$O(\log(n))$
Linear	$O(n)$
Near-Linear	$O(n\log(n))$
Polynomial	$O(n^k), k \geq 1$
Exponential	$O(a^n), a > 1$

1.4.1　Complexity classes

Most of the problems we will encounter are hard to solve. It means that we cannot expect to find efficient algorithms for any of them. In practical terms, this also means that any exact algorithm or method will have a running time which is exponential in the input size n, and we will never find any polynomial upper bound to the worst case running time. We will formalize that now.

Most problems in network design are optimization problems, where we search for an optimal value, such as cost. For the discussion on computational complexity, however, it is useful to think of the optimization problems as decision problems, that is, problems for which the output is either true or false only.

We can convert an optimization problem into a decision problem by introducing a parameter k and asking if the optimal value is as most k (or at least k). If we can show that a decision problem is hard, then its related optimization formulation must be hard too.

Example 1.4.1.　Suppose we have a network and want to find a shortest path from a node i to a node j. Formulating this as a decision problem, we introduce a constant k and ask: "is there a path with length at most k?"

In order to define some important complexity classes, we refer to the class of decision problems as a language L. An algorithm A is said to accept a language L, if, for each $x \in L$, it outputs the value true, and false otherwise. We assume that if x is in an improper syntax, then the algorithm given x will output false.

The polynomial time complexity class $\mathcal{P}$ is the set of all decision problems L that can be accepted in worst case polynomial time $p(n)$, where $p(n)$ is a polynomial and n is the size of x. The non-deterministic polynomial time complexity class $\mathcal{NP}$ is a larger class which includes the class $\mathcal{P}$ but also allow for languages (decision problems) that may not be in $\mathcal{P}$.

In the complexity class $\mathcal{NP}$, an algorithm can perform non-deterministic modifications of $x \in L$, such that it ultimately outputs true, where the verification is done in (polynomial) time $p(n)$, where $p(n)$ is a polynomial and n is the size of x. In other words, the time of asserting that x is true is polynomial, and the generation of such an x may require a polynomial number of non-deterministic modifications. Still there is no guarantee that such a solution will be found. We can only guess at a solution and verify it in polynomial time. If we try to explore all possible modifications of x in the

Table 1.3 The growth of the upper bound for the number of network configurations for some values of n

n	Bound	Time
5	1024	1 ms
10	$3.52 \cdot 10^{13}$	1.1 years
15	$4.06 \cdot 10^{31}$	1.29×10^{18} years[a]

[a] 283 million times the age of Earth.

algorithm, this procedure would become an exponential time computation, since the time required increases very rapidly with the size of the input.

Example 1.4.2. The total possible network configurations having n nodes grow very rapidly with n. The upper bound for the number of configurations is $2^{n(n-1)/2}$. This expression is given by the number of possible links $n(n-1)/2$ which can be present or not (and therefore the base is 2). Table 1.3 shows the upper bound for some values of n. Supposing that evaluation of each configuration takes 1 μs on a computer, the corresponding time to evaluate all configurations is listed in column 3.

No one knows for certain whether $\mathcal{P} = \mathcal{NP}$ or not. Most computer scientists believe that $\mathcal{P} \neq \mathcal{NP}$, which means that there are no efficient algorithms for solving any $\mathcal{NP}$ problem.

Example 1.4.3. In network design, there is no way to efficiently calculate how many links and which ones should be included in an optimal design. Selecting the links is therefore a non-deterministic procedure. Verification, that is, summing up edge costs and comparing them is, however, fast.

It should be noted that there may be little difference in the formulation of two problems, which nevertheless will happen to fall into different complexity classes.

Example 1.4.4. Finding the shortest path in a network is easy, while finding the longest path in a network is hard.

1.4.2 Showing problem hardness

Given a problem, how do we know whether an efficient algorithm exists for it or not? It is an important question, because if a problem belongs to the class of known hard problems, we would not have to spend time on trying to find an exact solution. This question is answered by the theory of $\mathcal{NP}$-completeness, which rests on a foundation of automata and language theory.

There are some problems that are at least as hard as every problem in $\mathcal{NP}$. The notion of hardness is based on the concept of polynomial time reducibility. A problem L is polynomial time reducible to another problem M if there is a function f, computable in polynomial time, such that $x \in L \iff f(x) \in M$. The problem M is said to be $\mathcal{NP}$-hard if every other problem L in $\mathcal{NP}$ is reducible to M in polynomial time. A problem is $\mathcal{NP}$-complete if it is $\mathcal{NP}$-hard and in the class $\mathcal{NP}$ itself. It is then one of the hardest problems in $\mathcal{NP}$. If one finds a deterministic polynomial time algorithm for even one $\mathcal{NP}$-complete problem, all $\mathcal{NP}$-complete problems can be solved in polynomial time. This would mean that $\mathcal{P} = \mathcal{NP}$. Note that there are problems believed to be in $\mathcal{NP}$ but that are not $\mathcal{NP}$-complete. Also, most $\mathcal{NP}$-hard problems are complete, but not all.

In order to show that a problem is $\mathcal{NP}$-complete, we need to have at least one $\mathcal{NP}$-complete problem. Such a problem is satisfiability (SAT) of a logical expression. It has been proven that satisfiability is $\mathcal{NP}$-complete. The proof is complicated, but it shows that satisfiability is at least as hard as any other problem in $\mathcal{NP}$. A variant of the satisfiability problem is the restricted *3-SAT*, restricted to clauses with three literals.

In SAT problems, we have a set of Boolean variables $\mathcal{V} = \{v_1, \ldots, v_n\}$ and a set of clauses (or sub-expressions) $\mathcal{C}$ over $\mathcal{V}$. The expression is a combination of the logical operations AND (denoted by $\cdot$) and OR (denoted by $+$). We denote by a bar the complement of a variable such that if v_i is true then $\bar{v}_i$ is false and vice versa. The problem is finding a combination of values of the variables in $\mathcal{V}$ such that each clause evaluates to true, and therefore the full expression evaluates to true.

The 3-SAT problem is a variant of SAT which takes three variables per clause. For example, the following formula could be an instance of 3-SAT. The clauses are the parentheses containing the three variables.

$$(\bar{v}_1 + v_2 + \bar{v}_7)(v_3 + \bar{v}_5 + v_6)(\bar{v}_2 + v_4 + \bar{v}_6)(v_1 + v_5 + \bar{v}_2).$$

We note that 3-SAT is in $\mathcal{NP}$, for we can construct a nondeterministic polynomial time algorithm that takes an expression with three variables per clause, guesses an assignment of Boolean values for these variables, and then evaluates it to see if it evaluates true by inserting the values into the clauses. 3-SAT is $\mathcal{NP}$-complete. An interesting fact is that the 2-SAT, expressions with only two variables in each clause, is in $\mathcal{P}$. We can use 3-SAT and reduction to show that some other problem in $\mathcal{NP}$ is $\mathcal{NP}$-complete, which is illustrated in Example 1.4.5.

Example 1.4.5.   Show that Integer Programming (IP) is $\mathcal{NP}$-complete. Consider the following integer programming example:

$$\max x_1 + 2x_2,$$
$$x_1 \geq 2,$$
$$x_2 \geq 0,$$
$$x_1 + x_2 \leq 4,$$
$$x_1, x_2 \quad \text{integers}.$$

First we formulate this problem as a decision problem. Introduce a constant to compare the objective function with, $k = 5$, say. The decision problem reads: is there a pair of values (x_1, x_2) such that the decision problem

$$x_1 + 2x_2 \geq k,$$
$$x_1 \geq 2,$$
$$x_2 \geq 0,$$
$$x_1 + x_2 \leq 4,$$
$$x_1, x_2 \quad \text{integers},$$

outputs true?

The IP problem is in $\mathcal{NP}$, because we can guess a pair of values (x_1, x_2) and verify that the side conditions are satisfied, and if so calculate the value of the objective function.

To show that IP is $\mathcal{NP}$-complete we use reduction from 3-SAT. Recall the form of 3-SAT,

$$(v_1 + v_2 + \bar{v}_3)(\bar{v}_1 + v_4 + v_5) \ldots$$

If we could solve IP in polynomial time, and 3-SAT can be formulated as an IP in polynomial time, then we can also solve 3-SAT in polynomial time, and as a consequence, $\mathcal{P} = \mathcal{NP}$.

Make the integer variables correspond to Boolean variables and have the constraints serving the same role as the clauses. The IP will have twice as many variables as the SAT instance, one for each variable and one for its complement.

A Boolean variable v_i can be expressed, letting $x_i = v_i$ and $y_i = \bar{v}_i$, as

$$1 \geq x_i \geq 0,$$
$$1 \geq y_i \geq 0,$$
$$x_i + y_i = 1.$$

A clause (v_1, v_2, v_3) is then expressed as $x_1 + x_2 + x_3 \geq 1$. The objective function is unimportant, we can simply let $k = 0$. Thus, IP is $\mathcal{NP}$-complete.

When one fails to prove that a problem is hard, there is a good chance that an efficient algorithm can be found. We will not prove that problems in this book are $\mathcal{NP}$-hard or $\mathcal{NP}$-complete, but just satisfy ourselves that most problems are and select solution methods based on this conviction.

1.5 Models and algorithms

In telecommunications, we consider two types of models: graphs describing topology, and probability distributions describing traffic and its impact on nodes. Therefore, most problems are linked to combinatorial optimization or probability distributions. Combinatorial optimization problems tend to be highly irregular and therefore difficult to solve. Probability models, on the other hand, involve the problem of randomness.

As an example of the complexity of network design, we may assume that the cost of a given network is dependent on the number of links used. However, for any fixed number of links in the network, there are many possible configurations, leading to different costs. Combining this problem with a random traffic pattern, the problem becomes even harder.

Most of the time, an *exact* solution cannot be found by reasonable effort, even if we know what we mean by a solution being exact. Instead, we have to resort to clever heuristics or approximations. For networks of realistic sizes, the choice of solution method may prove to be as important as solving the actual problem at hand. This text is to a large degree about actual algorithms as well as the problems they are intended to solve. Selection of algorithms is usually a trade-off between computational effort and accuracy. The additional time required to find a more 'accurate' solution may not be worthwhile spending. The effect of statistical variations and measurement errors of traffic may far outweight the accuracy of determining a configuration that the network designer tries to achieve.

Another issue is the verification of a result. Since it is usually impossible to find an exact solution in reasonable time, verification of a network design should be paid attention to. A good approach is to solve a problem using different algorithms and compare the results. If the results are similar, we may assume that we are on the right track, provided that the design criteria are reasonably accurately formulated.

Although most algorithms are best suited to be implemented on a computer, the discussion of selection of basic data structures (such as trees, graphs, stacks, or queues), as well as basic algorithms on these, such as searching and sorting, is omitted.

Communication network problems are usually some kind of optimization problem. We should keep in mind that we should not expect to find an absolute optimum, both due to problem hardness and statistical influence. Some general reflections on problem solving by Polya, 1988 are instructive.

What is the unknown? What are the data? What is the condition?

There are a number of very general principles that may be useful. Fundamentally, these are knowledge of the problem and its conditions. There may be methods devised for similar or related problems that can be useful. Secondly, it is important to devise a plan. Many times the network has to be designed with slightly different parameters repeatedly. It is useful to make a sensitivity analysis, changing some parameter slightly and observing how the solution is changing. In some cases, it may be possible to work backwards as well, assuming a solution and from it, retrieve input parameters.

Restricting a problem to a special case or making a generalization is often useful. Such specializations may lead to asymptotic limits. Generalizations, such as the analogy of communication networks with electrical networks, can also be also instructive. Since many of the problems discussed are hard, we cannot expect to prove the correctness of a solution. It may, therefore, be a good idea to solve the same problem using different methods, for example, a combinatorial approximation and a randomized method.

Most algorithms presented are described on high level and by examples. The examples are designed to show important characteristics of the algorithms rather than reflecting realistic real-world problems.

1.5.1 Classification of algorithms

We may define an algorithm as a step-by-step procedure for calculating some entity or processing data. Loosely speaking, an algorithm is a set of rules that defines a sequence of operations. It is desirable that an algorithm is efficient and well defined, so that any input data lead to a defined state, and that the algorithm does not 'hang' in an infinite loop. We therefore require that the number of steps the algorithm needs to complete a task is finite.

An algorithm can have deterministically controlled or random transition between states and have fixed or random input data. When the state transitions are deterministic, the algorithm is called deterministic, whilst an algorithm with stochastic state transitions, is called a randomized algorithm.

Algorithms may be classified in various ways, which also reflect the way they may be implemented on a computer (and even the choice of programming language). An algorithm may evaluate an expression directly; iteratively, by calling itself repeatedly with a value previously determined; or recursively where successive calls to the algorithm determine successively smaller problems. A recursive algorithm must have a base case which can be determined without a recursive function call, and a recursive part which specifies how the algorithm calls itself with successively smaller problem sizes. A simple example is the recursive evaluation of the factorial of an integer $f(n) = n!$. If $n \leq 1$ then $f(n) = 1$, otherwise $f(n) = nf(n-1)$. Since we are usually interested in evaluating $n!$ for $n > 1$, we simply take $f(1) = 1! = 1$. Example: $(4!) = 4 \cdot (3!) = (4 \cdot 3) \cdot (2!) = (4 \cdot 3 \cdot 2) \cdot (1!) = (4 \cdot 3 \cdot 2 \cdot 1)$. The factorial can also be viewed as an iteration. Starting from $f(1) = 1$, we get $f(2) = 2f(1)$, $f(3) = 3f(2)$, and $f(4) = 4f(3) = 4 \cdot 3 \cdot 2 \cdot 1$.

There are a number of important approaches for constructing an algorithms, of which we mention only a few. This is not intended to be an exhaustive or mutually exclusive set of classes, but rather a list of principles that are useful for hard problems.

Brute force. The method of sequentially listing and evaluating all possible configurations does certainly determine an optimal solution—eventually. It is only a reasonable method for very small problems, but is also useful in testing algorithm correctness.

Analytical methods. An analytical method can be evaluated exactly or numerically (that is, approximately with arbitrarily small error). The main analytical methods used are mathematical programming (analytical optimization), combinatorics and probability calculus. Such methods are not available for a general $\mathcal{NP}$-hard problem.

Sometimes, however, hard problems can be made analytically tractable by letting some group of parameters tend to infinity or zero (or any other suitable limit). For example, the flow in some networks can be described analytically when the size of the network grows to infinity. These bounds are very useful in analyzing hard problems.

Approximations. An approximative method gives a result close to the exact value and with a bound on its error—or equivalently, the distance between the approximation and the exact value. There are many ways to construct approximations. A problem is often simplified so that it becomes tractable by other methods. Limits and asymptotics are often used.

Heuristics. Heuristic methods differ from approximate in that there is no guarantee on the error limit. Another view is that a heuristic method is relying on a principle which may serve as a first 'best guess,' 'common sense', or 'intuition.' The optimality of the result remains unprovable, but heuristics often give reasonable results and are easy to program. One of the best known heuristics is the greedy principle (or the greedy heuristic). It is used in many various contexts, in exact methods as well as in approximations.

The greedy method is applicable to optimization problems, that is, problems that involve searching through a set of configurations to find one that minimizes or maximizes an objective function defined on these configurations. The general formula of the greedy method could not be simpler. In order to solve a given optimization problem, we make a sequence of choices. The sequence starts from some well-understood starting configuration, and then iteratively makes the decision that seems best from all of those that are currently possible.

The greedy approach does not always lead to an optimal solution. But there are several problems that it does work optimally for, and such problems are said to possess the greedy-choice property. This is the property that a global optimal configuration can be reached by a series of locally optimal choices (that is, those that are the best available choices at that time), starting from a well-defined configuration.

Problem restriction. An important way of making problems tractable is by restriction of the search space. Possibly some parameters are kept fixed while others are allowed to vary. Reduction of the state space is the method for deriving various bounds. In combinatorial optimization we may use a technique called branch-and-bound, where an integer optimization problem is replaced by its continuous counterpart, which is relatively easy to solve, and parameter values are successively refined based on these.

Divide and conquer. The divide and conquer principle amounts to solving a particular complex problem by dividing it into subproblems of smaller size, recursively solving each subproblem, and then merging the solutions to the subproblems to produce a solution to the original one. Decompositional methods can be classified as relying on this principle. Another related technique is dynamic programming. It is similar to divide and conquer in that it is very general, and it often produces efficient algorithms for otherwise hard problems.

1.5.2 Randomization

Randomization is a very general and important class of algorithms. By randomization algorithms we include simulation, Monte Carlo methods and meta-heuristics, or any method that is dependent on random numbers. The method class relies on random numbers of high quality, and from the author's experience it is worthwhile to study and implement algorithms for reproduceable and uniformly distributed random numbers.

Meta-heuristics are numerical optimization methods that mimic a physical system or biological evolution. Two commonly used methods are simulated annealing and evolutionary algorithms. The two methods represent two groups of techniques: local search and population-based search. The strength of these methods is that they are very general and often easy to formulate.

Monte Carlo methods and simulation are also indispensable tools for estimation and verification. Simulation should be used with care, however. It is not a panacea for solving hard problems. It can be misleading if some parameters are incorrectly set or conditions improperly formulated.

Local search.   Local search is an optimization method which is based on the principle that, given a candidate solution, the optimum may be found in the neighborhood of that solution if the initial configuration is close enough to the optimum. The search space is reduced by this location restriction. It has to be possible to evaluate the optimal value, or fitness value, which is the term used to describe optimality in some sense (such as low cost or high reliability).

We may consider a problem P, which is assumed to be $\mathcal{NP}$-hard so that it has a huge set S of potential solutions. Also, let there be a way to generate candidate solutions $s_1, s_2, \ldots$ from an initial configuration s_0. The principle is to reduce the search space, since we know that a brute force approach would not finish in reasonable time. The fitness value is given by the function $f(s)$, which gives a higher score for a better solution.

The generation of new candidate solutions is created from previous candidates by a neighborhood operator, say $g(\cdot)$. The function takes a candidate solution s and produces a new one: s', slightly different from s. One simple operator in the context of network topologies is letting links swap end points. The result is that two existing links are removed and two new ones are added. The total number of links remains constant.

We may also have a mutation operator, which in the generation of a new candidate with small probability changes a fundamental characteristic (such as the total number of links) of the initial solution. One of the simplest local search methods is hillclimbing which can be described as follows:

(1) Generate an initial candidate solution—possibly at random. Call this the current solution s. Calculate its fitness $f(s)$.
(2) Use the neighborhood operator $g(\cdot)$ to generate a new solution s', and calculate its fitness.

(3) If $f(s')$ is better than or equal to $f(s)$, then replace s with s'. Repeat from step 2 until a termination criterion is reached.

As the name suggests, the idea of hillclimbing is a kind of greedy principle. Whenever possible, move to a better or at least different solution. The method is searching the neighborhood of a current solution. The danger here is that the set of possible candidates may be stuck at a local optimum which is not, a global optimum. A modification to local search methods has ways of moving from a local optimum in the hope of finding a global one.

Population-based search. A population-based search method uses a set of initial configurations, rather than one. This approach enhances the probability of finding a good solution as compared to local search. The method is based on the paradigm of the survival of the fittest. The parent population enables search in many neighborhoods simultaneously. It can also be equipped with a mutation operator that mitigates the risk of trapping in the neighborhood of a non-global, local optimum. Some time will be spent, statistically, searching neighborhoods of only moderate or even poor solutions, in the hope that a global optimum may be found.

Another strength of the population-based algorithm is that the search direction may be described as the recombination of two solutions, giving a "linear interpolation" of the two candidates as the next one. This facilitates moves in directions, which with higher probability, contain an optimum. This also allows for much greater jumps than in local search, since two parents may be completely different. The population-based algorithm may be summarized as

(1) Generate an initial parent population of candidate solutions. Calculate the fitness of each solution.
(2) Select a subset of the population to be parents. This probability is proportional to the fitness value.
(3) Apply recombination to have children mutate (with low probability).
(4) Keeping the population size constant, replace some candidates of the previous generation with the children.
(5) Repeat from step 2 onwards until a termination criterion is reached.

The main principle of the method is the mimicking of a survival-of-the-fittest strategy. Candidates with higher fitness value have greater probability to be selected as parents. The recombination creates a child randomly from the chromosomes of the parents, with an additional mutation operator present, that allows for random moves. The selection of candidates that should be replaced by the children can be done in many ways. As indicated, the algorithm is powerful and flexible, but requires determination of control variables that may not be easy to determine.

1.5.3 Monte Carlo Techniques

A Monte Carlo method is a method performing sampling of random variates X from a set or probability distribution $X \in A$ and evaluates a function $f(X)$ on these variates. Often these function values are either used in optimization or estimation of expectations. When estimating expectation (mean), the approximation

$$\mathbf{E}(f(X)) \approx \frac{1}{N}\sum_{i=1}^{N} f(X_i),$$

is used. The population mean of $f(X)$ is estimated by a sample mean. When the samples $\{X_i\}$ are independent, the law of large numbers ensures that the approximation can be made as accurately as desired by increasing the sample size N.

A Monte Carlo algorithm may not produce an optimal result, but a very useful property of the algorithm is that it can be run repeatedly with independent random values each time, and then the probability of obtaining an optimal solution can be estimated and be made arbitrarily small, at the expense of increased running time.

The optimization variant of Monte Carlo is when the function $f(X)$ is the cost of candidates in X is to be minimized. Throughout the optimization, the pair $(X, f(X))$ is kept updated. A candidate $(Y, f(Y))$ is replaced by $(X, f(X))$ whenever $f(X) < f(Y)$. The output result is the candidate X representing the minimum cost configuration found thus far. If the successive costs $f(X)$ can be saved in an array, and after N iterations counting the number of minimum cost configurations, let it be called s, then s/N estimates the probability that a current solution or a better one exists.

1.6 Organization of this book

Network design is an iterative process, where different levels of initial data determine where in the iteration the process starts, and so it is somewhat arbitrary how the topics should be ordered.

The text can roughly be described as consisting of three parts. The first part consists of graph theoretical problems and algorithms, leading to design of cost optimal network topologies.

The second part focuses on stochastic traffic and performance evaluation of network, where we most of the time assume that a suitable topology has been found.

The third part discusses issues related to topological design which are subject to operational reliability constraints.

Chapters 2 and 3 discuss the main graph theoretical algorithms in network design, that constitute an important part of the toolbox for the subsequent chapters. Chapter 4 is dedicated to the design of network topology. This chapter is intended to serve as a starting point in the network design process. Chapters 5–11 specialize in more technology-oriented design problems: loss networks, general packet networks, flow-controlled packet networks and multiservice networks. Chapter 12 is dedicated to network resilience and related design problems.

1.7 **Summary**

This chapter barely skims the surface of complexity theory and algorithm analysis. It is intended to provide an understanding of the difficulties inherent in network design, and to present an overview of the methods and principles at hand.

$\mathcal{NP}$-hardness is discussed in most literature on algorithms, for example Goodrich and Tamassia (2002). Two references on heuristical methods are Polya (1988) and Skiena (1998), which may serve as an introduction to algorithms for hard problems (whether $\mathcal{NP}$-hard, harder, or not so hard). All basic principles are used throughout the text.

Networks and Flows

2

This chapter briefly introduces some terminology from graph theory related to networks and flows. A network can be thought of as a structure for transportation of some commodity between points in the network. A commodity is a general term; in this text it is some type of exchange of data. When a commodity is transported through a network, it constitutes a flow. The natural model of the structure of a network—its topology—is the graph. It consists of a set of vertices (or nodes), which are connected by a set of edges (or links). Depending on the problem we wish to solve, we may associate numbers with edges and vertices, such as link costs, flow or capacity limits, demands, and, when appropriate, more complex performance measures such as blocking or delay. These numbers are collectively referred to as weights.

We have adopted the convention that, when discussing results from graph theory, to use the terms graph, vertex, and edge, which represent the model. In situations where a planned or existing physical construction can be assumed, we use the terms network, node, and link. The word network may refer both to a kind of graph and to a physical entity. It should be possible to distinguish what the word is referring to from the context. These two meanings of the word are closely related, but usually we allow a physical network to be a wider concept containing more information, for example, technological details.

One of the main applications of graph theory is the study of network flows. Flow problems span from quite straightforward to rather intricate. The algorithms presented in this and the next chapter are therefore of increasing complexity. It turns out, however, that more complex flow problems often can be solved by reducing them to a sequence of simpler ones.

A network flow is called constant if it does not change over time. Throughout the chapter, we assume that a network topology is given, represented by a graph. The graph shows (by its vertices and edges) how the nodes are connected by links in the physical network. The graph may also include different types of weights. In most cases, we also have a matrix of demands between the vertices which induce flows in the network. A network flow is usually controlled by some routing principle that decides how the flows are transported throughout the network. That is, a routing principle determines one or more suitable routes (or paths) that a flow may follow. Since the flows in this chapter are assumed to be constant, the routing principles are fairly simple.

The content of this chapter and the next serve as an introduction to the rest of the text, where particular design objectives and network types are treated. In Section 2.1 we discuss various flow concepts and some preliminaries on graph theory. Section 2.2 shows some representations of graphs used in algorithms. Graph connectivity, the simplest flow-related problem, is the topic of Section 2.3. Section 2.4 is dedicated to algorithms for the shortest path problem which is a fundamental problem in routing of flows. A number of methods to find shortest paths are described: linear programming and a primal-dual formulation, dynamic programming and the greedy principle. These approaches will often be used throughout the text. The maximum flow problem is discussed in Section 2.5, where the objective is to find the maximum possible flow between two vertices in a graph. Maximum flows can be found by a primal-dual algorithm. An approximate algorithm based on the theory of electrical networks is also discussed.

2.1 Preliminaries

We now define some basic concepts of networks and flows, which may be rather intuitive, but nevertheless need a precise formulation to be the foundation for a rigorous network flow theory. In this section, proofs of the theorems are omitted. Comprehensive introductions to networks and flows including proofs can be found in many texts such as Biggs (1985), Papadimitriou and Steiglitz (1982), or Goldberg et al. (1990).

2.1.1 Basic flow concepts

Any network topology can be represented by a graph $G = (V, E)$, which has a set of vertices, V, and a set of edges, E. If the variables i and j of non-negative integers represent two distinct vertices, then the pair (i, j) represents an edge, if it exists. In algorithms, we usually let the total number of vertices $|V|$ be denoted by n, which is called the order of the graph. Similarly, we denote the number of edges $|E|$ by m, called the size of the graph. In this section we introduce some formal definitions and concepts, which will be useful in describing the algorithms that follow. Depending on the problem, edges may have associated with them one or more weights, for example, a cost of using that edge for transportation of a unit of flow, or capacity constraints on the amount of flow allowed to pass through the edge, or both. We distinguish between flow limits and capacities as follows. A flow limit is a maximum (or minimum) amount of flow that can be admitted onto an edge. A capacity is an upper bound of flow on an edge, which is not necessarily tight. For constant flows, however, it is tight and equals the upper flow limit. Vertices may also be assigned weights. A general weight is denoted by w_{ij} for an edge (i, j) and w_i for a vertex i.

The edges in a graph may be undirected, allowing flow in both directions of the edge, or directed, where only flow in a specified direction is permitted. Directed graphs are generalizations of undirected graphs—any undirected graph can be formulated as a directed graph. The converse is not true in general. The conversion of an undirected to a directed graph is straightforward; simply replace each undirected edge with two oppositely oriented directed edges, each having the same weight, that is $w_{ij} = w_{ji}$.

A directed edge is usually referred to as an arc in the literature. We will use the term directed edge since we are interested mainly in undirected networks, and just drop the word directed whenever possible. The reason behind this is that the formal definition of a network assumes a directed graph. In telecommunications, however, it is more natural to consider undirected graphs. An undirected edge (i, j) captures the operation of full duplex links, that is, a system that allows communication in both directions, simultaneously. (In a duplex system, callers may talk and listen at the same time.) One of the reasons for defining a network as a directed graph is that some algorithms are easier to formulate and prove for directed graphs, particularly graphs with single directed edges. This eliminates the need for preventing loops that may follow from using undirected edges. Similarly, routing protocols in physical networks need to be designed to avoid loops.

We will usually not allow multiple edges in G or self loops (i, i), that is, edges starting and ending at the same vertex. Therefore, any edge from a vertex i to a vertex j must be unique, and we denote it by (i, j) or e_{ij}. These two notations are used interchangeably, depending on the context, to make formulas readable. For example, if we have a function of an edge, we sometimes use the notation $f_{ij} = f(e_{ij})$ rather than $f(i, j)$ to stress that there is an edge present between i and j. The latter notation can be interpreted that it is a function of any two vertices—even two vertices without an edge present. For example, we denote the distance of an edge (i, j) by d_{ij} and the distance between any two vertices k and l by $d(k, l)$.

Figure 2.1 shows a graph with seven nodes and ten links. The order of the graph is $|V| = n = 7$ and its size is $|E| = m = 10$. Assigned to each link are flow limits and link costs. The numbers in brackets (l_{ij}, u_{ij}) denote the lower and upper flow limits

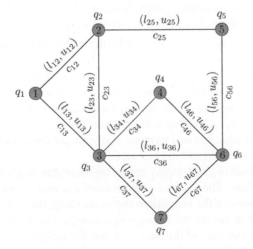

FIGURE 2.1

A graph representing an example network with edge and vertex weights. The edges are labeled with flow limits in brackets (l_{ij}, u_{ij}), and costs, c_{ij}. The vertices are labeled with demands q_i.

of link (i, j) and the number c_{ij} the cost of the link. The cost can be the distance between the vertices or a more general weight. A demand q_i is usually associated with each node. In case all demands are zero apart from, say, $q_s = -q_t > 0$, this demand generates a flow between the source s and the sink t. Here, we have adopted the convention that a demand is a deficit of some commodity, represented by a negative number. In other words, a demand is a negative flow out from a node. This slight abuse of terminology allows for positive flows in the direction from supplies to demands.

Suppose we have supplies and demands of some commodity between one or more vertex pairs across the network. Subject to the network topology and edge capacities, these induce the network flows. In this chapter, single commodity flows are discussed, where a flow f has a single source vertex s and a single sink vertex t only. It follows that, for a flow f, the inflow equals the outflow for all vertices except for the source and the sink.

Formally, a flow is a function $f : E \to \mathbb{R}$ that satisfies the following constraints:

$$f_{ij} \leq u_{ij}, \quad \text{for all } (i, j) \in E \quad \text{(capacity constraint)}, \tag{2.1}$$

$$f_{ij} = -f_{ji}, \quad \text{for all } (i, j) \in E \quad \text{(flow antisymmetry constraint)}, \tag{2.2}$$

$$\sum_{i \in V} (f_{ij} - f_{ji}) = 0 \quad \text{for } (i, j) \notin \{s, t\} \quad \text{(flow conservation constraint)}. \tag{2.3}$$

The capacity constraint states that no flow can exceed available edge capacity. The flow antisymmetry constraint means that if we send, say, two flow units from i to j and one unit from j to i, we have a net flow of one unit from i to j, so that oppositely directed flows cancel each other out. The flow conservation constraint implies that the flow into a vertex minus the flow out from the vertex is zero.

Given a flow f, the residual capacity function $u^f : E \to \mathbb{R}$ is defined by $u_{ij}^f = u_{ij} - f_{ij}$. The residual graph with respect to a flow f is given by $G_f = (V, E_f)$, where $E_f = \{(i, j) \in E : u_{ij}^f > 0\}$, that is, the edges which have spare capacity after the flow have been deducted (see Figure 2.2). The value $|f|$ of a flow f is the net flow into the sink t, $|\sum_{i \in V} f_{it}|$. It will be convenient to express edge capacities, edge costs, and flows as vectors of dimension m, the size of the graph.

When there is a flow in a graph, it must follow a path. A path is a sequence of distinct vertices $\{v_1, v_2, \ldots, v_k\}$ such that the vertices $(v_i, v_{i+1}), 1 \leq i \leq k - 1$ are adjacent. Thus, each edge has one terminal vertex in common with another edge of the path, apart from the start and the end vertices.

Two basic problems in graph theory are the topics of this chapter, the shortest path and the maximum flow. The shortest path between a source s and a sink t is a path in G for which the sum of the weights on the edges along the path is minimum. The edge weights are often simply the distance between the two terminal vertices of the edge, or some other estimate of the edge cost, but the weights may be arbitrary. For this reason, shortest paths algorithms turn out to be very useful.

In the maximum flow problem, we consider a graph $G = (V, E)$ where the edge weights are flow limits u_{ij} and we seek the maximum value of the flow between a

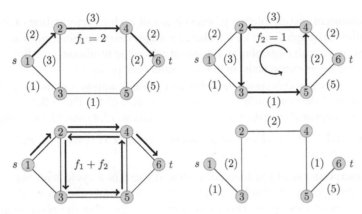

FIGURE 2.2

A graph with two flows (top), an $s - t$ flow and a circulation. The two superimposed flows (bottom) and the resulting residual graph $G_f(V, E_f)$ with respect to $f_1 + f_2$.

source s and a sink t. In illustrations, the capacity limit is shown as a number (u_{ij}) in brackets to distinguish it from edge costs, shown without brackets (see Figure 2.1). By convention, lower flow limits where $l_{ij} = 0$ are not shown. The maximum flow in this graph is the flow of maximum value that can be transported in the graph without violating the flow limits.

Given a flow f, we define an augmenting path to be an $s - t$ path in the residual graph. The following theorem, due to Ford and Fulkerson (1962), gives an optimality criterion for maximum flows.

Theorem 2.1.1. *A flow in a graph G is optimal if and only if its residual graph contains no augmenting path.*

A special type of flow is a circulation, a flow with $s = t$, so that $\sum_{i \in V} f_{ij} = 0$ for every vertex j. In effect, there is no distinguishable source and sink. The set of edges which the circulation follows is called a cycle, denoted by Γ. A circulation is shown in Figure 2.2.

Suppose that, in addition to edge capacities we also have a cost function on the directed edges $c : E \to \mathbb{R}$. In a network with both cost and capacity limits we can define a minimum cost flow, where a required flow of value $|f|$ is to be transported from the source s to the sink t in the cheapest possible way. We make the assumption that costs are antisymmetric:

$$c_{ij} = -c_{ji} \quad \forall (i, j) \in E \quad \text{(cost antisymmetric constraint).} \tag{2.4}$$

It follows that the cost of a circulation f is

$$c(\Gamma) = \sum_{(i,j) \in \Gamma : f_{ij} \geq 0} f_{ij} c_{ij}. \tag{2.5}$$

The minimum cost circulation problem is that of finding a circulation of minimum cost in a directed graph $G = (V, E)$ where the edges have capacities u_{ij} and costs c_{ij}. Minimum cost circulations play an important role in the minimum cost flow problem similar to the augmenting paths in the maximum flow problem.

Next we state two criteria for the optimality of a circulation. Let the cost of a cycle Γ be the sum of the costs of the edges along the cycle (2.5). The following theorem states the first optimality criterion. Based on Equations (2.4) and (2.5), we have:

Theorem 2.1.2. *A circulation in a directed graph G is optimal if and only if its residual graph contains no negative-cost cycle.*

The interpretation of the theorem is that, if there is a cycle with negative cost in the residual graph, we can lower the total cost by simply increasing the value of the circulation on this cycle. This follows from the fact that the residual graph by construction contains the unused capacity.

To state the second criterion, we need to introduce the notions of price function and reduced cost function. A price function is a vertex weight $\pi : V \to \mathbb{R}$. The reduced cost function with respect to a price function π is defined by $\bar{c}_{ij} = c_{ij} + \pi_i - \pi_j$. The concept of vertex prices originates in the theory of linear programming and is the base of many minimum cost flow algorithms. Being so-called dual variables in linear programming, vertex prices can be interpreted as the current 'market prices' of a commodity. The reduced cost $\bar{c}_{ij}$ can be seen as the cost of buying a unit of commodity at i, transporting it to j, and then selling it there. Due to the conservation constraints, the reduced costs define an equivalent problem. The following theorem states the optimality criterion of the dual problem, a problem involving vertex prices.

Theorem 2.1.3. *A circulation f is optimal for the minimum cost circulation problem on a graph $G = (V, E)$ with flow limits u_{ij} and costs c_{ij} if and only if it is optimal for the problem on G with the same edge flow limits u_{ij} and costs $\bar{c}_{ij}$ for every price function π.*

The second optimality criterion is as follows:

Theorem 2.1.4. *A circulation f is optimal if and only if there is a price function π such that, for each edge (i, j),*

$$\bar{c}_{ij} < 0 \text{ implies that } f_{ij} = u_{ij} \quad \text{(complementary slackness constraint)}. \quad (2.6)$$

The result can be interpreted as follows. If the reduced cost $\bar{c}_{ij}$ is negative, we get a net profit of buying a commodity at i, transporting it to j, and selling it. Therefore, it is desirable to transport as much as possible to maximize profit, that is, $f_{ij} = u_{ij}$.

A useful property of flows is the fact that they can be decomposed into a smaller number of 'primitive elements.' Flows can be decomposed into simple paths and circulations into simple cycles in the graph. Such a decomposition of a circulation is shown in Figure 2.3. We will only note here that such a composition is possible, and this fact is used in many flow algorithms.

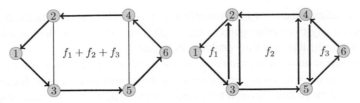

FIGURE 2.3

Decomposition of a circulation. Since the same commodity amount flows in all edges, the sum of the flow elements make up the total flow (opposite directed flows cancel each other out).

2.1.2 Graph flow problems

Figure 2.4 shows the relation between the flow problems addressed in this and the next chapter. The simplest flow problem in a graph is that of determining whether two vertices s and t are connected or not. If it is possible to send a (non-zero) flow f from s to t, the vertices are connected. This problem can be solved by performing a search on the graph, which is discussed in Section 2.3. Shortest path is a generalization of connectivity, where weights of cost-type (such as distances) on the edges are taken into account. Algorithms for shortest path are the topic of Section 2.4.

When the edge weights are capacities, we can formulate the problem of finding the maximum flow between two vertices s and t. This is also a generalization of the connectivity problem. In principle, we can try to send a unit flow from s to t, which is equivalent to the connectivity problem. If we record the path found from s to t, we can form the residual graph by deducting a unit of capacity from each edge along this path and iterate the procedure until s and t are disconnected. This is not particularly efficient, however, and better algorithms are discussed in Section 2.5.

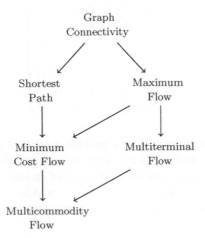

FIGURE 2.4

Interdependence of network flow problems.

By generalizing the $s - t$ maximum flow to the problem of finding the maximum flow between any two vertex pairs in the graph, considering one flow at a time, we arrive at the multi-terminal flow problem in Figure 2.4.

A generalization of the maximum flow and the shortest path, where we have both cost c_{ij} and capacity limits u_{ij} on the edges, is the minimum cost flow. It determines how to route a given flow through the graph so that capacity constraints are satisfied, at minimum cost.

When there are demands between more than two vertex pairs simultaneously, it is a multi-commodity flow problem. In this case, flows between several vertex pairs compete for the same network resources, and the objective is to meeting the vertex demands as closely as possible while satisfying the capacity limits, at minimum cost. The last three problems are discussed in Chapter 3.

2.2 **Network representations**

There are various ways to represent a graph as a matrix that can be used in equations. This is a direct consequence of the formulation of flow problems as linear programs. One such possibility is the $n \times n$ *adjacency matrix* A. For an undirected graph it is defined as a matrix with a positive one in the positions (i, j) and (j, i) whenever there is an edge between vertices i and j, and zeros elsewhere. The adjacency matrix is in this case symmetric. This is the most commonly used matrix representations in applications.

Example 2.2.1.   The adjacency matrix of the undirected network (that is, ignoring the edge directions) in Figure 2.1 is

$$A = \begin{pmatrix} 0 & 1 & 1 & 0 & 0 & 0 & 0 \\ 1 & 0 & 1 & 0 & 1 & 0 & 0 \\ 1 & 1 & 0 & 1 & 0 & 1 & 1 \\ 0 & 0 & 1 & 0 & 0 & 1 & 0 \\ 0 & 1 & 0 & 0 & 0 & 1 & 0 \\ 0 & 0 & 1 & 1 & 1 & 0 & 1 \\ 0 & 0 & 1 & 0 & 0 & 1 & 0 \end{pmatrix}. \tag{2.7}$$

Another representation, suitable for directed graphs, is the vertex-edge $n \times m$ incidence matrix where an entry is a positive one in the position of the origin of the directed edge, and a negative one in the position of its termination, and zeros elsewhere. The vertex-edge incidence matrix can also be defined for undirected graphs, where all non-zero entries are positive ones. In this text, we denote the vertex-edge incidence matrix by J.

Example 2.2.2.   The vertex-edge incidence matrix of the network in Figure 2.1 taking edge directions into account, is

$$J = \begin{pmatrix} 1 & 1 & 0 & 0 & 0 & 0 & 0 & 0 & 0 & 0 \\ -1 & 0 & 1 & 1 & 0 & 0 & 0 & 0 & 0 & 0 \\ 0 & -1 & -1 & 0 & 1 & 1 & 1 & 0 & 0 & 0 \\ 0 & 0 & 0 & 0 & -1 & 0 & 0 & 1 & 0 & 0 \\ 0 & 0 & 0 & -1 & 0 & 0 & 0 & 0 & 1 & 0 \\ 0 & 0 & 0 & 0 & 0 & -1 & 0 & -1 & -1 & 1 \\ 0 & 0 & 0 & 0 & 0 & 0 & -1 & 0 & 0 & -1 \end{pmatrix}. \tag{2.8}$$

This matrix is useful in representing many flow problems. Each row represents the incoming and outgoing edges. When the matrix is multiplied by a flow vector, the result is the inflow minus outflow, to and from each vertex, respectively. When this sum is set to zero, the equation represents a flow conservation constraint (2.3).

By indexing all the $s - t$ paths in the network, it is also possible to map paths between s and t against edges, with a positive one in the position corresponding to the position where an edge is part of the path, and zeros elsewhere. This representation is called the edge-path (or edge-chain) incidence matrix. This matrix, denoted by K, tends to be large and the adjacency or vertex-edge incidence matrices are therefore preferable representations.

Example 2.2.3. In the graph in Figure 2.1, we can identify the possible paths from s to t. The direction of the edges does not affect the matrix entries; either an edge is in a path or it is not. The paths $p_1, \ldots, p_7$ are given in Table 2.1.

$$K = \begin{pmatrix} 0 & 1 & 0 & 1 & 1 & 0 & 1 \\ 1 & 0 & 1 & 0 & 0 & 1 & 0 \\ 0 & 1 & 0 & 1 & 0 & 0 & 1 \\ 0 & 0 & 0 & 0 & 1 & 0 & 0 \\ 0 & 0 & 0 & 0 & 0 & 1 & 1 \\ 0 & 0 & 1 & 1 & 0 & 0 & 0 \\ 1 & 1 & 0 & 0 & 0 & 0 & 0 \\ 0 & 0 & 0 & 0 & 0 & 1 & 1 \\ 0 & 0 & 0 & 0 & 1 & 0 & 0 \\ 0 & 0 & 1 & 1 & 1 & 1 & 1 \end{pmatrix}. \tag{2.9}$$

Table 2.1 List of paths in the example network in Figure 2.1

Index	Path
1	1-3-7
2	1-2-3-7
3	1-3-6-7
4	1-2-3-6-7
5	1-2-5-6-7
6	1-3-4-6-7
7	1-2-3-4-6-7

So far, edges have been identified by a pair (i, j) of vertices. In computer implementations of most algorithms, however, it is useful to have a single parameter indexing an edge. Suppose that n vertices are indexed by the natural numbers. We can have a maximum of $n(n - 1)/2$ edges, so we can form a vector of this length. A simple mapping between a pair of vertices and a single edge index is obtained by introducing two variables l and s, the largest and the smallest of the two vertex indices, respectively, that is

$$l = \max\{i, j\}, \tag{2.10}$$

$$s = \min\{i, j\}. \tag{2.11}$$

Then, we can assign a single number k to any edge e_{ij} by $k = l \cdot (l - 1)/2 + s$. This indexing simplifies storing and retrieving data in vectors.

The converse transformation, given an edge index k, is performed by searching for the largest number l such that

$$k \leq l \cdot (l - 1)/2. \tag{2.12}$$

Then the largest of the vertex indices is l and the smallest is $s = k - l(l - 1)/2$. It is often convenient to remove nonexisting edges and adjust the index to represent only existing edges. Such an indexing representing the edges in Figure 2.1 is shown in Table 2.1.

2.3 Graph connectivity

One of the simplest but also one of the most fundamental problems in graph theory is to determine whether two vertices in a graph are connected or not. The problem may seem trivial enough in a small graph, but in larger graphs, a systematic approach becomes necessary. We define connectivity formally as follows.

Definition 2.3.1 (Connectivity).   In an undirected graph $G = (V, E)$, two vertices $i, j \in V$ are said to be connected if G contains a path from i to j. Otherwise, they are said to be disconnected. A graph is called connected if every pair of distinct vertices in V can be connected through some path. Otherwise it is called disconnected.

The connectivity problem can be solved by simply traversing all possible paths in the graph until either the destination vertex has been found, or all paths have been traversed without finding the destination vertex. There are two strategies for searching through a graph: depth-first search and breadth-first search. The problem of determining vertex connectivity can be solved efficiently using any of these strategies. A simple algorithm to ascertain the connectivity of a graph G is as follows:

Algorithm 2.3.1 (Graph Connectivity).   Given a graph $G = (V, E)$.

STEP 0: Start at an arbitrary node s in the graph G.
STEP 1: Proceed from node s using either depth-first or breadth-first search, counting all nodes reached.

STEP 2: Once the graph has been entirely traversed, if the number of nodes counted is equal to the number of nodes of G, the graph is connected; otherwise it is disconnected.

Output TRUE or FALSE.

2.3.1 Depth-first search

Suppose we are given a starting vertex s and we would like to know which other vertices s is connected to. The depth-first search (DFS) method can be used to determine the set of connected vertices. We label vertices as visited and unvisited to be able to keep track of where we are in the search. Initially, we let all vertices except s be marked as unvisited. Now, s is referred to as the current vertex. Whenever we find an unvisited neighbor vertex, we continue 'forward as far as we can' by marking the unvisited vertex as visited, and set the new vertex as the current vertex. When arriving at a point where no more unvisited neighbors can be found, we backtrack along the path we can until an unvisited vertex can be found again. If we arrive back to s without finding any such neighbors, the algorithm terminates; all vertices that can be reached have been visited. Of course, we can stop the search earlier if we are looking for a particular vertex t that has been visited before backtracking s. We can note that each edge is being traversed twice, once in the forward direction, once when backtracking.

Example 2.3.1.   If we perform a depth-first search on the network in Figure 2.1 starting from s, we would, for the first probe, get the sequence of vertices $s, 2, 3, 4, 6, 7$, if the directions of the edges are observed. Next, we backtrack to vertex 2 from where we reach vertex 5, omitting vertex 6, which has already been visited. Therefore, we backtrack again, but this time all the way to s, since we have, in fact, visited all vertices.

2.3.2 Breadth-first search

The alternative to DFS is breadth-first search (BFS) in which all the neighbors of the current one are visited before moving on to the next level. Consequently, there is no need for backtracking. Again, we start from a vertex s. When the current vertex has unvisited neighbors, these vertices are visited in turn. If no further unvisited vertices can be found, we move onto the next vertex in the order they have been visited. Continuing in this manner, we will eventually reach a vertex which has no unvisited neighbors and there is no vertex to move on to, and then the algorithm terminates.

Example 2.3.2.   Using BFS on the example in Figure 2.1, starting from s, the first scan gives 2, 3. Moving onto vertex 2, we reach vertex 5. Proceeding to vertex 3, we reach vertices 4, 6, and 7. Now there is no visited vertex with unvisited neighbors, so the search is finished.

2.4 Shortest paths

Finding the shortest path in a graph is a fundamental problem of immense importance in graph theory as well as in traffic engineering of communication networks. In graph theory, more complex problems can often be reformulated as a sequence of shortest path problems. The shortest path distances or costs can be defined arbitrarily to model various problems that can be solved by a shortest path algorithm. It is therefore important to have efficient algorithms for solving this problem. In communication networks, the shortest path is a basic routing principle used, for example, in the Open Shortest Path First (OSPF) and the Intermediate System to Intermediate System (IS-IS) protocols. An effective routing principle leads to decreasing transport distance which lowers propagation delay in the network, and also the transport cost per flow unit, which can be seen as a measure of the cost effectiveness of the network. Being a problem of such versatility, we will study the problem in some detail and solve it in various ways, thereby illustrating different solution techniques.

The shortest path can be found using different methods, and the problem is therefore suitable for illustration of important solution techniques such as the primal-dual approach in linear programming and dynamic programming as well as the greedy heuristic.

Shortest path algorithms can be formulated both for undirected and directed graphs. In the following subsections, examples of both formulations are presented.

It should be noted that some algorithms require that the edge weights are non-negative for the algorithms to work properly. This can be a limitation in cases where negative weights result from a higher-level algorithm. Therefore, an algorithm without this restriction is presented as well.

The shortest path problem has two variants: the (single source) shortest path and the all-pairs shortest path. In the former, the shortest path between a specific node pair $s - t$ is sought. In the latter, we are looking for the shortest paths between all vertex pairs at once.

2.4.1 Shortest path as a linear program

Consider a directed graph $G = (V, E)$ with some weight function w_{ij} defined on the edges $(i, j) \in E$ which we assume takes on positive values. The edge weights can be anything we define it to be, but for the time being we may think of the weights as the distances along the edges, using the notation d_{ij} for the distance of an edge (i, j). The shortest path between two vertices s and t is the path for which the sum of the distances along this path assumes its minimum.

Thus, the shortest path problem can be thought of as an optimization problem over the set of paths connecting s and t. It bears some similarities with the connectivity problem of Section 2.3, but the search algorithms discussed there cannot handle edge weights. We begin by formulating the shortest path as a linear program.

Firstly, we need to define a path in the network. If it is possible to send a unit flow from s to t, then the flow clearly must follow a path. In general, a shortest path need not be unique. Had it been, then a flow from s to t would send all the flow on this path. That implies that no flow is split in any vertex on the path. (We can prevent flow splitting when there are multiple shortest paths by requiring that the flow can take on values zero and one only.) By assumption, no flow can be added or absorbed along the way. We must therefore have flow conservation along this path, as expressed in Equation (2.3). This leads to the set of equations

$$\sum_j (f_{ij} - f_{ji}) = \begin{cases} 1 & i = s, \\ 0 & i \neq s, t, \\ -1 & i = t, \end{cases} \tag{2.13}$$

where $\sum(f_{ij} - f_{ji})$ means flow out from i minus flow into j. The path is now a set of edges carrying a unit flow; no other edges carry any flow. Thus, the linear program representation of the shortest path problem is

$$\min \quad \sum d_{ij} f_{ij},$$
$$\sum_j (f_{ij} - f_{ji}) = \begin{cases} 1 & i = s, \\ 0 & i \neq s, t, \\ -1 & i = t, \end{cases} \tag{2.14}$$
$$f_{ij} \geq 0 \quad \text{for all } (i, j) \in E.$$

The difference $f_{ij} - f_{ji}$ defines the direction of the flows, and since all flows must be positive, we must represent the graph as a directed graph to be able to formulate the flow conservation constraints. We know that all directed edges in the network should point outwards from s and inwards to t, but for edges which have neither s nor t as a terminal vertex, the orientation of the edges is not so clear. In vector representation, and slightly modified, the optimization problem (2.14) can be written compactly as

$$\min_{\mathbf{f}} \quad \mathbf{d}^T \cdot \mathbf{f},$$
$$J_{n-1}\mathbf{f} = \mathbf{b}_{n-1}, \tag{2.15}$$
$$f_k \geq 0, \quad \text{for all } 1 \leq k \leq m,$$

where J_{n-1} is the reduced $(n-1) \times m$ node-edge incidence matrix for $G = (V, E)$ with n nodes and m edges, $\mathbf{f}$ is the vector of flows, $\mathbf{d}$ is the vector of distances, and $\mathbf{b}_{n-1}$ is a reduced vector of flow constraints from the right-hand side in (2.13). The matrix J_{n-1} does not contain all n constraints in the system (2.13), because one is redundant. We therefore remove the last equation where $i = t$ (The flow is completely determined by the flow conservation conditions—it prevents any flow from being absorbed before reaching t.). We can also drop the double subscript from f, since the edge orientations are given by J_{n-1}. We illustrate the linear program in an example:

Example 2.4.1.   The reduced vertex-edge incidence matrix, the flow constraints, and the distance vector for the network shown in Figure 2.5, with the edge directions

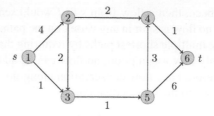

FIGURE 2.5
Example of a shortest path problem in a graph.

and distances as shown, are

$$A = \begin{pmatrix} 1 & 1 & 0 & 0 & 0 & 0 & 0 & 0 \\ -1 & 0 & 1 & 1 & 0 & 0 & 0 & 0 \\ 0 & -1 & -1 & 0 & 1 & 0 & 0 & 0 \\ 0 & 0 & 0 & -1 & 0 & 1 & -1 & 0 \\ 0 & 0 & 0 & 0 & -1 & 0 & 1 & 1 \end{pmatrix}, \quad \mathbf{b} = \begin{pmatrix} 1 \\ 0 \\ 0 \\ 0 \\ 0 \end{pmatrix},$$

$$\mathbf{d}^{\mathrm{T}} = (4, 1, 2, 2, 1, 1, 3, 6).$$

The first column has positive one for the origin (s) of the edge and a minus one for its terminal at vertex 2, and similarly for the rest of the columns. The last constraint, that is, the last row in the vertex-edge incidence matrix, has been removed. Using the simplex method to solve this linear program, gives the solution

$$\mathbf{f}^{\mathrm{T}} = (0, 1, 0, 0, 1, 1, 1, 0),$$

with optimum value $\mathbf{d}^{\mathrm{T}} \cdot \mathbf{f} = 6$. The result is shown in Figure 2.6 to the left. Note that reversing the direction of edge $(4, 5)$ gives a different shortest path, now with a cost of 7. This case is depicted on the right-hand side in Figure 2.6.

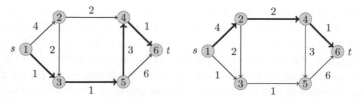

FIGURE 2.6
The shortest path problem formulation is dependent on the edge directions. The linear program solution in Example 2.4.1 has a minimum value of 6 (left), but on reversing the edge $(4, 5)$ the shortest path has a value of 7 (right).

By introducing vertex prices, we can formulate the dual linear program of (2.14),

$$\max \quad \{\pi_s - \pi_t\}, \tag{2.16}$$
$$\pi_i - \pi_j \le d_{ij} \quad \forall (i, j) \in E.$$

Instead of using flows as variables, we use the prices associated with the vertices. In the dual program, we optimize an n-vector π of prices rather than the m-vector $\mathbf{f}$ of flows. From (2.16), we can see that we have to use the transpose of the vertex-edge incidence matrix. Equivalent to deleting the last Equation in (2.14), in the dual we let $\pi_t = 0$.

In matrix form, the dual problem can be expressed as

$$\max_{\pi} \quad \mathbf{g}^{\mathsf{T}} \pi$$
$$J^{\mathsf{T}} \pi \le \mathbf{d} \tag{2.17}$$
$$\pi_k \ne 0, \quad 0 \le k \le t - 1$$
$$\pi_t = 0.$$

Here, we seek the value of π_s, so we set $\mathbf{g} = (1, 0, \ldots, 0)$ in the objective function. It is obvious from Figure 2.7 that we cannot find prices π_i such that $\pi_i - \pi_j = d_{ij}$ for all (i, j), because there are only $n = 6$ prices but $m = 8$ edge distances. In other words, the corresponding system of equations is overdetermined. It is easily seen, however, that prices can be found so that their differences satisfy the inequality constraints in (2.17). By maximizing the prices, we arrive at the same value as in the primal program, that is, the cost of the shortest path equals the price associated with the source s. Formally, this can be stated as follows:

Theorem 2.4.1 (Duality Theorem of Linear Programming). If either (2.14) or (2.17) has a finite optimal solution, so does the other, and the corresponding values of the objective functions are equal. If either problem has an unbounded objective, the other problem has no feasible solution.

Example 2.4.2. Using (2.17) and solving the dual problem with the simplex method, we obtain the solution $\pi = (6, 2, 5, 1, 4, 0)$, where the positions in the vector correspond to the vertices. It is easily seen that the difference in prices are less than or equal to the distances in $\mathbf{d}$.

The problem can also be solved by iteratively assigning the largest possible vertex prices so that the distance constraints are fulfilled. This is illustrated in Figure 2.7. Initially, let $\pi_t = 0$ and the set of 'saturated' edges S. The set S thus contains edges for which the price differences match the distance constraints with equality. Working backwards through the network starting with the shortest edge to any neighbor, we successively increase the prices. In the first iteration, we obtain $\pi_4 = 1$ since $\pi_4 - \pi_t = 1 = d_{4t}$. This price is propagated to all vertices that have not yet been visited, which guarantees that all price differences between such vertices remain zero. The edge e_{4t} is now added to S. In the next iteration, vertex 2 can be assigned the price 3, and e_{24} is added to S. Continuing in this way, the final price for the source s is 6 and the shortest path is the subset of edges in S that connects s with t.

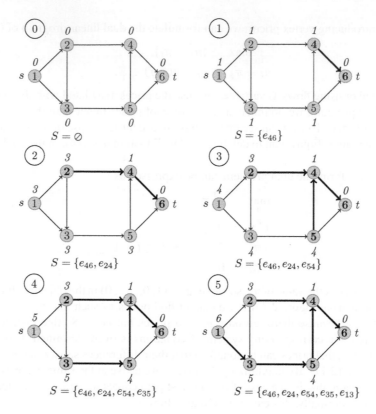

FIGURE 2.7
The dual problem can be solved by iteratively increasing the prices associated with the vertices so that the distance constraints are satisfied.

Note that the price for vertex 2 can be chosen so that $2 \leq \pi_2 \leq 3$, where the lower bound gives that the price difference with s matches the distance d_{12}, and the upper bound matches the distance d_{24}.

The primal-dual approach turns out to very useful, also for more complex problems. By using vertex prices—the dual variables—the number of variables and the number of constraints in the problem are interchanged. The approach often leads to a sequence of simpler subproblems that can be solved iteratively. In the case of shortest path, the minimization problem is broken down into $n - 1$ maximization problems with fewer variables involved.

2.4.2 Dijkstra's algorithm

One of the most widely used algorithms for shortest paths in applications is Dijkstra's algorithm. It works on undirected graphs but the weights (distances) need to be non-negative for the algorithm to work. Note that even if we only want the shortest path between the vertices s and t, there is no better way (in the worst case) than to determine the shortest paths from s to all other vertices.

Starting from a source s, the algorithm scans its neighbors to find the distances to them. Next, the neighbor that is closest is chosen and is scanned for its neighbors' distances in turn. It continues selecting scanned neighbors at increasing distances, and progresses throughout the network until the destination t has been reached.

The manner in which it progresses throughout the network is referred to as the greedy principle, in that it always choses the closest vertex for subsequent scanning.

The algorithm works by using a distance 'best estimate' $l(j)$ from the source s to every other vertex j in the network, initially set to ∞. The estimate $l(j)$ is successively updated as the algorithm progresses. Let $d(s, j)$ be a function representing the actual distance from s to some vertex j. The algorithm keeps track of vertices to which the actual distance has been determined in a set S of visited vertices. In determining the actual distance to a yet unvisited vertex i neighboring to S, the relation

$$d(s, j) = \min_{i \in S}(d(s, i) + d_{ij}) \tag{2.18}$$

is used, where d_{jx} is the direct distance from j to x. This means that, knowing the actual distances from s to all vertices $j \in S$, the actual distance to a new vertex x is the smallest sum of the distance from s to any vertex j in S and the direct distance from j to x. When scanning for neighbors from S, the algorithm chooses the closest unvisited neighbor to S in each step. Therefore, the relation (2.18) is guaranteed to work; there cannot be a shorter way to reach x.

Algorithm 2.4.2 (Dijkstra's Algorithm).   Given an (undirected) graph $G = (V, E)$, non-negative edge costs $c(\cdot)$, and a starting vertex $s \in V$.

    STEP 0:
        Set $S := \{s\}, l(s) := 0, l(i) = \infty, i \in V, i \neq s$

    STEP 1 to $|V| - 1$:
        while $S \neq V$ **do**
        find x such that $l(x) = \min\{l(y) : y \neq S\}$
        set $S := S \cup \{x\}$
        for all $y \in V - S$ **do**
        $l(y) := \min\{l(y), l(x) + d_{xy}\}$
        end (while)

Output the shortest distances from s to all $i \in V$ in the vector $\mathbf{l}$.

Algorithm 2.4.2 shows the steps in Dijkstra's algorithm. In its basic form, the algorithm does not return the shortest path, only its distance. It is, however, easy to reconstruct the path by adding the preceding vertex to the estimated distance label when it is updated via (2.18). Every time a vertex is added to S, the edge corresponding to the vertex and its preceding vertex is recorded as part of the path. When the destination t has been found and processed, we can backtrack to s following the recorded vertices to find the actual path.

We can estimate the running time of the algorithm as follows. Each iteration requires processing of vertices that have not yet been visited, which is, at most, $|V|$. Since there are $|V|$ iterations (including the initialization), the algorithm has a running time of $O(|V|^2)$. This running time refers to a 'standard' implementation. By using clever data structures, faster implementations can be achieved.

Example 2.4.3. The application of Dijkstra's algorithm to the same problem as in Example 2.4.1 is summarized in Figure 2.8. Initially, the set S consists of the source s only and all labels are set to ∞. The two neighbors of s, vertices 2 and 3, are labeled with their distances from s. The closest vertex 3 is then added to S. Reaching vertex 2 from vertex 3 gives a total distance of 4, so the previous label of vertex 2 remains unchanged. Then, vertex 2 is added to S. The new closest neighbor of S, vertex 5, can be reached from s via vertex 3 at a distance of 2. The only unvisited neighbor of S is vertex 4 at distance 5 from s. In the following iteration, vertex 5 is added to S. Its neighbors, 4 and t, can now be labeled with the estimated distances 4 and 7,

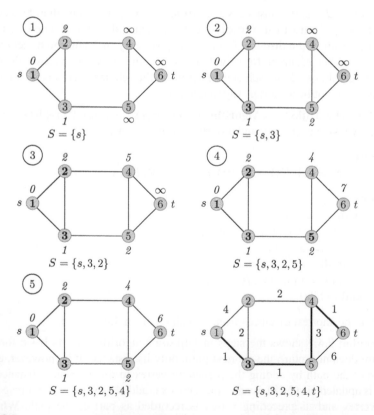

FIGURE 2.8

Dijkstra's algorithm applied to the shortest path problem.

respectively. Next, vertex 4 is added to S, and its neighbor t can be reached at a total distance of 6 from s. The sink t can now be added to S, and the algorithm terminates.

Note that in the third iteration, we have a tie between the distance to vertices 2 and 5. It does not matter which of the two vertices we visit; the final result is the same. Unlike the solution to the linear program in Example 2.4.1, Dijkstra's algorithm does not produce the actual shortest path. In order to find this, we can add the previous vertex to the label and backtrack from t once the algorithm terminates.

Dijkstra's algorithm can be classified in different ways, depending on which of its features we are looking at. The successive improvement of distance estimates suggests that it is a primal-dual algorithm. The way neighbors are selected for scanning makes it a greedy algorithm as well. Finally, the equation used in the updating process is closely related to dynamic programming, which is presented in the next section. The protocols OSPF and IS-IS both use Dijkstra's algorithm for shortest path calculations.

2.4.3 The Bellman-Ford algorithm

Another algorithm for finding shortest paths is the Bellman-Ford algorithm. It is similar to Dijkstra's algorithm in that it computes successive refinements of the vertex distances, but contrary to Dijkstra's algorithm, it processes all vertices in each iteration. It is slower than Dijkstra's algorithm, but has the advantage of being capable of handling negative edge weights.

The algorithm is based on the following principle, called Bellman's equations:

$$v_s = 0,$$
$$v_j = \min_i\{v_i + d_{ij}\}, \quad j \neq s, \tag{2.19}$$

where d_{ij} is the distance (or, more generally, the weight) of the edge (i, j). For a source s, the distance estimates v_s are initially set to zero and all other estimates are set to ∞. The estimates are successively improved by iterating (2.19) until no further improvement is achieved. It turns out that in the worst case $|V| - 1$ iterations are sufficient to reach the final result. Equation (2.19) is also called a dynamic programming equation. It is called the necessary optimality condition of dynamic programming.

This so-called principle of optimality can be summarized as follows. Suppose we know that the shortest path from s to t is known to pass through vertex i. Then the path from i to t must be the shortest path from i to t. The principle works 'backwards' in that, starting from vertex i closer to t, the shortest path from i to t is 'unaware' of how to get form s to i. Should it happen that the shortest path from s to t passes through i, then the $i - t$ shortest path must be a segment of the $s - t$ shortest path.

The algorithm uses successive approximation, starting from an initial approximation for v. After a sufficient number of iterations using (2.19), a point is reached where no further improvements can be made. Such a point is called a fixed point. As an initial approximation we take $v_s = 0$ and $v_i = \infty, i \in V/\{s\}$.

Bellman's equations (2.19) are nonlinear (because they involve minimization) and cannot be solved as a system of linear equations. They can, however, be solved iteratively. By successively refining the values in each iteration k taking the best value of the current and the new estimate, we arrive at the iterative equation

$$v_j^{k+1} = \min\{v_j^k, \min_{i \neq j}\{v_i^k + d_{ij}\}\}, \tag{2.20}$$

for all j and $k = 1, 2, \ldots, n - 1$.

Clearly, $v_j^k \leq v_j^{k-1}$ for all j and k. We can regard v_j^k as the distance of a shortest path from s to j using k or fewer edges. After a maximum number of $n - 1$ iterations, Equation (2.20) it is guaranteed to converge to the fixed point. If $v_j^{k+1} = v_j^k$ for all j and some $k < n - 1$, then these values do not change further, and since they solve Bellman's equations they are the values of the shortest paths.

In the same way as in Dijkstra's algorithm, we can record the immediate predecessor when updating the distances and, when reaching the fixed point, backtrack to get the actual path.

Even if the algorithm can handle negative distances, it will fail in the presence of a negative distance cycle (a cycle whose distances sum to a negative value).

The Bellman-Ford algorithm runs in $O(|V| \cdot |E|)$ time, or since $O(|E|) = O(|V|^2)$, in $O(|V|^3)$ time. This can be seen from (2.20), since we have n vertices, minimize over maximum $n - 1$ neighbors, and iterate a maximum of $n - 1$ times. Compared to Dijkstra's algorithm it is therefore slower, but being capable of handling negative weights, more robust.

Example 2.4.4.   Using the Bellman-Ford algorithm on the same problem as in Examples 2.4.1 and 2.4.3, summarized in Table 2.2, we begin by letting $u_s^0 = 0$ and all $u_i^0 = \infty$ for $i \neq s$. In the first iteration, we have only vertex s with $u_s < \infty$, so $u_2^1 = \min\{\infty, 0 + 2\} = 2$ and $u_3^1 = \min\{\infty, 0 + 1\} = 1$. For all other vertices, the distances are trivially $u_i = \min\{\infty, \min_{k \neq j}\{\infty + d_{ki}\}\}$. The second iteration gives $u_2^2 = \min\{2, 2\}$ with the second argument equal to $2 = \min\{2, 1 + 3, \infty\}$ for $k = s$ ($k = 3$ gives $1 + 3 = 4$ and all other vertices give ∞). Continuing with vertices 4 and 5 gives the distances 5 and 2, respectively. Continued iteration successively improves the distance estimations, until no improvement can be made (after iteration 5).

The fixed point is reached after four iterations, equaling the number of nodes traversed to reach t from s: $\{s, 3, 5, 4, t\}$. In the worst case, the algorithm converges after up to $n - 1$ iterations. The first vertex s is the starting point, so a maximum of $n - 1$ vertices can be traversed by the shortest path.

The Bellman-Ford algorithm is used for example in the Routing Information Protocol (RIP) in autonomous systems (subsystems like IP networks owned by an operator). Each node in the network calculates the distances to all neighboring nodes within the autonomous system and sends its tables to the other nodes. When a node receives a table from a neighbor, it recalculates its distance table based on the information received.

Table 2.2 Three iterations of the Bellman-Ford algorithm for the shortest path problem. After the third iteration, no further improvements are achieved. The length of the shortest path is the value of u_6^4

i	u_i^0	u_j^0	d_{ij}	u_i^1
1	0	−	−	0
2	∞	0	2	2
3	∞	0	1	1
4	∞	∞	−	∞
5	∞	∞	−	∞
6	∞	∞	−	∞

i	u_i^1	u_j^1	d_{ij}	u_i^2
1	0	−	−	0
2	2	0	2	2
3	1	0	1	1
4	∞	2	3	5
5	∞	1	1	2
6	∞	∞	−	∞

i	u_i^2	u_j^2	d_{ij}	u_i^3
1	0	−	−	0
2	2	0	2	2
3	1	0	1	1
4	5	2	2	4
5	2	1	1	2
6	∞	5	2	7

i	u_i^3	u_j^3	d_{ij}	u_i^4
1	0	−	−	0
2	2	0	2	2
3	1	0	1	1
4	4	2	2	4
5	2	1	1	2
6	7	4	2	6

2.4.4 The Floyd-Warshall algorithm

The Floyd-Warshall algorithm is an efficient method to find the all-pairs shortest paths, that is, the shortest paths between all vertices, at once. Let the distance between any two vertices i and j be denoted by $d(i, j)$ and the edge distances (weights) be d_{kl} for $(k, l) \in E$.

The algorithm uses an $n \times n$ matrix $D = (d(i, j))$ of distances that is successively updated. The distances are updated using the triangle operation:

$$d(i, k) = \min\{d(i, k), d(i, j) + d(j, k)\}, \quad \text{for all } i, j, k = 1 \ldots, n; i, k \neq j. \tag{2.21}$$

The algorithm is based on the following theorem. The inductive proof of the theorem gives an illustration of how the algorithm works.

Theorem 2.4.3.   *If we perform a triangle operation for successive values $j = 1, 2, \ldots, n$, each entry $\mathrm{d}(i, k)$ becomes equal to the length of the shortest path from i to k, provided that the edge lengths are non-negative.*

Proof.   We use induction to show that after the triangle operation for j has been performed, $\mathrm{d}(i, k)$ is the distance of the shortest path from i to k with intermediate vertices $v \leq j$, for all i and k. In each step, fix j. For $j = 1$ (letting s be the first vertex), the shortest distances between neighbors of s are computed by (2.21), so the shortest of the direct edges and paths via s is chosen as distances.

Suppose the statement is true for $j - 1$ and consider the triangle operation for j. If the shortest path from i to k does not pass through j, $\mathrm{d}(i, k)$ remains unchanged and is optimal for intermediate vertices up to j. If, on the other hand, the shortest path from i to k passes through j, $\mathrm{d}(i, k)$ is replaced by $\mathrm{d}(i, j) + \mathrm{d}(j, k)$, which is shorter. Since both $\mathrm{d}(i, j)$ and $\mathrm{d}(j, k)$ are assumed to be optimal up to $j - 1$, the resulting distance must be optimal up to j.   □

The algorithm requires a total of $n(n - 1)^2$ comparisons, so its running time is $O(|V|^3)$, which is slower than Dijkstra's algorithm.

We can keep track of the shortest paths by constructing another $n \times n$ matrix, say $H = (h_{ij})$, where we let h_{ij} be the highest numbered intermediate node on the shortest path from i to j, if there is one, and zero otherwise. This matrix is updated during each iteration j. Whenever a better distance estimate $\mathrm{d}(i, k)$ is found, H is updated with $h_{ik} = j$. More formally, let all $h_{ij} = 0$, initially, and when performing the triangle operations set,

$$h_{ij} = \begin{cases} j & \text{if } d_{ik} > d_{ij} + d_{jk}. \\ h_{ij} & \text{otherwise.} \end{cases} \tag{2.22}$$

The shortest path from i to j can then be reconstructed easily from the final H matrix. By looking up an (i, j) entry, we either find the highest numbered intermediary vertex, say k, or zero if a direct edge exists. The (i, k) entry gives the next vertex 'backwards' in the path, and so on.

Algorithm 2.4.4 (The Floyd-Warshall Algorithm).   Given an $n \times n$ matrix $\{d_{ij}\}$ with $d_{ij} \geq 0$.

STEP 0:

> **for all** $i \neq j$ **do** $\mathrm{d}(i, j) := d_{ij}$ when i and j are neighbors and $\mathrm{d}(i, j) = \infty$ otherwise
> **for** $i = 1, \ldots, n$ **do** $\mathrm{d}(i, i) := \infty$

STEP 1 to $|V| - 1$:

> **for** $j = 1, \ldots, n$ **do**
> **for** $i = 1, \ldots, n, i \neq j$ **do**
> **for** $k = 1, \ldots, n, k \neq j$ **do**
> $\mathrm{d}(i, j) := \min\{\mathrm{d}(i, j), \mathrm{d}(i, k) + \mathrm{d}(k, j)\}$

Table 2.3   Initial values for the Floyd-Warshall algorithm in Example 2.4.5

s	2	3	4	5	t
∞	4	1	∞	∞	∞
4	∞	2	2	∞	∞
1	2	∞	∞	1	∞
∞	2	∞	∞	3	1
∞	∞	1	3	∞	6
∞	∞	∞	1	6	∞

Table 2.4   All-pairs shortest paths in Example 2.4.5

s	2	3	4	5	t
2	4	1	5	2	6
4	4	2	2	3	3
1	2	2	4	1	5
5	2	4	2	3	1
2	3	1	3	2	4
6	3	5	1	4	2

Output the $n \times n$ matrix $D = (\mathrm{d}(i, j))$ where $\mathrm{d}(i, j)$ is the shortest distance from i to all j under $\{d_{ij}\}$.

Example 2.4.5.   Consider the all-pairs shortest path problem for the network in Figure 2.5. Initially, D has the entries given in Table 2.3.

Now, fix $j = s$. The row and column corresponding to s will not be changed, but all entries are modified by applying Equation (2.21). Note that also $\mathrm{d}(i, i)$ is updated; in this iteration taking the path via s. For example, the distance of $2 - s - 2$ is 8 in this step. After all n vertices have been processed in this way, we have the all-pairs shortest paths as given in Table 2.4. The $s - t$ entry gives the distance of the shortest $s - t$ path.

2.5 Maximum flows

The maximum flow problem is to find the maximum possible flow between a source s and a sink t in a capacitated network. In contrast to the shortest path problem in which the edge weights can be thought of as distances or costs, the edge weights in the maximum flow problem are capacity limits which cannot be exceeded by the flow. The resulting maximum flow solution is referred to as a single commodity flow, as opposed to multi-commodity flows where there are flows between more than one source-sink pair. The more general multi-commodity flow problem is discussed in Chapter 3.

First, the maximum flow problem is formulated as a linear program. The second approach is the *labeling algorithm* by Ford and Fulkerson. Finally, an analogy between maximum flows and electric currents is shown to lead to an approximate method with interesting properties.

2.5.1 Maximum flow as a linear program

Consider a graph $G = (V, E)$ with a weight function $w : E \to \mathbb{R}$ defined for each edge. This number will be referred to as the capacity of the edge. The capacity is thus the maximum amount of flow that a particular edge can carry. We denote this quantity by u_{ij} for an edge (i, j).

The maximum flow problem amounts to finding the maximum possible flow for a single commodity between a source node s, and a sink node t in a network with a given topology $G = (V, E)$.

The maximum flow problem can be stated as follows. Given a directed network $G = (V, E)$, a source vertex s and a sink vertex t, and positive (integer) numbers u_{ij} representing the capacity of each edge $(i, j) \in E$, maximize the total flow from s to t.

Let f_{ij} be the flow on edge $(i, j) \in E$ and $\sum f_{ij}$ be the total flow across the network. Then we have the linear program

$$\max \sum_{(i,j) \in E} f_{ij},$$

$$\sum_j (f_{ij} - f_{ji}) = \begin{cases} \sum_j f_{ij} & i = s, \\ 0 & i \neq s, t, \\ -\sum_j f_{ij} & j = t, \end{cases} \quad (2.23)$$

$$0 \leq f_{ij} \leq u_{ij} \quad \forall (i, j) \in E.$$

This linear program (slightly modified) can in principle be solved by the simplex method. However, there are more efficient methods for obtaining the maximum flow.

Example 2.5.1. Consider a network $G = (V, E)$ where the edges have a capacity constraint u_{ij}. In Figure 2.9 the capacity limits are presented in brackets. Let $s = 1$ and $t = 7$.

Note that the capacity constraints are m-dimensional and the flow constraints are n-dimensional. The solution—the flow—is of course an m-dimensional vector.

In order to formulate the problem as a proper linear program, the redundant equations have to be removed. We need to delete the two equations which contain the value of the objective function from the system. The set of conservation equations forces the net flow into t to be $q_t = -q_s$, so this equation can be omitted. We can also omit the equation for s, since the flow is the quantity we wish to maximize in the objective function. Thus, we have the following formulation:

$$\max_{\mathbf{f}} \quad \mathbf{x}^{\mathsf{T}} \cdot \mathbf{f},$$

$$0 \leq I_m \mathbf{f} \leq \mathbf{u},$$

$$J_{n-2} \mathbf{f} = \mathbf{0},$$

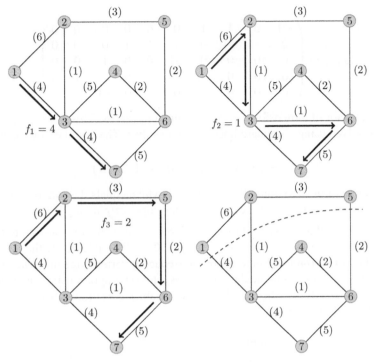

FIGURE 2.9

A maximum flow problem: the numbers in brackets by the edges show link capacities. The dashed line shows the minimum cut.

where

$$\mathbf{x}^T = \begin{pmatrix} 0 & 0 & 0 & 0 & 0 & 0 & 1 & 0 & 0 & 1 \end{pmatrix},$$

$$I_m = \begin{pmatrix} 1 & 0 & 0 & 0 & 0 & 0 & 0 & 0 & 0 & 0 \\ 0 & 1 & 0 & 0 & 0 & 0 & 0 & 0 & 0 & 0 \\ 0 & 0 & 1 & 0 & 0 & 0 & 0 & 0 & 0 & 0 \\ 0 & 0 & 0 & 1 & 0 & 0 & 0 & 0 & 0 & 0 \\ 0 & 0 & 0 & 0 & 1 & 0 & 0 & 0 & 0 & 0 \\ 0 & 0 & 0 & 0 & 0 & 1 & 0 & 0 & 0 & 0 \\ 0 & 0 & 0 & 0 & 0 & 0 & 1 & 0 & 0 & 0 \\ 0 & 0 & 0 & 0 & 0 & 0 & 0 & 1 & 0 & 0 \\ 0 & 0 & 0 & 0 & 0 & 0 & 0 & 0 & 1 & 0 \\ 0 & 0 & 0 & 0 & 0 & 0 & 0 & 0 & 0 & 1 \end{pmatrix},$$

$$\mathbf{u}^T = \begin{pmatrix} 6 & 4 & 1 & 3 & 5 & 1 & 4 & 2 & 2 & 5 \end{pmatrix},$$

$$J_{n-2} = \begin{pmatrix} -1 & 0 & 1 & 1 & 0 & 0 & 0 & 0 & 0 & 0 \\ 0 & -1 & -1 & 0 & 1 & 1 & 1 & 0 & 0 & 0 \\ 0 & 0 & 0 & 0 & -1 & 0 & 0 & 1 & 0 & 0 \\ 0 & 0 & 0 & -1 & 0 & 0 & 0 & 0 & 1 & 0 \\ 0 & 0 & 0 & 0 & 0 & -1 & 0 & -1 & -1 & 1 \end{pmatrix},$$

and $\mathbf{0}$ is a vector of dimension $n - 2$ containing only zeros. This linear program can easily be solved using a standard linear program solver. The optimal flow vector thus obtained is

$$\mathbf{f}^T = \begin{pmatrix} 3 & 4 & 1 & 2 & 1 & 0 & 4 & 1 & 2 & 3 \end{pmatrix}.$$

The value of the maximum flow is $|f| = 7$, which equals the sum of the flow on the edges $(3, t)$ and $(6, t)$ (or the corresponding edges originating from s).

The problem thus consists of two sets of constraints, capacity constraints $I_m = \mathbf{u}$ and the flow conservation constraints $J_{n-2} = \mathbf{0}$. In Example 2.5.1, the maximum flow problem is formulated in terms of the (reduced) vertex-edge incidence matrix. It is also possible to formulate the problem as a linear program using the edge-path incidence matrix. To see this, recall that the variables constitute a path flow vector. We must therefore have

$$\begin{aligned} \min \quad & \mathbf{e}^T \cdot \mathbf{f}, \\ & K\mathbf{f} \leq \mathbf{d}, \\ & f_k \geq 0, \quad k = 1, 2, \ldots, p, \end{aligned} \tag{2.24}$$

where p is the number of paths. In the objective function, $\mathbf{e} = (1, 1, \ldots, 1)$, so we are minimizing the sum of all path flows.

Example 2.5.2. Equation (2.24) can be solved using the simplex method. The matrix K is the same as (2.9) in Example 2.2.3. The resulting vector of path flows is $\mathbf{f}_p^T = \{3, 1, 1, 0, 2, 0, 0\}$. Table 2.1 lists the paths in the graph, and by making a mapping from the path flows to edge flows, we arrive at the edge flows $\mathbf{f}_e^T = \{3, 4, 1, 2, 0, 1, 4, 0, 2, 3\}$. We obtain this mapping by summation of the contributions by each path flow to individual edges. The only difference between the solution obtained in Example 2.5.1 and this one is the path of f_2 in Figure 2.9. The former solution has the path $p_2 = \{1, 2, 3, 4, 6, 7\}$, whereas this solution gives $p_2 = \{1, 2, 3, 6, 7\}$. The value of the flow is of course the same. This example shows that a maximum flow need not be unique.

2.5.2 The Ford-Fulkerson labeling algorithm

The Ford-Fulkerson algorithm is an efficient primal-dual algorithm for finding single commodity maximum flows. We assume integer capacities on the edges in order to have the algorithm converge in a finite number of steps. Note that any rational number can be multiplied by its denominator to yield an integer, so in practical situations, this assumption is not a serious restriction.

A maximum flow must satisfy both the capacity constraints and the flow conservation constraints. The algorithm starts with a feasible flow, that is, a flow satisfying the capacity constraints, and searches for augmenting paths as defined in Section 2.1. Such a path must, by construction, satisfy the flow conservation constraints. The augmenting paths (and hence, the increase in flow) are constructed so that the flow never violates the capacity constraints.

The algorithm is based on Theorem 2.1.1 and can be described as follows. Suppose we have a network with edge capacities, a source s, and a sink t. Let the initial flow be the zero flow, which must be feasible. Starting from s, we perform a breadth-first search, recording the order the vertices are visited and labeling the visited vertices with the maximum capacity available along a given path. When t has been reached, an augmenting path has been found. We increase the flow on this path, form the residual graph and repeat the procedure.

Algorithm 2.5.1 (The Ford-Fulkerson Algorithm).   Given an undirected graph $G = (V, E)$, positive integer edge capacities u_{ij}, a source $s \in V$, and a sink $t \in V$.

STEP 0:
>   Let $\mathbf{f} = (0, 0, \ldots, 0)$ be an m-dimensional flow vector, $f_{max} = 0$,
>   set vertex labels $l(s) := 0, l(i) = \infty, i \in V, i \neq s$, mark all vertices as UNVISITED

STEP 1 to N:
>   **while** G is connected **do**
>   insert s into Q and mark s as VISITED
>   **while** $Q \neq \varnothing$ **do**
> $v \leftarrow first(Q)$
>   **if** x is adjacent to an UNVISITED vertex y (BFS search)
>   set $l(y) := \min\{l(x), u_{xy}\}$
>   insert y into Q, mark y as VISITED
>   **if** $y = t$
>   augment f: $f_{max} := f_{max} + l(y)$; set $Q = \varnothing$
>   backtrack and set $f_{ij} := l(y)$ where the edge (i, j) is in the augmenting path
>   **else** remove x from Q
>   form residual graph with $u_{ij} := u_{ij} - f_{if}$, set $\mathbf{f} = \mathbf{0}$

Output the maximum flow f_{max}.

If the algorithm does not reach t from s, let S be the set of vertices s is connected to, and let $\bar{S}$ be the complement of S. Then the flow f_{max} given by the algorithm is a maximum flow and $(S, \bar{S})$ is a minimum cut. The maximum flow and the capacity of the minimum cut are in fact equal. This result, referred to as the max-flow min-cut theorem, was stated by Ford and Fulkerson (see Table 2.5).

Theorem 2.5.2.   *The maximum value of a flow from s to t in a network is equal to the minimum capacity of any cut separating s and t.*

Table 2.5   The ford-fulkerson algorithm for maximum flows

Q	Scan	Arr.	Label	Dep.
s	s	2	6	–
s2	s	3	4	–
s23	s	–	–	s
23	2	5	3	–
235	2	–	–	2
35	3	4	4	–
354	3	6	1	–
3546	3	7	4	–
s	s	2	6	–
s2	s	–	–	s
2	2	3	1	–
23	2	5	3	–
235	2	–	–	2
35	3	4	1	–
354	3	6	1	–
3546	3	–	–	3
546	5	–	–	5
46	4	–	–	4
6	6	7	1	–
s	s	2	5	–
s2	s	–	–	s
2	2	5	3	–
25	2	–	–	2
5	5	6	2	–
56	5	–	–	5
6	6	7	2	–
s	s	2	3	–
s2	s	–	–	s
2	2	5	1	–
25	2	–	–	2
5	5	–	–	5

Example 2.5.3.   In the network shown in Figure 2.9, let $s = 1$ and $t = 7$. The tables show the progress of the Ford-Fulkerson algorithm. We form a queue (denoted by Q in the tables) of vertices that have been visited. The first vertex in the queue is the one being scanned, and the vertices found by the breadth-first search are vertices in the arrival (Arr.) column. The label is the smallest available capacity along a path. The available capacity is thus dependent on the capacity label of the vertex that is being scanned. When there are no more reachable neighbors, the vertex is put in the departure (Dep.) column, and the algorithm moves on to the next vertex in the queue. Backtracking is done by going from the current vertex in the arrival column to the

vertex that has been scanned, then finding the scanned vertex in the Arr. column, and tracking back to the vertex that was scanned, and so on.

We find three paths, $p_1 = \{1, 3, 7\}$ with capacity 4; $p_2 = \{1, 2, 3, 6, 7\}$ with capacity 1; and $p_3 = \{1, 2, 5, 6, 7\}$ with capacity 2. The total flow is $f_1 + f_2 + f_3 = 7$. Furthermore, with this flow, the connected subgraph is $S = \{1, 2, 5\}$. Its complement is $\bar{S} = \{3, 4, 6, 7\}$, and the minimum cut is therefore $(S, \bar{S})$, which also have a capacity of 7. The minimum cut is shown in Figure 2.9.

2.5.3 Approximate maximum flow

Approximate methods are mainly used for two reasons: either no exact method exists for the problem in question, or the approximation algorithm is substantially faster than an exact algorithm. If the speed of computation is of importance, this can often be traded for a slight inaccuracy in the result. This is typically the case when the calculations have to be repeated a large number of times on different instances.

We illustrate an approximate solution method for the maximum flow problem, presented in Christiano et al. (2010). The approximation is based on the theory for electrical networks. If we let the edges be assigned resistors in an electrical network and send a current through the network, the electrical flow will represent the maximum flow in our original problem. The electrical flow satisfies the flow conservation constraints, but the capacity constraints may not be satisfied for the resistances assigned to the edges. Therefore, the procedure successively adjusts the resistances so that the capacity constraints are satisfied too. After a sufficient number of iterations, we have an electrical flow in the network which approximates the maximum flow.

The essence of the approximation algorithm is to reformulate the optimization problem as a sequence of systems of linear equations. The method presented here is a simplified version of the algorithm proposed in Christiano et al. (2010). The algorithm described is built on techniques that render a fast algorithm. Here, we will only discuss the analogy to electrical circuits and formulate the approximation, omitting the computational aspects. The procedure gives a $(1 - \varepsilon)$ approximation to the maximum flow, so that, if $f_{\max}$ is the value of the true maximum flow and f is the maximum flow given by the algorithm, then $f \geq (1 - \varepsilon) f_{\max}$.

Let $G = (V, E)$ be an undirected graph with integer edge capacities u_{ij} for $(i, j) \in E$, a source s, and a sink t. It is convenient to index with respect to edges with a subscript u_e rather than u_{ij} when summing over edges in the expressions below.

Next, we give an orientation to each edge. This is completely arbitrary. The orientation of the edges is chosen in accordance with the direction of the fictitious electrical flows, when possible. The resulting network is shown in Figure 2.10.

An $s - t$ flow f is feasible if $|f(e)| \leq u_e$ for each edge e, that is, when the amount of flow routed through any edge does not exceed its capacity. The maximum $s - t$ flow is the problem of finding a feasible $s - t$ flow in G of maximum value. We denote a maximum flow in G (with the given capacities) by f^*, and we denote its value by $|f^*|$. We say that f is a $(1 - \epsilon)$-approximation of the maximum flow if it is a feasible $s - t$ flow of value at least $(1 - \epsilon)|f^*|$.

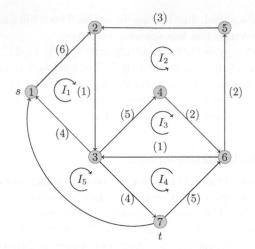

FIGURE 2.10

Fictitious currents in the maximum flow problem. The edges are labeled with capacity limits u_{ij} in brackets.

2.5.3.1 *Electrical flows*

Viewing the network as an electrical network, we can use Kirchhoff's Voltage Law and Ohm's Law to determine the electrical currents and potentials in the network. (An account of electrical circuit theory can be found in, for example Kuphaldt, 2006.) Let each edge have a resistance $R_{ij} > 0$ associated with it, denote by ϕ_i the electrical potential associated with each vertex (where the difference in potentials between two vertices equals the voltage across them) and I the current circulating in any closed loop in the network.

We introduce fictitious currents in the network. These currents can be directed arbitrarily, but we must note of the sign of the flows. Ohm's Law states that the difference in electrical potential between two nodes i and j is the product of the resistance R_{ij} and the current I_{ij}, so that $(\phi_i - \phi_j)/R_{ij} = I_{ij}$ is the current flowing through an edge (i, j).

By Kirchhoff's Current Law, all currents into and out from a node i, sum to zero, that is

$$\sum_{j:(i,j)\in E} I_{ij} = 0.$$

This is just the flow conservation constraint. Furthermore, Kirchhoff's Voltage Law states that the sum of electric potential differences in any closed loop is zero, that is

$$\sum_{(i,j)\in\Gamma} (\phi_i - \phi_j) = 0, \tag{2.25}$$

where Γ is a loop (or cycle). This law can be used to formulate a system of equations. First, we show an example of electrical circuit analysis.

Example 2.5.4.   Consider the maximum flow problem in Figure 2.10. Let the resistances on every edge (i, j) be $R_{ij} = 1$ and suppose a power source (that is, a battery) is connected to the source s and the sink t. Initially, let the power source have a unit voltage. We can then identify five loops and introduce a circulating current in each loop, as shown in the figure. Using Kirchhoff's Voltage Law (2.25) and the unit resistances, we arrive at the following system of equations,

$$I_1 + (I_1 + I_2) + (I_1 - I_5) = 0,$$
$$(I_1 + I_2) + 2(I_2 + I_3) + 2I_2 = 0,$$
$$2(I_2 + I_3) + (I_3 + I_4) = 0,$$
$$(I_3 + I_4) + (I_4 + I_5) + I_4 = 0,$$
$$(-I_1 + I_5) + (I_4 + I_5) - 1 = 0,$$

where we take current as positive in the direction indicated by the arrows in the figure. The last equation for the loop in which I_5 is circulating, also contains the power source, hence, the -1 in the last equation. Moving out the voltage term to the right-hand side, we can write the system in matrix form as

$$\begin{pmatrix} 3 & 1 & 0 & 0 & -1 \\ 1 & 5 & 2 & 0 & 0 \\ 0 & 2 & 3 & 1 & 0 \\ 0 & 0 & 1 & 3 & 1 \\ -1 & 0 & 0 & 1 & 2 \end{pmatrix} \begin{pmatrix} I_1 \\ I_2 \\ I_3 \\ I_4 \\ I_5 \end{pmatrix} = \begin{pmatrix} 0 \\ 0 \\ 0 \\ 0 \\ 1 \end{pmatrix}. \tag{2.26}$$

The system of equations can be solved using, for example, a standard mathematical package. The result is $\mathbf{I} = (0.333, -0.156, 0.222, -0.356, 0.844)$. The last component I_5 represents the current flowing into s and out from t. The current is less than unity, because we have a network of resistors whose total resistance is greater than unity. We know from Example 2.5.1 that the maximum flow is 7. Dividing this value by 0.844 should give us the voltage that has to be applied to the network in order to have a current of 7 (since Ohm's Law is linear). Adjusting the last component in the right-hand side vector in (2.26) to 8.29 (=7/0.844) gives the result $\mathbf{I} = (2.76, -1.29, 1.84, -2.95, 7.00)$. The edge flows are easily found by combining the currents as indicated in Figure 2.10. Thus, we have the flows (with unit resistances)

$$f_1 = I_1 = 2.76,$$
$$f_2 = -I_1 + I_5 = 4.24,$$
$$f_3 = I_1 + I_2 = 1.47,$$
$$f_4 = -I_2 = 1.29,$$
$$f_5 = I_2 + I_3 = 0.55,$$

$$f_6 = -I_3 - I_4 = 1.11,$$
$$f_7 = I_4 + I_5 = 4.05,$$
$$f_8 = I_2 + I_3 = 0.55,$$
$$f_9 = -I_2 = 1.29,$$
$$f_{10} = -I_4 = 2.95.$$

When solving the maximum flow problem, we usually do not know the result beforehand. Next we will present an algorithm that uses edge capacities and the related congestion rates as flow constraints.

Example 2.5.4 shows the main principles behind the approximation of a maximum flow by using electrical network analysis. This can be formulated as a system of linear equations. Let the m-dimensional vector of resistances be denoted by $\mathbf{R}$. For a vector of flows $\mathbf{f}$ from s to t, its energy with respect to $\mathbf{R}$ is defined as

$$\mathcal{E}_{\mathbf{R}}(\mathbf{f}) \equiv \sum_{ij} R_{ij} f_{ij}^2.$$

The electrical flow of value $|f|$ (with respect to $\mathbf{R}$) is the flow that minimizes $\mathcal{E}_{\mathbf{R}}(\mathbf{f})$ among all $s - t$ flows f having this value.

2.5.3.2 *The algorithm*

We have formulated the maximum $s - t$ flow problem as a linear program, but in view of Example 2.5.4 we may find an approximate solution to the same problem by solving a sequence of linear equations, provided we can assure that the capacity constraints are not violated.

We start from the vertex-edge incidence matrix J, the $n \times m$-matrix with rows representing the vertices and the columns representing the edges. We construct J from the arbitrary orientation of the edges. We assign a positive one for the origin and negative one for the terminal of each edge. (This may have different edge orientations than the representation in (2.8).)

Let the network flow be an m-dimensional vector $\mathbf{f} \in \mathbb{R}^m$ and the voltage applied to each node be denoted by $\mathbf{b}_{st}$. Then we have

$$J\mathbf{f} = \mathbf{b}_{st}.$$

Next, we define the Laplacian matrix as the $n \times n$-matrix L (with respect to the resistances R_{ij} assigned to the edges) as

$$L = JCJ^{\mathrm{T}},$$

where C (the conductance) is the $m \times m$ diagonal matrix with entries $C_{e,e} = 1/R_e$, $e = (i, j) \in E$, using the single parameter indexing of edges. Let R be the diagonal matrix with $R_{e,e} = R_e$. The energy of a flow $\mathbf{f}$ can then be written as

$$\mathcal{E}_R(\mathbf{f}) = \sum_e r_e f_e^2 = \mathbf{f}^T R\mathbf{f} = ||R^{1/2}\mathbf{f}||^2.$$

A unit electrical flow thus corresponds to the vector $\mathbf{f}$ that minimizes $||R^{1/2}\mathbf{f}||^2$ subject to $J\mathbf{f} = \mathbf{b}_{st}$. If $\mathbf{f}$ is an electrical flow, then by Ohm's Law it is a potential flow, which means that there is a vector $\boldsymbol{\phi} \in \mathbb{R}^V$, such that

$$f_{ij} = \frac{\phi_j - \phi_i}{R_{ij}}.$$

That is, in matrix form,

$$\mathbf{f} = CJ^T\boldsymbol{\phi} = R^{-1}J^T\boldsymbol{\phi}.$$

Using $J\mathbf{f} = \mathbf{b}_{st}$, we have $J\mathbf{f} = JCJ^T\boldsymbol{\phi} = L\boldsymbol{\phi} = \mathbf{b}_{st}$, and hence the electrical potentials of the vertices are given by

$$\boldsymbol{\phi} = L^\dagger\mathbf{b}_{st},$$

where $L^\dagger$ denotes the Moore-Penrose pseudoinverse of L. (The pseudo inverse is a generalization of the matrix inverse to matrices which are non-invertible.) Thus, the electrical flow $\mathbf{f}$ is given by the expression

$$\mathbf{f} = CJ^T L^\dagger \mathbf{b}_{st}. \tag{2.27}$$

The algorithm also needs to adjust the resistances so that the capacity constraints are satisfied. For this purpose, we define the congestion of an edge $e = (i, j)$ to be the ratio

$$\nu \equiv \frac{|f_e|}{u_e}$$

of the flow on an edge to its capacity. In particular, an $s - t$ flow is feasible if and only if $\nu \leq 1$ for all $e \in E$. In Christiano et al. (2010), two auxiliary parameters are used: the precision of the approximation $0 < \varepsilon \leq 0.5$ and a multiplicative factor ρ, such that $f_{ij} < \rho u_{ij}$ for all f_{ij}. The factor ρ is set to $\rho = 3\sqrt{m/\varepsilon}$. In Christiano et al. (2010), the authors show that their algorithm converges in N iterations, where

$$N = \frac{2\rho \ln m}{\varepsilon^2}. \tag{2.28}$$

This N is therefore the number of iterations after which the convergence of the algorithm is guaranteed. In practice, however, this may be a rather pessimistic estimate.

When a flow has been computed from Equation (2.27), we can calculate the congestion for each edge and use these figures to improve the estimate by adjusting the resistances. Flows causing high congestion are penalized by a higher resistance. The new resistances are calculated in two steps, by calculating edge weights and then by adjusting the resistances. The weights, reflecting the deviation from edge capacities as measured by the congestion, are calculated as

$$w_e^{(k)} = w_e^{(k-1)} \left(1 + \frac{\varepsilon}{\rho} \frac{|f_e|}{u_e}\right), \tag{2.29}$$

where k denotes the kth iteration. Finally, the resistances are updated according to

$$R_e^{(k)} = \frac{1}{u_e}\left(w_e^{(k-1)} + \frac{\varepsilon|\mathbf{w}^{(k)}|_1}{3m}\right), \tag{2.30}$$

where $|\mathbf{w}|_1 = \sum_{i=1}^n |w_i|$ is the taxicab norm. Initially, $\mathbf{w}^{(0)} = (1, 1, \ldots, 1)$ and $\mathbf{R}^{(0)} = (1, 1, \ldots, 1)$. The details of the choice of weights and resistances are beyond the scope of this text and the interested reader is referred to Christiano et al. (2010).

We can now summarize the algorithm based on the equations in this section. First, we select a number $0 < \varepsilon \le 0.5$, set $\rho = 3\sqrt{m/\varepsilon}$, assign unit resistances to each edge, and form the matrix $C = 1/R_e$. Then, we form vertex-edge incidence matrix J and the Laplacian $L = JCJ^T$ from C and J. The approximate network flow is $\mathbf{f} = CJ^T L^\dagger \mathbf{b}_{st}$, where $L^\dagger$ is the Moore-Penrose pseudoinverse of L.

Next, we compute the auxiliary edge weights according to Equations (2.29) and (2.30). Iterating this procedure gives successively better approximations to the maximum flow.

The algorithm described so far gives the edge flows $\mathbf{f}$, given the total $s - t$ flow in the load vector $\mathbf{b}_{st}$. However, this total flow is the quantity that we want to compute. A straight forward way of calculating the maximum flow is to adjust the load vector based on the maximum congestion in the network after the approximation has been carried out. Since a feasible flow must be such that $f_e \le u_e$, we can simply divide the non-zero loads in $\mathbf{b}$ by v_{max}, which we take as $v_{max} = \max_e\{|f_e|/u_e\}$. Using this principle and starting with a unit total flow gives an approximately correct result after a few iterations (as shown in Example 2.5.5).

Example 2.5.5. To formulate the problem, construct the vertex-edge incidence matrix from the fictitious currents induced in the network, as shown in Figure 2.10. This gives

$$J = \begin{pmatrix} 1 & -1 & 0 & 0 & 0 & 0 & 0 & 0 & 0 & 0 \\ -1 & 0 & 1 & -1 & 0 & 0 & 0 & 0 & 0 & 0 \\ 0 & 1 & -1 & 0 & 1 & -1 & 1 & 0 & 0 & 0 \\ 0 & 0 & 0 & 0 & -1 & 0 & 0 & 1 & 0 & 0 \\ 0 & 0 & 0 & 1 & 0 & 0 & 0 & 0 & -1 & 0 \\ 0 & 0 & 0 & 0 & 0 & 1 & 0 & -1 & 1 & -1 \\ 0 & 0 & 0 & 0 & 0 & 0 & -1 & 0 & 0 & 1 \end{pmatrix}.$$

We also have the vectors $\mathbf{w}^T = (1, 1, 1, 1, 1, 1, 1, 1, 1, 1)$, $\mathbf{b}_{st}^T = (1, 0, 0, 0, 0, 0, -1)$ and $\mathbf{u}^T = (6, 4, 1, 3, 5, 1, 4, 2, 2, 5)$. We use the single parameter edge index in the following expressions.

Now, we compute in order $|\mathbf{w}^{(k-1)}|_1$, $C_{e,e}^{(k)} = (1/u_e^2 \cdot w_e^{(k-1)} \varepsilon|\mathbf{w}^{(k-1)}|_1/3m)^{-1}$, $L^{(k)} = JC^{(k)}J^T$, $\mathbf{f}^{(k)} = C^{(k)}J^T(L^{(k)})^\dagger \mathbf{b}_{st}$, $\gamma_e^{(k)} = |f_e^{(k)}|/u_e$ and $w_e^{(k)} = w_e^{(k-1)} \cdot (1 + \gamma^{(k)}\varepsilon/\rho)$, starting from $k = 1$. This is the inner iteration.

With the $(1+\varepsilon)$-approximate flow determined by the inner iteration, we determine the maximum congestion γ_{max} and modify the load vector. Thus, in the outer iteration, the load vector $\mathbf{b}_{st}$ is set to $\mathbf{b}_{st} := \mathbf{b}_{st}/\gamma_{max}$, and the inner iteration is run again.

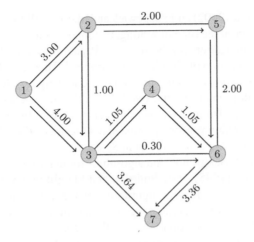

FIGURE 2.11

Approximate solution of the maximum flow problem. An effect of the approximation algorithm is that flows are distributed to lower the energy, whenever possible.

The signs of the flows are dependent on how we have defined their directions, but these are irrelevant since we are seeking the value of the maximum flow. By iterating the procedure, we get the result depicted in Figure 2.11.

It is worth noting that the algorithm performs load balancing as compared to the methods discussed earlier. This is due to the minimization of energy, which is dependent on the resistance, which in turn is a function of the congestion on the links. To summarize, the presented approximation algorithm relaxes the edge capacities, and fictitious resistances are used in place of the capacities, and these resistances are then successively modified to force the flow to satisfy the capacity constraints. An algorithm in a similar spirit is discussed for the multi-commodity flow problem of Section 3.3.

In the approximation algorithms, the electrical potentials play a similar role as the prices in the linear program formulation. It also illustrates the principle of successive approximation, which is very useful in many contexts.

2.6 **Summary**

Two of the most important mathematical disciplines in the context of network design are graph theory and probability theory. The present chapter introduces fundamental algorithms in graph theory that are useful in many contexts. By 'divide-and-conquer' or 'decomposition,' many more complex problems can be formulated in these simpler ones. In addition, several algorithmic approaches are illustrated, such as dynamic programming, the greedy principle, and the primal-dual optimization approach.

For an account of graph theory, see, for example Biggs (1985). The algorithms used are described in Skiena (1998) and Shaffer (1997), which also contain useful C code,

Goodrich and Tamassia (2002) and Sedgewick and Flajolet (1996). The original paper by Dijkstra (1959) can, at the time of writing, be found on the Internet. Fundamentals of combinatorial optimization can be found in Papadimitriou and Steiglitz (1982), from which the classification of algorithms is borrowed. Theoretical fundamentals on linear programming are provided in, for example, Luenberger (1989).

The shortest path problem is discussed from different angles in Papadimitriou and Steiglitz (1982) and from an engineering perspective in Bertsekas and Gallager (1987). Dijkstra's algorithm was first presented in Dijkstra (1959). Some analysis and extensions to this algorithm can be found in Sniedovich (2006).

General theoretical background and algorithms for maximum flows can be found in, for example, Biggs (1985), Papadimitriou and Steiglitz (1982) and Goldberg et al. (1990). One of the first efficient algorithms for maximum flows is the labeling algorithm (also known as the augmenting path algorithm) by Ford and Fulkerson. There exist more efficient algorithms, such as the algorithms by Edmonds and Karp (1972) and Dinitz (1970), and, more recently, the push-relabel algorithm by Goldberg and Tarjan (1986).

Advanced Flow Theory

This chapter discusses three problems which play crucial roles in network design: multi-terminal flows, minimum cost flows, and multi-commodity flows. The multi-terminal flow problem is a natural simplification of multi-commodity flows and can be seen as a step toward a structured method of network design. The minimum cost problem is used in network optimization problems where a network topology with both capacity limits and edge costs is given. Finally, multi-commodity flows allow for several commodities simultaneously—competing for the same resources. Efficient algorithms for these problems form the foundation of any network analysis method. In this chapter, we will gradually relax the constraints imposed previously to arrive at an algorithm for the multi-commodity flow problem, which is an idealized formulation of the routing problem in communication networks.

3.1 Multi-terminal flows

When there are several flow sources and destinations in an undirected network, we can define a multi-terminal flow—the maximum flows between all pairs of vertices in the network. Note, however, that each flow is considered in isolation, so this is still a case of single commodity flow. It turns out that each network with edges representing the flow requirements between multiple source-destination pairs can be represented by a special tree. This fact will prove useful in topological design, for example, when looking for capacity bottlenecks or when we wish to find the global minimum cut in a network. First, we introduce some characteristics of trees.

Definition 3.1.1. A tree T is a graph such that:

(1) T is connected; and
(2) there are no cycles in T.

Two immediate properties of a tree are that:

(1) for every pair of vertices i, j there is a unique path from i to j;
(2) $|E_T| = |V_T| - 1$, that is, the number of edges is one less than the number of vertices.

Given a graph $G = (V, E)$. If $T = (V_T, E_T)$ is a tree with $V_T = V$ and $E_T \subset E, T$ it is called a spanning tree of G. Thus, a spanning tree T consists of all vertices in G and a subset of edges in G, such that T satisfies the conditions in Definition 3.1.1. If the edges in G have weights $w_{ij}, (i, j) \in E$, we may form a minimum spanning tree, which is a spanning tree such that the total weight

$$w(T) = \sum_{(i,j) \in E_T} w_{ij} \tag{3.1}$$

is minimized. In contrast, a maximum spanning tree is a tree such that expression (3.1) is maximized. Algorithms for finding such trees are discussed in Chapter 4. For the present discussion, however, we will just note that such spanning trees exist and can be constructed efficiently.

Let $G = (V, E)$ be an undirected graph with upper bounds on flows u_{ij} and $f_{ij} = f_{ji}$ the maximal flows on edge (i, j) in G. Since the network is undirected, the function $f : (i, j) \mapsto \mathbb{R}^+$ is symmetric. For the sake of completeness, let $f_{ii} = \infty$ for all $i \in V$. We state a theorem on the necessary and sufficient conditions for a symmetric function f to be the maximum multi-terminal flow in a network. A flow function satisfying these conditions is said to be realizable. If a function f fails to be realizable, then it cannot be the maximum multi-terminal flow in the network.

Theorem 3.1.1. *A necessary and sufficient condition for a non-negative symmetric function f to be realizable is that*

$$f_{ik} \geq \min\{f_{ij}, f_{jk}\} \quad \text{for all } i, j, k \in V. \tag{3.2}$$

Further, if f is realizable, then it is realizable by a tree network.

Proof.   To show necessity, let $(S, \bar{S})$ be a minimal cut separating vertices i and k (with $i \in S$ and $k \in \bar{S}$) and having capacity $u(S, \bar{S})$. Hence, $f_{ik} = u(S, \bar{S})$. Now, vertex j can be either in S or in $\bar{S}$. In the first case, $f_{jk} \leq u(S, \bar{S})$ since the cut separates j and k, and in the second case, $f_{ij} \leq u(S, \bar{S})$ since this cut separates i and j. Using this relation recursively, we have

$$f_{iq} \geq \min\{f_{ij}, f_{jk}, f_{kl}, \ldots, f_{pq}\}. \tag{3.3}$$

In order to show sufficiency, we construct a network that achieves this flow. Using the given f_{ij} as edge weights in G, create a maximum spanning tree. Now consider the tree with f_{ij} interpreted as edge capacities. We show that this is the required network. Since this is a maximum spanning tree, we must have

$$f_{iq} \leq \min\{f_{ij}, f_{jk}, f_{kl}, \ldots, f_{pq}\},$$

where (i, q) is an edge outside the tree and the other edges are parts of the tree. But by assumption, Equation (3.3) holds for any edge (i, q). This implies the equality $f_{iq} = \min\{f_{ij}, f_{jk}, f_{kl}, \ldots, f_{pq}\}$, so the flow f_{iq} can be accommodated by the constructed tree, which proves sufficiency.   □

The proof also shows that there are no more than $n - 1$ distinct values for maximum flows, since the network is represented by a tree.

Example 3.1.1. In Example 2.5.3, it is shown that the maximum $(1, 7)$ flow is 7, and the minimum cut is $(S, \bar{S})$ with $S = \{1, 2, 5\}$ and $\bar{S} = \{3, 4, 6, 7\}$. Now, let $j = 5$, which is in S. We can immediately see that the maximum flow to or from vertex 5 cannot exceed the value 5, the sum of the capacities of the edges connecting the vertex to the rest of the network. With $i = 1$ and $k = 7$ in Equation (3.2), we therefore have $7 = f_{17} \geq \min\{f_{15}, f_{57}\} = \min\{5, 5\} = 5$. If $j = 2$, then $7 = f_{17} \geq \min\{f_{12}, f_{27}\} = \min\{9, 7\} = 7$. Thus, the conditions of Theorem 3.1.1 are satisfied.

3.1.1 The Gomory-Hu algorithm

Suppose we would like to find the maximum flows between all vertex pairs in a network. These can obviously be found by determining all $n(n-1)/2$ maximum flows. There is, however, a much more efficient algorithm requiring only $n - 1$ maximum flow calculations—known as the Gomory-Hu algorithm.

Two n-vertex networks are said to be flow equivalent if they have the same flow function f. By Theorem 3.1.1, every network is flow equivalent to a tree, called the Gomory-Hu tree. The Gomory-Hu algorithm identifies the maximum flows and the flow equivalent tree. Since we work with single commodity flows, these values are also the minimum cut capacities. Note that the algorithm gives the maximum flows, not the actual edge flows.

The algorithm builds on some observations which we now state. It computes the maximum flows between clusters of vertices formed by condensing them, that is, grouping them together.

Lemma 3.1.2. *Let $(S, \bar{S})$ be a minimal cut separating an origin s and destination t in G. Let $p, q \in S$. Then there is a minimum cut $(X, \bar{X})$ separating p and q in the original network with either $\bar{S} \subset X$ or $\bar{S} \subset \bar{X}$. Thus, condensing the vertices in $\bar{S}$ does not change the value of minimal cuts separating p and q and, hence, the maximal flow between them.*

Proof. Let $(Y, \bar{Y})$ be a minimal cut separating p and q. Since p and q are both in S, we may assume (without loss of generality) that $p \in S \cap Y$ and $q \in S \cap \bar{Y}$. Since s is in S, we may also assume (again without loss of generality) that $s \in S \cap Y$. Now there are two possibilities for t. Either $t \in \bar{S} \cap Y$, or $t \in \bar{S} \cap \bar{Y}$. Let $t \in \bar{S} \cap Y$ (the other case follows from symmetry).

Let $A = S \cap Y$, $B = \bar{S} \cap Y$, $\bar{A} = S \cap \bar{Y}$, and $\bar{B} = \bar{S} \cap \bar{Y}$. Since $(S, \bar{S})$ is a minimal cut separating s and t and $(A \cup \bar{A} \cup \bar{B}, B)$ is another cut separating s and t, we have

$$u(S, \bar{S}) = u(A, B) + u(\bar{A}, B) + u(A, \bar{B}) + u(\bar{A}, \bar{B})$$
$$\leq u(A \cup \bar{A} \cup \bar{B}, B) = u(A, B) + u(\bar{A}, B) + u(\bar{B}, B),$$

which implies

$$u(A, \bar{B}) + u(\bar{A}, \bar{B}) - u(\bar{B}, B) \leq 0. \qquad (3.4)$$

Similarly, since $(Y, \bar{Y})$ is a minimum cut separating p and q and $(A \cup B \cup \bar{B}, \bar{A})$ is another cut separating p and q, we have

$$u(Y, \bar{Y}) = u(A, \bar{A}) + u(A, \bar{B}) + u(B, \bar{A}) + u(B, \bar{B})$$
$$\leq u(A \cup B \cup \bar{B}) = u(A, \bar{A}) + u(B, \bar{A}) + u(\bar{B}, \bar{A}),$$

which implies

$$u(A, \bar{B}) + u(B, \bar{B}) - u(\bar{A}, \bar{B}) \leq 0. \tag{3.5}$$

Combining (3.4) and (3.5) yields $u(A, \bar{B}) \leq 0$, but since the capacities are positive, we must have $u(A, \bar{B}) = 0$. This implies that

$$u(\bar{A}, \bar{B}) \leq u(\bar{B}, B),$$

$$u(B, \bar{B}) \leq u(\bar{A}, \bar{B}),$$

so that $u(B, \bar{B}) = u(\bar{A}, \bar{B})$; and therefore

$$u(Y, \bar{Y}) = u(A, \bar{A}) + u(a, \bar{B}) + u(B, \bar{A}) + u(B, \bar{B}) = u(A \cup B \cup \bar{B}, \bar{A}).$$

This proves the lemma. The procedure used in this proof is called uncrossing.   □

This lemma gives us the basis for the Gomory-Hu algorithm. First, we select two vertices, say s and t, arbitrarily, and solve a maximal flow problem with s and t, as origin and destination, respectively. Let the minimal cut separating s and t be denoted by $(S, \bar{S})$ with $u(S, \bar{S}) = f_1$. Now choose two other vertices, say p and q, both of which are in one of these sets as the next origin-destination vertex pair. If both are in S, then condense $\bar{S}$ (else condense S) and solve the maximal flow problem in the condensed network between p and q. This gives a minimum cut $(X, \bar{X})$ separating p and q with S being entirely on one side, say $S \subset \bar{X}$, and the value of this cut is $u(X, \bar{X}) = f_2$. (If $S \subset X$, the roles of X and $\bar{X}$ would be exchanged.)

This process is repeated exactly $n - 1$ times, at the end of which each vertex of the tree will be a singleton. The resultant tree T is then flow equivalent to the original network.

Lemma 3.1.3. *The maximal flow f_{ij} between i and j in the original network is given by*

$$f_{ij} = \min\{f_{ik}, f_{kl}, \ldots, f_{pj}\},$$

where the edges $(i, k), (k, l), \ldots, (p, j)$ form the path in the tree T between i and j.

Proof.   Since each edge in the tree represents a cut separating the same set of vertices in the tree as in the original network, it follows that $f_{ij} \leq \min\{f_{ik}, f_{kl}, \ldots, f_{pj}\}$. To show the reverse inequality, we show that at each stage of the tree construction, where two condensed vertex sets are connected by an edge in the tree, there is at least one vertex in each set such that the maximum flow in the original network between them is equal to the flow value of the edge in the tree. By using induction, we note that in the first step of building the tree, this is true by construction. Suppose at some stage there are two sets, X and Y, in the tree connected by an edge with flow value f. By the induction hypothesis there are vertices $i \in X$ and $j \in Y$ such that $f_{ij} = f$. Let Y be split at the next stage into Y_1 and Y_2, and let X be connected to Y_1 in the new tree. Suppose that the split is the result of solving a maximum flow problem between s and t in Y with $s \in Y_1$ and $t \in Y_2$, and let the capacity of this new cut be f'. Clearly, the new edge thus created satisfies the hypothesis. However, we must show that the flow on the existing edge remains unaffected. If the edge value of the new edge is f', and the value of the old edge is f, we show that there exist nodes $k \in X$ and $l \in Y$ with $f_{kl} = f$. We have two cases to consider:

Case 1: If $l \in Y_1$, then $f_{kl} = f$ and since $k \in X$ and $l \in Y_1$ by assumption, and we are done.

Case 2: If $l \in Y_2$, then k and s are one side of the (s, t)-mincut and l and t are on the other side, and this cut has a capacity f'. Hence, from the lemma, when computing the maximum flow f_{ks}, we can condense all the nodes of Y_2 into a single node. Denoting the maximum flows in the original network by f_* (without bars) and those in the condensed network (Y_2 condensed) by $\bar{f}_*$ (with bars), we have

$$\bar{f}_{ks} = f_{ks},$$

$$\bar{f}_{kl} \geq f_{kl} = f,$$

$$\bar{f}_{lt} = \infty,$$

$$\bar{f}_{ts} \geq f_{ts} = f'.$$

Using the inequality of the lemma, we have

$$\bar{f}_{ks} \geq \min\{\bar{f}_{kl}, \bar{f}_{lt}, \bar{f}_{ts}\} = \min\{\bar{f}_{kl}, \bar{f}_{ts}\}.$$

Hence, $f_{ks} = \bar{f}_{ks} \geq \min\{f, f'\}$. There is an (s, t) minimum cut with a value f' that separates k and l and, hence, $f' \geq f_{kl} = f$. By the inequality, $f_{ks} \geq f$. But there is a cut that separates k and s whose value is f and hence $f_{ks} = f$, which is the desired result.   □

Algorithm 3.1.4 (Gomory-Hu Algorithm).   Given an undirected graph $G = (V, E)$ with edge capacities $u_{ij} > 0$.

STEP 0:   Set $V_T = V$ and $E_T = \emptyset$.

STEP 1:   Choose some $X \in V_T$ with $|X| \geq 2$ if such X exists. Otherwise, go to step 5.

STEP 2:   Let $T \backslash X$ be the components of T that are not in X; contract the components of $T \backslash X$ to a single node v (when $T \backslash X \neq \emptyset$). Form the graph $G' = (v \cup X, E_X \cup_{x \in X} (v, x))$. Then G' consists of v and the vertices in X, together with the edges within X and an edge between v and X with weight equal to the value of the minimum cut $(T \backslash X, X)$ (known from an earlier iteration).

STEP 3: Choose two vertices $s, t \in X$ and find a minimum $s - t$ cut $(S, \bar{S})$ in G'.
Then $S \cap X$ and $\bar{S} \cap X$ represent a split of X.

STEP 4: Set $V_T = (V_T \backslash X) \cup \{S \cap X, \bar{S} \cap X\}$.

To E_T, add the new edge between $S \cap X$ and $\bar{S} \cap X$ with weight equal to
the minimum $s - t$ cut.
Reconnect the edges between $T \backslash X$ and X so that if one end is in $T \backslash X \cap S$,
connect to $S \cap X$,
else connect to $\bar{S} \cap X$. Go to step 2.

STEP 5: Replace each singleton set $\{i\} \in V_T$ by i and the each edge $(\{i\}, \{j\}) \in E_T$
by (i, j). Output $T = (V_T, E_T)$.

In step 2, forming the graph G' reduces the size of the network, but the maximum
flow can still be computed in the original network, since this value does not change by
the operation of condensing the nodes. Step 5 can be seen as a mere formality—the
algorithm works with sets of vertices, and these are converted to ordinary vertices
and edges in this last step.

Example 3.1.2.   Consider the network in Example 2.5.3. In order to construct the
Gomory-Hu tree, first condense all vertices into one cluster S_0. The steps of Algorithm 3.1.4 are shown in Figure 3.1. Let $X = S_0$ and randomly select two vertices,
for example $s = 4$ and $t = 7$. Using the Ford-Fulkerson algorithm, we determine
the maximum flow to $f_{47} = 7$ and the minimum cut $(S_1, \bar{S}_1)$ to $S_1 = \{4\}$ and
$\bar{S}_1 = \{1, 2, 3, 5, 6, 7\}$. Separate S_1 and $\bar{S}_1$ and introduce an edge between the sets
with weight 7. Select a set X with $|X| \geq 2$, in this case $X = \bar{S}_1$, and choose two ver-
tices from X, say $s = 2$ and $t = 7$. The maximum flow is $f_{27} = 7$, $S_2 = \{3, 4, 6, 7\}$
and $\bar{S}_2 = \{1, 2, 5\}$. Now we have the sets $\{4\}$, found earlier, $S_2 \cap X = \{3, 6, 7\}$
and $\bar{S}_2 \cap X = \{1, 2, 5\}$. The edge between $\{4\}$ and $\{3, 6, 7\}$ remains (since the
second set includes vertex 7, for which this maximum flow was computed), and
a new edge with weight 7 is introduced between $\{3, 6, 7\}$ and $\{1, 2, 5\}$. Now, let
$X = \{3, 6, 7\}$ and select two new vertices, $s = 6$ and $t = 7$, say. The maximum flow is $f_{67} = 9$, $S_3 = \{1, 2, 3, 4, 5, 6\}$ and $\bar{S}_3 = \{7\}$. The set $\{3, 6, 7\}$ is
therefore split into $S_3 \cap X = \{3, 6\}$ and $\bar{S}_3 \cap X = \{7\}$. The edge between the
two sets has weight 9. Continuing in this way, let $X = \{3, 6\}$, $s = 3$ and $t = 6$.
Then $f_{36} = 9$, $S_4 = \{1, 2, 3, 4, 5\}$ and $\bar{S}_4 = \{6, 7\}$, so that we have the sets
$\{\{4\}, \{3\}, \{1, 2, 5\}, \{6\}, \{7\}\}$. The new edge between vertices 3 and 6 has weight 9.
Letting $X = \{1, 2, 5\}$, $s = 2$ and $t = 5$ gives $f_{25} = 5$, $S_5 = \{1, 2, 3, 4, 6, 7\}$
and $\bar{S}_5$. The edge between $\{1, 2\}$ and $\{5\}$ has weight 5. Finally, $X = \{1, 2\}$ gives
$f_{12} = 9$, $S_6 = \{1, 3, 4, 6, 7\}$ and $\bar{S}_6 = \{2, 5\}$, and the final edge $(1, 2)$ is assigned
weight 9. Now, all sets are singletons and are simply interpreted as the vertices themselves. The Gomory-Hu tree is given by these vertices and the edges introduced in the
process.

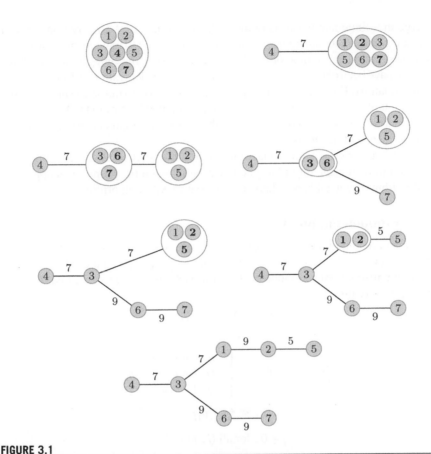

FIGURE 3.1

Construction of the Gomory-Hu tree in Example 3.1.2; the vertex numbers marked in boldface represent the next two vertices to be analyzed in the algorithm.

3.2 Minimum-cost flows

We now turn to problems defined on networks with both flow limits and costs defined on its edges. Such a problem is finding the minimum cost flow. The minimum cost flow generalizes both the maximum flow and the shortest path.

The minimum cost flow shows the cheapest way to transport a flow f of known value $|f|$ from an origin vertex s to a destination vertex t through a network with both flow limits and costs assigned to the links. Just as with the shortest path, the link costs can be defined arbitrarily, and that means that the minimum cost flows can be used as integral parts of algorithms for multi-commodity flow problems.

We assume throughout this section that all costs, flows, and flow limits are integers. The rationale for this is that under these assumptions the algorithm is guaranteed to converge in a finite number of iterations. This is not much of a restriction, since if values are given as rational numbers, they can be converted to integers by multiplying by the least common multiple of the denominators. After the (integer) results of the algorithm have been obtained, these are easily rescaled back to rationals by dividing by the same quantity. For all practical purposes, the restriction to rationals suffices. (Any irrational number can be arbitrarily closely approximated by a rational. Furthermore, computer arithmetics is built on rational numbers, since a computer can only represent a finite sequence of decimals.)

It is straightforward to formulate the minimum cost flow as a linear program. There are more efficient algorithms, however, building on the primal-dual approach. The algorithm discussed here is known as the out-of-kilter algorithm.

3.2.1   Problem formulation

The problem definition in linear program formulation is as follows. Given a directed graph $G = (V, E)$, lower bound on the edge flow l_{ij}, upper bound on the edge flow u_{ij}, and a unit flow transportation cost c_{ij} for each edge $(i, j) \in E$, find the edge flows f_{ij}, satisfying

$$\min \sum_{(i,j) \in E} c_{ij} f_{ij},$$

$$\sum_j (f_{ij} - f_{ji}) = \begin{cases} q & \text{for } i = s, \\ 0 & \text{for } i \neq s, t, \\ -q & \text{for } i = t, \end{cases}$$

$$l_{ij} \leq f_{ij} \leq u_{ij},$$

$$f_{ij} \geq 0 \quad \text{for all } (i, j) \in E,$$

where q is the required flow for which the minimum cost routes are sought. Note that a demand typically has to be split onto several routes. Optimally, it starts shipping as much flow as possible on the cheapest route, then onto the next cheapest, etc. A necessary condition for feasibility is that $l_{ij} \leq u_{ij}$ for each $(i, j) \in E$ and this will be assumed valid throughout this discussion.

To see that the shortest path problem is a special case of the minimum cost flow, let $c_{ij} = d_{ij}, l_{ij} = 0, u_{ij} = 1$, and apply a unit flow from s to t. Then the linear program becomes

$$\min \sum_{(i,j) \in E} d_{ij} f_{ij}$$

$$\sum_j (f_{ij} - f_{ji}) = \begin{cases} 1 & \text{for } i = s \\ 0 & \text{for } i \neq s, t \\ -1 & \text{for } i = t \end{cases}$$

$$f_{ij} \geq 0 \quad \text{for all } (i, j) \in E,$$

with $f_{ij} = 1$ on all edges (i, j) along a shortest route.

To show that maximum flow problem also is a special case, expand the network by adding the edge (t, s) where t is the origin and s is the destination. Let the lower bound be $-\infty$ and the upper bound be ∞ for this new edge. Let $c_{ij} = 0$ for all original edges and -1 for the new edge. The artificial edge now carries the backwards flow from t to s. Thus, we have:

$$\min - \sum_{(i,j)\in E} f_{ij},$$

$$\sum_j (f_{ij} - f_{ji}) = \begin{cases} \sum_j f_{ij} & \text{for } i = s, \\ 0 & \text{for } i \neq s, t. \\ -\sum_j f_{ij} & \text{for } i = t, \end{cases}$$

$$f_{ij} \leq u_{ij},$$

$$f_{ij} \geq 0 \quad \text{for all } (i, j) \in E.$$

We return to these comparisons in the contexts of the out-of-kilter algorithm. In its most general form, the algorithm is defined for directed graphs. In this presentation, however, we assume an undirected graph, being the most natural representation of a communication network. This also leads to some simplifications in the algorithm. Even if the minimum-cost flow can be found by solving a linear program, the out-of-kilter algorithm is far more efficient.

Consider a graph $G = (V, E)$. Let c_{ij} denote the cost of transporting a unit of commodity on edge (i, j) from node i to node j. The edges typically have an upper flow bound u_{ij}, which cannot be exceeded by any flow (or sum of flows). It is also convenient to introduce a lower flow bound l_{ij} on the edges. A lower flow bound is seldom used as a functional limit, since in practical applications, the edges are allowed to carry zero flow. It is rather used as a control parameter for artificial edges, as will be clear in the following discussion. Note also that we refer to u_{ij} as a flow bound, rather than capacity in this section in order to conform with the notion of l_{ij}—the author finds the term 'lower capacity' ambiguous. It is clear that we can set $c_{ij} = 0$ for links with no cost, $l_{ij} = 0$ if there is no lower flow limit, and $u_{ij} = \infty$ if there is no upper flow limit.

The out-of-kilter algorithm, first published in Fulkerson (1961), works by finding optimal circulations in a modified network. An edge flow f_{ij} is called feasible if

$$l_{ij} \leq f_{ij} \leq u_{ij}. \tag{3.6}$$

In a final solution, we require all flows to be feasible. This condition is also present in the linear program formulation. Since the algorithm works with circulations, the flow conservation constraints

$$\sum_j (f_{ij} - f_{ji}) = 0 \tag{3.7}$$

have to be satisfied at all vertices. This differs from the linear program formulation, because the out-of-kilter algorithm works with a modified problem. Suppose we want to find the minimum cost flow f from an origin s to a destination t. In order to have a circulation, an artificial edge is added back from t to s. The artificial (t, s)-edge is then labeled with the lower and upper flow bounds $(-|f|, |f|)$, where $|f|$ is the value of the flow, and zero cost. Note that this labeling ensures to force the flow from s to t to have value $|f|$, and that no additional cost is added to the problem by the artificial link.

The algorithm classifies edges as in kilter (in order) or out-of-kilter. For an edge to be in kilter, the flow has to be feasible (satisfy Equation (3.6)) and in addition satisfy certain conditions on its reduced cost. Otherwise the edge is out-of-kilter. This notion is defined below.

Similarly to the maximum flow algorithm by Ford and Fulkerson, the out-of-kilter algorithm starts with an initial flow satisfying the flow conservation constraint (called a nominal flow). The vertices are also labeled by the algorithm. In addition to these labels, so-called prices are assigned to each vertex. In principle, a vertex price π_j at j can be interpreted as the price of a commodity unit, bought at another vertex i at price π_i, plus the cost of transportation from i to j. Note, however, that the prices π are related to the entire route structure and not only to the transportation cost from a vertex i to another vertex j. The vertex prices are therefore dependent on the flow distribution throughout the network and may be changed as the algorithm progresses. The augmenting paths in the maximum flow algorithm are corresponded by minimum cost circulations in the minimum cost problem.

When an optimal circulation does not yet exist, some edges are out-of-kilter. The algorithm selects an arbitrary out-of-kilter edge and tries to rearrange flows to bring that edge into kilter without forcing any other edges out-of-kilter or farther out-of-kilter. If the edge can be brought into kilter, the algorithm modifies the flow accordingly, and looks for another out-of-kilter edge. If there is an edge which cannot be brought into kilter in such a way, the problem cannot be solved. Since there is only a finite number of edges, the algorithm finds an optimal solution or determines that there is no solution after a finite number of iterations.

3.2.2 The out-of-kilter algorithm

Given a set of vertex prices π, define the reduced cost of edge (i, j) as

$$\bar{c}_{ij} = c_{ij} + \pi_i - \pi_j. \tag{3.8}$$

The modified cost represents the unit transportation cost from i to j with respect to existing flows (which affect the prices). Now, the optimality conditions of the edge flows f_{ij} are

$$\text{if } \bar{c}_{ij} < 0, \text{ then } f_{ij} = u_{ij}, \tag{3.9}$$

$$\text{if } \bar{c}_{ij} = 0, \text{ then } l_{ij} \leq f_{ij} \leq u_{ij}, \tag{3.10}$$

$$\text{if } \bar{c}_{ij} > 0, \text{ then } f_{ij} = l_{ij}. \tag{3.11}$$

In the the first case, the negative reduced transportation cost gives a minimum by sending as much of the commodity as possible from i to j. In the second case, the flow volume does not affect the total flow cost, but has to be within the flow limits (3.6). Finally, if $\bar{c}_{ij}$ is positive, the flow should be at a minimum. An edge which meets these three conditions is said to be in-kilter. Edges which do not are termed out-of-kilter.

There are, therefore, two cases when an edge can be out-of-kilter. Either the flow is feasible, but not optimal according to Equations (3.9)–(3.11), or the flow is infeasible.

For edges with feasible but suboptimal flows we must have either

$$\bar{c}_{ij} < 0 \quad \text{and} \quad f_{ij} < u_{ij}, \quad \text{Case I,}$$

or

$$\bar{c}_{ij} > 0 \quad \text{and} \quad f_{ij} > l_{ij}, \quad \text{Case II.}$$

For edges with infeasible flows, we must have one of the following cases:

$$\bar{c}_{ij} > 0 \quad \text{and} \quad f_{ij} < l_{ij}, \quad \text{Case III,}$$

$$\bar{c}_{ij} = 0 \quad \text{and} \quad f_{ij} < l_{ij}, \quad \text{Case IV,}$$

$$\bar{c}_{ij} = 0 \quad \text{and} \quad f_{ij} > u_{ij}, \quad \text{Case V, or}$$

$$\bar{c}_{ij} < 0 \quad \text{and} \quad f_{ij} > u_{ij}, \quad \text{Case VI.}$$

Based on these relations, define a kilter number k_{ij}, for each edge (i, j) as

$$k_{ij} = \begin{cases} |f_{ij} - u_{ij}| & \text{if } \bar{c}_{ij} < 0, \\ \max\{0, \, f_{ij} - u_{ij}, l_{ij} - f_{ij}\} & \text{if } \bar{c}_{ij} = 0, \\ |f_{ij} - l_{ij}| & \text{if } \bar{c}_{ij} > 0. \end{cases} \tag{3.12}$$

The kilter number is always positive or zero. It indicates how much out-of-kilter an edge is. For feasible cases, the kilter number measures the degree of suboptimality; for infeasible cases, the kilter number indicates how much the flow is infeasible. An edge which is in-kilter has kilter number zero as a consequence of the optimality conditions in Equation (3.9).

The out-of-kilter algorithm arbitrarily selects an out-of-kilter edge and attempts to decrease the corresponding kilter number to zero. During an iteration, the kilter numbers of other edges cannot increase, but may decrease. The algorithm terminates returning an optimal solution when the kilter numbers of all edges are zero.

Once the problem has been formulated correctly by adding artificial edges where necessary, the out-of-kilter algorithm can be initiated with any set of vertex prices and any circulation which satisfies the flow conservation constraint. The prices and the circulation can conveniently be set to zero, assuming that all $l_{ij} = 0$ on the original edges (but not on the artificial edge).

Arbitrarily chose an out-of-kilter edge (i, j) for which the flow should be reduced. In order to have a circulation, another path in the network from j to i must be found, along which the flow values can be changed.

Labeling procedure.   When trying to find a circulation, a label is created at each visited vertex. The label has two components. The first component is the identity of the previous vertex in the path, and the second the amount of flow Δf that could be changed. The labels are used to create the path flow back from j to i.

The second component at vertex j, Δf_{ij}, is computed by careful inspection of the optimality criteria (3.9)–(3.11). The capacity can, however, never exceed that component of the label in the previous node. Suppose that we have a partial path $\{s, i, j\}$ and vertex i has been labeled $[s, \Delta f_{si}]$. If, on edge (i, j), there is free capacity of, say, u'_{ij}, at j we can nevertheless only have, at most,

$$\Delta f_{ij} = \min\{\Delta f_{si}, u'_{ij}\}. \tag{3.13}$$

This procedure is necessary to guarantee that feasibility is not violated, or that any kilter number is increased on any edge. The label of the first vertex is not critical. In the current scenario it is usually labeled $[t, u_{ts} - f_{ts}]$, that is, the label contains the remaining flow required to bring edge (t, s) into kilter.

If such a path is not found with the current vertex prices, the algorithm determines new vertex prices π for the unlabeled vertices. By systematically doing so, the reduced costs on some edges can be brought to zero or negative. When this happens, the algorithm allows flow on these edges thus creating more possibilities of edges to include in the path. By greedily modifying the prices, the costs of additional flow are guaranteed to follow the cheapest path. It is important that, at each step, the prices are increased as little as possible while bringing the reduced cost of some edge (i, j) to zero—an edge with i labeled and j unlabeled. The situation when a path is found is called a breakthrough.

Breakthrough.   Consider a minimum cost flow from s to t. After breaking through, the optimal path is given by backtracking through the first components of the labels, starting from t. The flow that is allowed on the path is indicated by the second component of the label at $[t, \Delta f_{ts}]$, and the circulation is augmented by this amount.

All labels are then erased, and we look for another out-of-kilter edge (possibly the same edge again) and the procedure is repeated. The same path will not be used again, since at least one edge along the path will be at its optimum, having reached its upper (or lower) bound. If all edges are in kilter, the optimality and feasibility conditions are satisfied, and the algorithm terminates with an optimal solution.

Non-breakthrough.   Given an out-of-kilter edge (i, j). If all nodes that can be labeled have been labeled, but no path consisting of labeled nodes includes j, we have a non-breakthrough. There can be two reasons for this. The first is when the reduced edge costs are such that increasing the flow would increase the overall cost. The second is when on each possible path, there is an edge which is saturated (with flow equaling the flow limit). In the first case, a new set of vertex prices can be computed, and the algorithm run again. If the second case occurs, the problem must be deemed infeasible. This must be so, since any change of flow will force at least one edge into an infeasible state.

In the first case, when only optimality conditions fail to be satisfied, the repricing of π_i will eventually bring $\bar{c}_{ij}$ to zero and hence bring the edge into kilter. The problem is infeasible, however, if the flow f_{ij} is outside the bounds (l_{ij}, u_{ij}) and there is no path through the network to bring the flow within the bounds.

Let us examine the case of non-breakthrough caused by a suboptimal cost structure. In this case, the set of labeled vertices has been handled by the algorithm already. In order to proceed and try to find a potential new path, we must try to connect a labeled vertex with an unlabeled vertex. For a potentially useful edge (x, y), we must therefore have $l_{xy} \leq f_{xy} \leq u_{xy}$ and $\bar{c}_{xy} > 0$. Otherwise, the vertex y would have received a label already. We can therefore try to decrease the reduced cost on this edge by increasing vertex price at the unlabeled vertex y. The price should, however, be increased as little as possible. If y has a current price of π_y, we try to find a minimum $b > 0$ giving a new reduced cost $\bar{c}'_{xy} = c_{xy} + \pi_x - (\pi_y + b)$. The constant b should be the smallest possible number to give some $\bar{c}'_{xy} = 0$. The algorithm can now be applied with these new prices. This can be repeated as long as there are potential edges on which it is possible to increase the flow.

Algorithm 3.2.1 (The Out-of-Kilter Algorithm).   Given an undirected graph $G = (V, E)$, integer upper u_{ij} and lower l_{ij} edge flow limits, integer edge costs c_{ij}, an origin vertex $s \in V$, and a destination vertex $t \in V$ for a required flow f_{st}.

STEP 0:

  Let $\mathbf{f} = (0, 0, \ldots, 0)$, $\mathbf{k} = (0, 0, \ldots, 0)$, and $\bar{\mathbf{c}} = \{c_{ij}\}$ be
  m-dimensional and $\boldsymbol{\pi} = (0, 0, \ldots, 0)$ n-dimensional.
  Connect t with s by an artificial edge and let $l_{ts} = u_{ts} = |f_{st}|$.
  Introduce the variables $f_{ij} = 0$ and $\bar{c}_{ij} = c_{ij}$ for each edge
  and $\pi_i = 0$ for each vertex. Calculate kilter numbers k_{ij} for all edges.
  Label s by $[t, |f_{st}|]$.

STEP 1 to N:

  while some $k_{ij} > 0$ and the problem is not infeasible **do**
    while t has not yet been reached and there is not a non-breakthrough
    do   Increase vertex prices π_i with the smallest amount so that some
      modified edge cost becomes $\bar{c}_{ij} \leq 0$.
      Label vertices with $\bar{c}_{ij} \leq 0$ by the preceding vertex and maximum
      possible flow. If no vertices can be labeled, there is a non-breakthrough.
    if t has been labeled (breakthrough) **then** increase flows on the path
      given by the labels. Denote the flow by f^r, where r is the current iteration.
      Recalculate the kilter numbers k_{ij}. Delete all labels and repeat.
    elseif Some edge from a labeled vertex to an unlabeled vertex has $f_{ij} < u_{ij}$,
      modify costs according to $\bar{c}'_{xy} = c_{xy} + \pi_x - (\pi_y + b)$ and repeat.

else STOP: the problem is infeasible.
if all $k_{ij} = 0$, an optimal solution has been reached.

Output the flows $f^1, f^2, \ldots, f^N$ and their respective paths.

Example 3.2.1.   The out-of-kilter algorithm applied to a rather simple problem for a graph with undirected edges is illustrated in Figure 3.2. The edge costs are here taken to be the (rounded) distances. Let s be at vertex 1 and t at vertex 7, and that we want the optimal flow of value 7 through the network. First, we modify the problem by adding the edge $(7, 1)$, that is, from t back to s. To this artificial edge we assign cost zero and flow limits $l_{71} = 7$ and $u_{71} = 7$, the value of the flow. The lower limit will force the network to transport this amount, if possible. The upper limits can in principle be set at

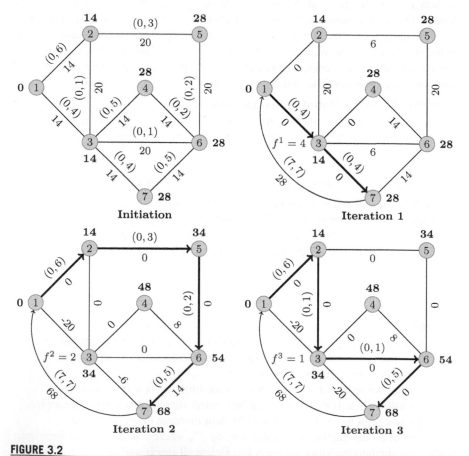

FIGURE 3.2

Progress of the Out-of-Kilter algorithm in Example 3.2.1.

any value greater than the lower value, but this choice is convenient here. Initially, all $\pi_i = 0$ and we take the flow $f_{ij} = 0$ which satisfies the flow conservation constraint trivially. This flow is feasible since all $l_{ij} = 0$. With this choice of π_i, all $\bar{c}_{ij} = c_{ij}$. It remains to calculate the kilter numbers using Equation (3.12). It is clear that the kilter numbers are zero for all edges apart from the artificial edge with $k_{ts} = 7$. This also illustrates why the problem has to be modified in the first place; the algorithm needs somewhere to start. In the first iteration, s is labeled $[t, 7]$. Initially, all $\bar{c}_{ij} > 0$, so the prices need to be adjusted. The smallest amount with which we can increase the prices and make a change in some reduced edge costs is 14. This amount is added to the prices at all unlabeled vertices, that is, all but s. Now, $\pi = (0, 14, 14, 14, 14, 14, 14)$, and so the reduced cost of edges $(1, 2)$ and $(1, 3)$ drops to zero and vertices 2 and 3 can be labeled $[s, 6]$ and $[s, 4]$, respectively. The second component in the respective label is the available edge capacity (since the flow is still zero). To progress, the prices need to be increased further. With $\pi = (0, 14, 14, 28, 28, 28, 28)$, another increase of 14 to prices of unlabeled vertices, vertices 4 and 7 are labeled $[3, 4]$ and $[3, 4]$, respectively. Since t (vertex 7) has been labeled, we have reached a breakthrough. Tracking back gives the path $t - 3 - s$, the circulation $s - 3 - t - s$ is augmented with the flow $f^1 = 4$, the labels are erased, and the reduced costs and kilter numbers are recalculated. Note that the prices remain as they are to the next iteration. In the second iteration, s is labeled $[t, 3]$. Vertex 2 can immediately be labeled $[s, 3]$. The lowest reduced cost is now 6, for edge $(2, 5)$. By increasing the prices with this amount, vertex 5 can be reached and labeled $[2, 3]$. Next, with an increase of 14, vertices 3 and 4 can be labeled $[2, 1]$ and $[3, 1]$. Adding 6 reaches vertex 6 labeled $[5, 2]$. Here, we have a tie—the label could also be $[3, 1]$, but we chose the larger flow increment (although there is no guarantee in general that such a choice will yield a larger flow). By incrementing the price further with 14, we reach a breakthrough. The label of t is $[6, 2]$, the circulation path is $s - 2 - 5 - 6 - t - s$ which is augmented by $f^2 = 2$. In the third iteration, with s labeled $[t, 1]$, we can actually label all nodes (since we reached a tie in iteration 2), and the final circulation path is $s - 2 - 3 - 6 - t - s$ with $\Delta f = 1$. Now, all edges, including the artificial, are in kilter, and we are done.

This example is similar to the maximum flow problem 2.5.4, but now costs have been taken into account. We could have started with any other flow. For example, with a flow of 6, we would have reached the optimal solution after the second iteration.

To see how the out-of-kilter algorithm can be used to solve a maximum flow problem, let all $l_{ij} = 0$ and $c_{ij} = 0$. On the artificial (t, s) edge, set $c_{ts} = -M$ and $u_{ts} = M$, where M is a large number (for example, the sum of all edge flow limits). The algorithm will then force as much flow as possible through the network in order to minimize c_{ts}.

The algorithm can also be used to find shortest paths. In this case, let $u_{ij} = 1$ and $c_{ij} > 0$ for all but the artificial edge. For the latter, set $l_{ts} = 0$, $u_{ts} = 1$, and $c_{ts} = -M$, with M large. The algorithm then sends a unit flow along the shortest path after which all edges will be in kilter.

3.2.3 **Variants and modifications**

The out-of-kilter algorithm is usually stated for directed networks. In that case, there may be negative flows and consequently more cases need to be considered in the algorithm. In particular, since in a standard formulation $l_{ij} = 0$, we need to consider cases where $f_{ij} < l_{ij}$ too. These are represented by the cases III and IV listed earlier. Therefore, for any directed edge with

$$\bar{c}_{ij} > 0 \quad \text{and} \quad f_{ij} < l_{ij},$$

$$\bar{c}_{ij} \leq 0 \quad \text{and} \quad f_{ij} < u_{ij},$$

its terminal vertex can be labeled; in the first case with $\Delta f_{ij} = f_{ij} - l_{ij}$, and in the second with $\Delta f_{ij} = u_{ij} - f_{ij}$. A net increase of a flow on an edge may be achieved by decreasing the flow in the opposite (negative) direction.

The algorithm can be modified to handle piecewise constant and monotone increasing edge costs. Let c_0 be an arbitrary "start-up" cost. Then if $c_1 \leq c_2 \leq c_3 \leq \cdots \leq c_r$ where

$$
\begin{aligned}
c_1: & \quad x_0 \leq x \leq x_1, \\
c_2: & \quad x_1 < x \leq x_2, \\
c_2: & \quad x_2 < x \leq x_3, \\
& \quad \vdots \qquad \vdots \\
c_r: & \quad x_{r-1} < x \leq x_r,
\end{aligned}
$$

we can define a piecewise linear cost function $c(x)$ with r segments, using the variable x to denote flow (this is a real variable, or 'generalized flow,' contrary to the assumptions in this section) as

$$
\begin{aligned}
x_0 \leq x \leq x_1, & \quad c(x) = c_1 x + c_0, \\
x_1 < x \leq x_2, & \quad c(x) = c(x_1) + c_2(x - x_1), \\
x_2 < x \leq x_3, & \quad c(x) = c(x_2) + c_3(x - x_2), \\
& \quad \vdots \qquad \vdots \\
x_{r-1} < x \leq x_r, & \quad c(x) = c(x_{r-1}) + c_r(x - x_{r-1}).
\end{aligned}
$$

In the network, this is represented by r parallel edges (with the same origin and destination vertices) with

$$
\begin{aligned}
l^{(1)} &= 0 & u^{(1)} &= f^{(1)}, \\
l^{(2)} &= 0 & u^{(2)} &= x^{(2)} - x^{(1)}, \\
l^{(3)} &= 0 & u^{(3)} &= x^{(3)} - x^{(2)}, \\
& \vdots & & \vdots \\
l^{(r)} &= 0 & u^{(r)} &= x^{(r)} - x^{(r-1)},
\end{aligned}
$$

where superscripts are denoting the edge number $1, 2, \ldots, r$ and the subscript has been dropped for simplicity. The total flow on this composite edge is then the sum of the flow on the individual edges, that is,

$$\sum_{i=1}^{r} f^{(i)}.$$

The vertices can be allowed to have net inflow or outflow, so that for a vertex i

$$\sum_{(i,j)\in E} f_{ij} = q_i.$$

Note that the problem is still handling a single commodity only. Any such q_i, called vertex flow, only represents local disturbances in the flow conservation. When all $q_i = 0$, the network is conservative. It is possible to make the network conservative by adding one more vertex, a super-vertex, with edges from each node that is not conservative (that is, with $q_i \neq 0$) to the super-vertex. The lower flow limit and the upper flow limit of each such edge are both set to be equal to the vertex flow.

A network where all lower flow limits l_{ij} for all edges (i, j) are zero is called simple. In case there are some $l_{ij} \neq 0$, the network can be translated to make it simple. To do so, we let

$$\tilde{u}_{ij} = u_{ij} - l_{ij}$$
$$\tilde{f}_{ij} = f_{ij} - l_{ij}$$

for all edges (i, j). It can easily be seen that all bounds are satisfied with this translation. The solution of the translated problem is easily translated back to yield the solution to the original problem.

3.3 Multi-commodity flows

In the previous section, we saw how to route a single commodity so as to minimize the transport cost to satisfy a certain demand. The problem in this section is generalized to the case with more than one commodity, leading to flows competing for network resources. In a multi-commodity flow problem there are thus demands between more than one origin-destination pair in a network. With each commodity is associated a demand, so that flows between different origins and destinations constitute different commodities. The notion of commodity is therefore a little confusing, but should be understood as demands that are mutually independent, apart from sharing the same network resources.

The multi-commodity flow problem is an important step in optimal network design. It appears in flow assignment (routing) and resource allocation tasks. In this section, we discuss a rather general approach. Slightly different methods are used in later chapters, taking technological and various performance aspects into account as well.

The multi-commodity flow problem with integer flows satisfying all demands is $\mathcal{NP}$-complete. If fractional flows are allowed, linear programming can be used to find a solution in polynomial time. However, the size of a problem instance grows rapidly, with the number of vertices, the number of edges, and the number of commodities, which makes linear programming impractical for solving all but small problems.

A more efficient approach is to use an approximation algorithm. It is in general much faster than linear programming approach and can in theory achieve a solution with arbitrary precision. The running time increases linearly with the number of commodities and inversely to the square of the chosen precision. The execution time is dominated by a number of minimum cost problems that must be solved in each iteration of the approximation algorithm.

3.3.1 Problem formulation

Let $G = (V, E)$ be an undirected graph with n nodes and m edges. Associated with each edge $(i, j) \in E$ there is a flow limit u_{ij} and possibly a cost c_{ij} (which, for example, may be proportional to the physical distance). Suppose there is a total of $K > 1$ commodities in the network. Each commodity k is specified by its origin s_k, destination t_k, and demand d_k. For each commodity, the flows may be split onto different paths. When the path flows of different commodities interact, the sum of the flows on any single edge is referred to as the edge flow. A flow of commodity k carried on edge (i, j) is denoted f_{ij}^k.

Only undirected graphs are considered here. An important consequence of having several commodities is that flow antisymmetry does not hold in general, that is, $f_{ij} \neq -f_{ji}$. It is true only for flows belonging to the same commodity $f_{ij}^k = -f_{ji}^k$. It is also not likely that flows of the same commodity flow in opposite directions in undirected networks in reality. Therefore, we will usually relax this condition here.

The multi-commodity flow problem can be expressed as a linear program as follows:

$$\max \Psi, \tag{3.14}$$

$$\sum_{(i,j) \in E} (f_{ij}^k - f_{ji}^k) = \begin{cases} d_k & i = s_k, \\ 0 & i \neq s_k, t_k, \\ -d_k & i = t_k, \end{cases} \tag{3.15}$$

$$0 \leq |f_{ij}^k| \leq u_{ij}^k, \tag{3.16}$$

$$0 \leq \sum_k |f_{ij}^k| \leq u_{ij}. \tag{3.17}$$

Just as for the minimum cost flow, we require that the multi-commodity flow satisfies flow conservation (3.15) for each commodity separately and that each flow is feasible by (3.16) and (3.17). In most cases the bound for individual commodities (3.16) is omitted. An example of where such bounds exist is in networks with trunk reservation, an important admission control technique that is discussed further in subsequent chapters.

Equation (3.15) also requires that $\sum_{(i,j) \in E} f_{ij}^k = d_i$ for $i = s_k$, that is, that the total flow out from s_k sums to the demand d_k of commodity k, and similarly for $i = t_k$. This condition is referred to as demand satisfaction.

There are several variants of the multi-commodity flow problem. The objective function Ψ above depends on the quantity we wish to optimize. In the minimum cost multi-commodity flow problem, there is a cost $c_{ij} f_{ij}$ of sending flow on edge (i, j).

In the linear program formulation we then maximize

$$\Psi = - \sum_{(i,j)\in E} c_{ij} \sum_{k=1}^{K} f_{ij}^{k}.$$

The minimum cost variant is a generalization of the minimum cost flow problem. In the maximum multi-commodity flow problem, there are no restrictions on individual commodities, but the total flow is maximized. The sum of the outflows from each origin s_k for all commodities k gives the total network flow, so that

$$\Psi = \sum_{k=1}^{K} \sum_{j:(s_k,j)\in E} f_{s_k,j}^{k}.$$

The objective may be to find a feasible solution, that is, a multi-commodity flow which satisfies all the demands and obeys the capacity constraints. More generally, however, we might want to know the maximum number z such that at least z percent of each demand can be transported without violating the capacity constraints. This problem is known as the concurrent flow problem. An equivalent problem is determining the minimum factor by which the capacities must be increased in order to transport all demands. In the maximum concurrent flow problem, the commodity with the poorest flow relative to its demand (referred to as its throughput) is maximized,

$$\Psi = \min_{k \leq K} \frac{\sum_{j\in V} f_{(s_k,j)}^{k}}{d_k}.$$

It is this variant of multi-commodity flow we will discuss here.

3.3.2 **An approximation algorithm**

The algorithm presented here, proposed by Leighton et al. (1993), is a $(1 + \epsilon)$-approximation algorithm for the maximum concurrent multi-commodity flow problem. The algorithm as originally presented is rather complex and we will therefore allow for some simplifications in line with the ones suggested in Leong et al. (1992).

Leighton et al. describe an algorithm for solving the general concurrent flow problem with arbitrary capacities and demands. It is initiated with an arbitrarily routed flow which is gradually improved by rerouting individual commodities from highly congested edges to lightly congested ones. Flows are successively rerouted along minimum cost flows computed in specially constructed auxiliary graphs.

The discussion focuses on the operation aspects of the algorithm and much of the underlying theoretical foundation has been simplified. Some of these modifications are justified by computational aspects—we want to avoid too large or too small numbers in the algorithm which may lead to numerical problems like overflow, underflow, or cancellations.

We note that, using the presented algorithm, approximately computing a k-commodity concurrent flow is about as difficult as computing k single commodity maximum flows.

To formulate the algorithm, we define the problem parameters as follows. Let a network be described by an undirected graph $G = (V, E)$ with positive upper flow limits (or capacities) u_{ij} assigned to each edge $(i, j) \in E$. We assume that G is connected and that it has no parallel edges. To simplify the discussion, we consider flow in only one direction, chosen arbitrarily, for each commodity. This can be justified by the fact that when we have a point-to-point flow in telecommunications, it is likely to be a duplex system. The role of origins and destinations in undirected networks can therefore be considered as symmetric. For any flow, some resources are reserved between the source and the sink, and from a planning point of view it does not matter in which direction the information flows, as long as the flow reflects the total demand between the origin-destination pair.

Let n, m, and K be the number of vertices, edges, and commodities, respectively. (We assume that the demands and the capacities have integer values.) A commodity k is defined by its origin-destination pair $s_k, t_k \in V$ and the corresponding demand $d_k > 0$. This can be expressed in the triple of integers (s_k, t_k, d_k).

In principle, due to the minimum cost problems that we need to solve—and these are usually solved exactly—we would assume integer demand and capacity values to guarantee convergence. Hence, when the capacities are modified in the course of the algorithm, these values would need to be rounded to the nearest integer values. In practice, however, the outcome of the minimum cost flow subproblems only yield alternative paths for rerouting and approximate solutions are therefore acceptable. We assume that we have a suitable termination condition for minimum cost flows, perhaps by properly rounding the input values. In any case, this rounding will not be considered as part of the core algorithm.

The optimization variables are the flows in the network. We distinguish between edge flows- the flow through an edge- and path flows, the fraction of a commodity flow following a specific path. Let $\mathcal{P}_k$ denote a collection of paths from s_k to t_k in G and let $f_p^k \geq 0$ for every path $p \in \mathcal{P}_k$. Then the value of the flow is the sum of all path flows along $p \in \mathcal{P}_k$,

$$f^k = \sum_{p \in \mathcal{P}_k} f_p^k.$$

The edge flow f_{ij}^k through an edge (i, j) is similarly defined as

$$f_{ij}^k = \sum_p \left\{ f_p^k : p \in \mathcal{P}_k \quad \text{and} \quad (i, j) \in p \right\}.$$

The total flow on edge (i, j) is the sum of flows from all commodities using edge (i, j),

$$f_{ij} = \sum_k f_{ij}^k.$$

In order to describe the optimization problem, we introduce a network scaling parameter $0 < z < 1$, called the throughput, whose value is such that at least z

percent of each demand can be transported over the network without violating the capacity constraints.

The throughput, z, should be maximized in the maximum concurrent multi-commodity flow problem. Since the algorithm is an $(1 + \epsilon)$-approximation, the best result we can expect is a solution for which $z \geq (1 - \epsilon)\hat{z}$, where $\hat{z}$ is the maximum possible throughput. However, rather than maximizing z, we will minimize the parameter $\lambda = 1/z$, called the congestion. The congestion on any edge is defined as

$$\lambda_{ij} = f_{ij}/u_{ij}, \quad \text{for any edge } (i, j) \in E.$$

Let the maximum edge congestion be $\lambda = \max_{(i,j) \in E} \lambda_{ij}$ in the network, and denote the smallest possible achievable congestion by $\hat{\lambda} = \min \lambda$. Without loss of generality, we assume that the multi-commodity problem is simple, where each commodity has a single source and a single sink. Otherwise, a commodity can be decomposed into a simple multi-commodity problem.

The problem of minimizing λ is equivalent to maximizing z, and $z \geq (1 - \epsilon)\hat{z}$ implies $\lambda \leq (1 + \epsilon)\hat{\lambda}$, where $\hat{\lambda}$ is the minimum possible λ. This is given by the fact that

$$\frac{1}{1 + \epsilon} \leq 1 - \epsilon,$$

by the first term of the Maclaurin expansion of the fraction. The reasons for using λ in the formulation are that congestion can be defined for any flow $0 \leq f < \infty$, which is convenient when forming derived quantities for solving the problem (since we want to scale the network resources rather than the demands). With this approach, a solution can be found to a problem for which no feasible solution exists without scaling.

The parameter ϵ is an input parameter to the algorithm. We shall assume implicitly throughout that $0 < \epsilon < 1/9$. It should be noted, however, that as ϵ decreases, the running time increases inversely proportional to its square. In practice, it should be possible to find a value of ϵ which is of the magnitude of the measurement errors of other input parameters.

Next, we define the weight of each edge as

$$w_{ij} = e^{\alpha \lambda_{ij}},$$

where α is a parameter defined as $\alpha = c \cdot s/\lambda$. The product of the two constants c and s scales the exponent and can be adjusted to increase the performance of the algorithm. In Leong et al. (1992) the values

$$c = 19.1 - \ln m$$

and

$$s = 0.25$$

are used. The weights are used as cost parameters in minimum cost flow problems that have to be solved in each iteration of the algorithm. With this choice of α, the same congestion level gives the same weight for any edge. Note that the value of α

must be chosen carefully. A too-small value does not guarantee any improvement, whereas a too-large value leads to very slow progress.

In addition, a potential Φ, defined as

$$\Phi_{ij} = u_{ij} \cdot w_{ij}$$

$$\Phi = \sum_{(i,j) \in E} \Phi_{ij}$$

serves as a measure of the convergence of the algorithm. By rerouting an amount of flow onto a less congested path results in a significant decrease in Φ. The potential is closely related to the overall cost of the solution with respect to the length w_{ij},

$$c_{ij} = f_{ij} \cdot w_{ij} = \lambda_{ij} \cdot \Phi_{ij}, \tag{3.18}$$

$$c = \sum_{(i,j) \in E} c_{ij}, \tag{3.19}$$

where f_{ij} are the edge flows. The cost improvement is used as a termination condition for the algorithm. Finally, the optimal amount of path flow to reroute is described by the parameter $0 < \sigma \leq 1$, whose value is found by minimizing Φ with respect to σ. This step ensures rapid convergence of the algorithm.

The algorithm terminates when the difference in costs between the old path p and the new path q satisfies the bound

$$c_p^k - c_q^k \leq \epsilon(c_p^k + \lambda \cdot \Phi/K) \tag{3.20}$$

for all path flows f_p^k.

The main principles of the algorithm are as follows. The routine identifies an edge with high congestion λ_{ij} and selects one of the paths flowing through this edge for rerouting. The fictitious weight w_{ij}, based on the congestion and assigned to each edge, is penalizing edges with high congestion and promoting edges with low congestion. To compute the possible benefits of rerouting a path flow, all of the selected path flow is used as required demand in a minimum cost flow problem. The solution to this problem is an alternative route for the path flow with a cost that can be compared to the cost of the existing path. Usually, only a fraction of the selected flow may need to be rerouted to achieve optimum. Therefore, an optimization problem is solved, where the minimization of the potential—again proportional to the edge weights—gives the fraction σ of flow to be rerouted. The new flow and its path are thereby completely specified, the rerouting is performed, and the congestion and other parameters are recomputed. The procedure is repeated until the cost improvement goes below the limit (3.20).

The algorithm starts with flows that satisfy all the demands, but not necessarily the capacity constraints. The algorithm then reroutes flow from heavily congested edges to edges with lower congestion in order to decrease the value of λ. In order to do so, it selects a heavily congested edge. This can be done by selecting the edge with the largest λ_{ij}, or, if there is more than one with the same congestion, any of these edges. We can formulate a lower bound on $\hat{\lambda}$.

Lemma 3.3.1.   *Suppose that there exists a multi-commodity flow satisfying capacities $\lambda \cdot u_{ij}$. Then for any weight function w_{ij}, the value $\sum_{k=1}^{K} \hat{c}^k(\lambda)/(\sum_{(i,j)\in E} w_{ij} u_{ij})$ is a lower bound on $\hat{\lambda}$.*

The goal of the algorithm is to find a multi-commodity flow f and a weight function w, such that this lower bound is within a $(1 + \epsilon)$ factor of optimal, that is

$$\lambda \leq (1 + \epsilon) \sum_{k=1}^{K} \hat{c}^k(\lambda) / \left(\sum_{(i,j)\in E} w_{ij} u_{ij} \right).$$

In this case, we say that f and w are ϵ-optimal. Note that we are using the term ϵ-optimal to refer both to a flow itself and a flow and weight function pair.

Selecting commodity to reroute.   Starting with any flows satisfying the demands but not necessarily the capacity constraints, calculate λ_{ij} for each edge. Let $\lambda = \max_{(i,j)} \lambda_{ij}$ and $\alpha = cs/\lambda$. Next, the edge weights w_{ij} are computed. For each commodity, solve a minimum-cost flow problem with the edge weights as costs, the scaled capacities $\bar{u} = \lambda u_{ij}$, and demands d_k. The result is a set of alternative paths for each commodity. The path flow cost c^k per commodity can now be calculated by using (3.18), summing over the edges that build up the path $p \in \mathcal{P}_k$ used by commodity k,

$$c_p^k = \sum_{(i,j)\in p: p\in \mathcal{P}_k} f_{ij}^k w_{ij}.$$

By comparing the costs of using the old paths, and the new paths just determined, we have an indication of which commodity to reroute—typically the commodity that gives the highest reduction in cost. Thus, if p is the old path and q the new path, the condition

$$c_p^k - c_q^k > \epsilon(c_p^k + \lambda \cdot \Phi/K)$$

shows that rerouting a proportion of commodity k from p to q would decrease the cost and the network congestion. The costs are computed for routing the entire path flow f_p^k of commodity k on paths p and q, respectively.

The algorithm selects the path with the largest flow through the most congested edge for which a new path has been found, or else, a path with the largest flow for which a path already exists. The selected path is the target path for rerouting flow. Compute costs for using the old and the new paths for routing the demand, respectively. The difference in cost shows the potential gain of rerouting. By inspecting the cost gain, a set of flows that can be rerouted can be identified. From this set, a flow can be selected deterministically according to the cost gain, or randomly.

Computing fraction to reroute.   In order to calculate σ, the fraction of flow to reroute, we solve an optimization problem, minimizing the difference in potential $\Delta\Phi$ between using the old path and the new path. It is actually sufficient to consider the difference in potential for these two paths only, since the rest of the terms in Φ remain unchanged. A simple way to implement this step is to calculate Φ_p and Φ_q

for $0 < \sigma \leq 1$ with step size, ϵ, say, and find the minimum of $\Phi_p - \Phi_q$, where p is the old path and q is the new path. This gives an approximate value for σ. It seems to be necessary to limit the value σ by a constant $\bar{\sigma}$, with $0.5 \leq \bar{\sigma} < 1$ to avoid oscillating behavior of the algorithm. Once σ has been determined, we reroute the flow accordingly, leaving $(1 - \sigma) f_p$ on the old path and σf_p on the new path. Also, to avoid wasting time routing small amounts of flow, we reroute a commodity only if σ_f is at least as large as $O(\epsilon/\alpha\lambda)$.

Stopping conditions. The stopping condition (3.20) measures the gain in cost achievable by rerouting a flow. In addition, it may be instructive to monitor how λ decreases. If λ does not decrease, either an approximately optimal solution has been found, or the path flow selected for rerouting does not improve the value. Also the potential function can be used to monitor the optimality of the flow; the algorithm yields a ϵ-optimal flow when the potential function becomes sufficiently small.

Example 3.3.1. Consider the network in Figure 3.3. Suppose we are given the demands in Table 3.1. Choose the constants $c = 19.1 - \ln m$ with $m = 5$, $s = 0.25$, and $\varepsilon = 0.1$. Also, for simplicity, let $\sigma = 0.5$ in this example. To find initial flows, use the shortest paths for each commodity in the network. The resulting flows are also shown in Figure 3.3. Note that the flows actually are feasible. Next, calculate the congestion λ_{ij}, weight w_{ij}, potential Φ_{ij}, and cost c_{ij} for each edge. These values are shown in Table 3.2. The maximum congestion is $\lambda = 1$. We may suspect that this is not an optimal flow, since other congestion values are much lower. The high congestion is caused by commodity 1, so we scale the capacities to $\lambda \cdot u_{ij}$ and solve a minimum cost flow problem with the scaled capacities and the weights w_{ij} as edge costs. The result is an alternative path for commodity 1, path $1-2-4-5$. At this point we may compute a proportion of the demand to reroute, represented by σ. Here we have just set $\sigma = 0.5$, so we reroute half of the demand onto the alternative path. This is illustrated in Figure 3.3 and the parameters with respect to this flow are tabulated

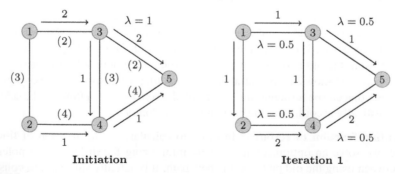

Initiation   **Iteration 1**

FIGURE 3.3

A multi-commodity flow problem: the initial solution (left), and final flow after rerouting (right).

Table 3.1 Flow demands for Example 3.3.1

k	s_k	t_k	d_k
1	1	5	2
2	2	5	1
3	3	4	1

Table 3.2 Initial flow parameters in Example 3.3.1

(i, j)	u_{ij}	f_{ij}	λ_{ij}	w_{ij}	Φ	c_{ij}
(1, 2)	3	0	0.00	1.0	3.0	0
(1, 3)	2	2	1.00	75.7	151.4	151.4
(2, 4)	4	1	0.25	3.0	11.8	3.0
(3, 4)	3	1	0.33	4.2	12.7	4.2
(3, 5)	2	2	1.00	75.7	151.4	151.4
(4, 5)	4	1	0.25	3.0	11.8	3.0

Table 3.3 Flow parameters after the first iteration in Example 3.3.1

(i, j)	u_{ij}	f_{ij}	λ_{ij}	w_{ij}	Φ	c_{ij}
(1, 2)	3	1	0.33	4.2	12.7	4.2
(1, 3)	2	1	0.50	8.7	17.4	8.7
(2, 4)	4	2	0.50	8.7	34.8	8.7
(3, 4)	3	1	0.33	4.2	12.7	4.2
(3, 5)	2	1	0.50	8.7	17.4	8.7
(4, 5)	4	2	0.50	8.7	34.8	8.7

in Table 3.3. Note, however, how the sum of potentials drops from 342 to 130 by this reassignment. Thus, the new multi-commodity flow is much closer to optimum.

Algorithm 3.3.2 (Approximation Algorithm for the Maximum Concurrent Multi-Commodity Flow Problem).   Given a graph $G = (V, E)$, non-negative edge costs $\{c_{ij}\}$, edge capacities u_{ij}, and K commodities (s_k, t_k, d_k). Set the precision ϵ.

STEP 0:
    Let $\mathbf{w}$ be the vector of $w_{ij} = c_{ij}$ and $\bar{\mathbf{u}}$ the vector of $\bar{u}_{ij} = \infty$.
    for all $k = 1, \ldots, K$ **do** solve a minimum cost problem
    MINCOST$(\bar{\mathbf{u}}, \mathbf{w}, s_k, t_k, d_k)$ and assign initial flows f_{ik}^k.

STEP 1 to N:
    for $i, j = 1, \ldots, n$ **do**
        Compute λ_{ij}, $\lambda = \max_{(i,j) \in E} \lambda_{ij}$, and α

Table 3.4   The specification of the commodities in Example 3.3.2

k	s_k	t_k	d_k
1	1	7	5
2	3	5	2
3	2	6	2

Compute weights w_{ij}, modified capacities $\bar{u}_{ij} = \lambda u_{ij}$,
and potential $\Phi = \sum_{(i,j) \in E} \Phi_{ij}$
for all $k = 1, \ldots, K$ **do** solve a minimum cost problem
MINCOST($\bar{\mathbf{u}}, \mathbf{w}, s_k, t_k, d_k$), giving a set $\mathcal{Q}_k$ of alternative paths.
Compute cost of path flows $c_p^k = \sum_{(i,j) \text{ in } p : p \in \mathcal{P}_k} c_{ij}^k$ and
$c_q^k = \sum_{(i,j) \text{ in } q : q \in \mathcal{Q}_k} c_{ij}^k$ for each commodity.
if $c_p^k - c_q^k < \epsilon(c_p^k + \lambda\Phi/K)$ for all k **then** STOP:
flows are approximately optimal
else
    Select a flow f_p^k such that $c_p^k - c_q^k \geq \epsilon(c_p^k + \lambda\Phi/K)$
    (typically the flow for which the difference is largest). Fix k.
    Calculate σ: let σ be such that $\Phi_p^k(\sigma) - \Phi_q^k(\sigma)$ is minimum,
    where $\Phi_p^k(\sigma)$ is the potential along $p \in \mathcal{P}_k$ with flow $(1 - \sigma)f_p^k$
    and $\Phi_q^k(\sigma)$ is the potential along $q \in \mathcal{Q}_k$ with flow σf_p^k.
    Reroute flow so that $f_q^k := \sigma f_p^k$ and $f_p^k := (1 - \sigma)f_p^k$

Output all path flows f_p^k, $k = 1, \ldots, K$, and $p \in \mathcal{P}_k$.

Example 3.3.2.   Consider the network depicted in Figure 3.4 and three commodities
(to keep the example easy to follow). The commodities are given in Table 3.4.

Let the parameters $m = |E| = 10$, $c = 19.1 - \ln(m)$, $s = 0.25$ be given as in the
description of the algorithm, and choose the shortest paths for the initial flows of each
commodity. We want to find an approximate solution to this multi-commodity flow
problem with tolerance $\epsilon = 1/10$. We calculate, in order, edge flows f_{ij}, congestion λ_{ij}, maximum congestion λ, α, edge weights w_{ij}, the potentials Φ_{ij}, and costs
c_{ij}.

3.4 Summary

Three problems are discussed in this chapter: the multi-terminal flow, minimum cost
flow and multi-commodity flow. The multi-terminal flow is used as a basis for the
formulation of a topological design method in Chapter 4. The minimum cost flow is
important as a subroutine in multi-commodity flow. The multi-commodity network
flow problem is fundamental to any network performance evaluation. It is related to a

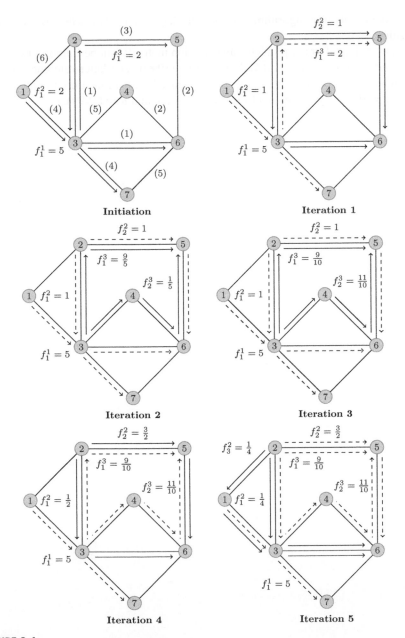

FIGURE 3.4

A multi-commodity flow problem with three flows. In the initial solution (Initiation) short-est paths are used to route the three flows. In Iteration 1, flow f^2 is split into two paths. In iterations 2 and 3, flow f^3 is split and rerouted. Iteration 4 shifts flow from f_1^2 to f_2^2 and in iteration 5, a fraction of flow f_1^2 is rerouted onto the third path, f_3^2.

multi-commodity flow algorithm discussed in Chapter 6, known as the flow deviation algorithm.

An overview of graph problems and solution methods can be found in Papadimitriou and Steiglitz (1982) and Goldberg et al. (1990). The latter reference is quite comprehensive in terms of graph-related problems. Proofs of the important theorems can be found in Leighton et al. (1993) and Leighton and Rao (1999).

Topological Design

A fundamental step toward a cost efficient and well performing network is the topological network design. In order to investigate properties of optimal topologies, we will make simplifying assumptions regarding the flows in the network.

There are two types of network design problems with different objectives: capacitated network design and survivable network design. The former is the topic of this chapter and will be expanded further with focus on different network technologies and flow types in Chapters 6–11. In capacitated network design, we seek a network topology that is capable of transporting the required flows at minimum cost. In contrast, the objective of survivable network design is finding a topology meeting some reliability criteria, again at minimum cost. That is the subject of Chapter 12. These two types of design problems will usually lead to different topologies.

Approximations, heuristics and meta-heuristics are used to find a minimum cost core (or backbone) network, given some node locations, a flow (traffic) demand matrix, and a cost structure of links interconnecting the nodes. Some common network configurations and their properties are presented. The design methods in this chapter are—to a large degree—independent of the flow characteristics and transport technology.

The chapter is organized as follows. The topological design problem is formalized in Section 4.1. Most topological design methods are built on the concept of the minimum spanning tree (MST). The MST, its properties and its relation to network topologies is discussed in Section 4.2. Methods for creating the common ring architecture are described in Section 4.3. We discuss a greedy algorithm for network design in Section 4.4. Section 4.5 introduces a design method based on the Gomory-Hu tree. The next two sections discuss randomized network design using Monte Carlo methods (Section 4.6) and a genetic algorithm (Section 4.7). Some particular problems related to resource allocation are addressed in Section 4.8.

4.1 Capacitated network design

Formally, the minimum cost capacitated network design is the problem of designing a network, given flow demands A between the node pairs, and link costs C, such that the network fulfills some pre-defined performance criteria of the carried flow, at minimum cost. Like in Chapter 2, the flows in this section are assumed to be constant, so in effect, the only performance criteria we use here are that the flows can be transported by the network and, in some cases, that a simple reliability constraint is fulfilled. The results of this chapter provide a starting point for successive refinements in subsequent chapters, where random flows and various related performance criteria are discussed. We use the term flow in Chapters 2,3 and in this chapter, and reserve the term traffic for the more realistic situation where the demands are generated by stochastic processes. It has been shown (Mansour and Peleg, 1998) that the capacitated network design problem is $\mathcal{NP}$-hard.

We assume that we are given n nodes with known locations, which form the vertex set V of the corresponding graph. The flow demands between vertex pairs are specified by an $n \times n$ flow demand matrix $A = (a_{ij})$, where a_{ij} is the amount of flow that is required to be transmitted between nodes i and j for $i \neq j$. When the flow is originating and terminating at the same node, it formally never enters the network (we are only considering link costs in this model).

The link costs are given by an $n \times n$ matrix $C = (c_{ij})$ where c_{ij} is the cost between nodes i and j for $i \neq j$. The diagonal elements can be set to zero, since by assumption, we do not have any flow originating and terminating at the same node. We also make the assumption that the network has to be connected (even if there may not be any demand between all node pairs). The outcome of the design is the number and capacity of links between network nodes, as well as some routing scheme for flow that is not transported on direct links between nodes. The collection of nodes and links, represented by a graph, is referred to as the network topology. The optimal topology is greatly dependent on the cost structure of the links and on the underlying transport technology.

Suppose we have a network topology candidate. A network candidate among all possible network topologies fulfilling the pre-defined performance criteria is called feasible. In other words, the given requirements filter out candidates of interest from those that are not.

4.1.1 Cost functions

The cost structure of the links, or their cost function, is a decisive factor in topological design. Usually, communication links are available in different capacities at different costs. The cost function is in general a nonlinear function of capacity. The task of optimally assigning capacity to links so that the performance objectives are met is referred to as resource allocation, and is discussed further in Section 4.8. By initially relaxing this step and assuming links of a single type with unit capacity, we obtain an approximate solution to the more complex problem with nonlinear capacity links.

A cost function can be continuous or discrete, linear or nonlinear. In addition, a link may or may not have an initial set-up cost. We make the assumption that the cost of flow paths $p \in \mathcal{P}$ obeys the triangle inequality, so that given three nodes i, j, and k, connected by links e_{ij}, e_{ik}, and e_{kj}, it is cheaper to send a flow from i to j over e_{ij} than over the path $\{e_{ik}, e_{kj}\}$.

A typical cost function $c(u, d)$ is a function of capacity u and link distance d. A common simplification is letting $c(u, d) = \alpha(u \cdot d)$ for some constant α, so that c is

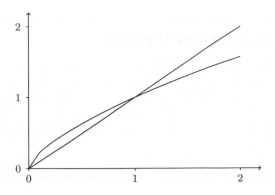

FIGURE 4.1

A concave and a linear cost function. The concave function is a power law function with $\beta = 0.65$.

a linear function of u and d. In reality, however, the cost is usually decreasing with capacity (and possibly also with distance), which leads to a concave cost function. This represents an economy-scale principle. This type of functions are defined as follows.

Definition 4.1.1.   A real-valued function f is called (downwards) concave if, for any two points x_1 and x_2 on an interval I and any $t \in [0, 1]$,

$$f(tx_1 + (1 - t)x_2) \geq tf(x_1) + (1 - t)f(x_2).$$

This is equivalent to a function with second derivative $f''(x) \leq 0$ and a decreasing first derivative $f'(x)$.

Most realistic cost functions are discrete and, using the continuous approximation, concave. A concave and linear cost function is illustrated in Figure 4.1.

4.1.2 Routing

Routing is the process of directing the flow from the source to the destination so that the performance requirements are met and the constraints set by the network fulfilled (such as capacity limits). In other words, when there are several possibilities of which path a flow can take, routing is the collective rules on how these paths should be selected. Routing is essential in maximizing network performance and minimizing the network cost.

There exist many routing methods, usually implemented as part of the network control in network protocols. One of the goals of network routing is robustness. A routing method should be fast, reasonably accurate, and not cause any degenerative behavior of the network, such as infinitely circulating messages. One of the simplest routing principles is the shortest path. The length of an edge can be defined arbitrarily, and then the shortest path will give different results depending on the definition used. For example, the length need not only be the physical distance of a link, but a function estimating link congestion. A shortest path may then be the path with the lowest congestion.

4.2 Important properties of graphs

Since we are using graphs to model network topologies, we can express desirable properties of networks in graph theoretical terms. In topological network design, we want to balance two—often conflicting—objectives. As the network cost is dependent on the number of links, we would like as few links as possible. The cost of transportation of a flow in terms of resources used in the network, however, is related to the number of links on the path of the flow, and we desire this number to be as low as possible. The second objective is minimized when there is a large number of links in the network. The most cost-efficient network topology is likely to be somewhere in between these two extremes.

4.2.1 Sparseness of graphs

In order for a network to have low cost, its corresponding graph should be *sparse*. Intuitively there should not be any 'unnecessary' links present, where the importance of a link is a balance between the conflicting goals of the cost of the network and the cost of routing a flow (the longer the path a flow has to take to reach its destination, the more expensive it is). The sparseness of a graph is measured by its size and its weight. We assume that the graphs have a fixed order, its number of vertices (or nodes) n. We state these properties as two definitions.

Definition 4.2.1. The size of a graph, $s(G)$ where $G = (V, E)$, is the number of edges in G, that is $s(G) = |E|$.

Definition 4.2.2. The weight of a graph, $w(G)$ where $G = (V, E)$ is the sum of the edge weights of G, that is

$$w(G) = \sum_{(i,j) \in E} w_{ij},$$

where i and j are vertices in G.

As a matter of fact, it is very straightforward to find the sparsest possible subgraph of a general weighted graph $G = (V, E)$. A tree T which contains every vertex of a connected graph $G = (V, E)$ is called a spanning tree for G. Suppose G has a weight function w_{ij} defined on its edges (for example, the respective node distances). The tree T which contains all vertices in G and minimizes the sum of the weights of the edges of T,

$$w(T_{\min}) \leq w(T) = \sum_{(i,j) \in T} w_{ij}, \quad \text{for all } T,$$

is called a minimum spanning tree (MST) for G, denoted $T_{\min}$. Note that the MST need not be unique. There are efficient algorithms to find the minimum spanning tree due to the following theorem.

Theorem 4.2.1. *Let $G = (V, E)$ be a weighted connected graph, and let V_1 and V_2 be a partition of the vertices of G into two disjoint nonempty sets. Consider the* *set E' of edges in G which have one end point in V_1 and the other in V_2. Let $e \in E'$ be such an edge with minimum weight. Then there is a minimum spanning tree $T_{\min}$ which has e as one of its edges.*

Proof. Let $T_{\min}$ be a minimum spanning tree of G. If $T_{\min}$ does not contain edge e, the addition of e to $T_{\min}$ must create a cycle. Denote this subgraph by $T_{\min} \cup \{e\}$. Then, there is some edge e' of this cycle that has one end point in V_1 and the other in V_2. Moreover, by the choice of e, $w(e) \leq w(e')$. Therefore, if we remove e' from $T_{\min} \cup \{e\}$, we obtain a spanning tree whose total weight is no more than before. Since $T_{\min}$ was a minimum spanning tree, the new tree must also be a minimum spanning tree of G.   □

Theorem 4.2.1 is important because it indicates how to construct a minimum spanning tree. In Kruskal's algorithm, this principle is used to build the minimum spanning tree by greedily forming clusters. Initially, each vertex constitutes its own cluster and the set of edges of the tree $E_T = \varnothing$. The algorithm then investigates each edge in turn, where the edges are ordered by increasing weight. If an edge (i, j) connects two different clusters, then (i, j) is added to the set of edges of the minimum spanning tree E_T, and the two clusters connected by (i, j) are merged into a single cluster. Any edge (i, j) connecting two vertices that are in the same cluster, is discarded. When all vertices are connected the algorithm terminates and the result is a minimum spanning tree.

Algorithm 4.2.2 (Kruskal's Algorithm for Finding an MST). Given a connected weighted graph $G = (V, E)$ with n vertices and m edges.

STEP 0: (initialize)
> **for** each vertex $i \in V$, **do**
>> define the clusters, $C(i) \leftarrow \{i\}$;
>
> **end**;
> Sort E with respect to weights in increasing order. Call the list L;
> Let $E_T \leftarrow \varnothing$;

STEP 1: (iterate)
> **while** E_T has fewer than $n - 1$ edges, **do**
>> starting from the top of L and progressing downwards,
>> get an edge (i, j) from L;
>> **if** $C(i) \neq C(j)$, **then**
>> add edge (i, j) to E_T;
>> merge $C(i)$ and $C(j)$ into one cluster;
>
> **end**;

Output $T_{\min} = (V, E_T)$, a minimum spanning tree of G.

Kruskal's algorithm can be implemented to run in $O(m \log(m))$ time.

In the *Prim-Jarník algorithm*, we grow a minimum spanning tree starting from an initial vertex. The main idea is similar to that of Dijkstra's algorithm. We begin with a vertex s, which defines the initial cluster of vertices C. Then, in each iteration, we greedily choose a minimum-weight edge (i, j) connecting a vertex i in the cluster C to a vertex j outside of C. The vertex j is then brought into the cluster C and the process is repeated until a spanning tree is formed. Since we are always choosing the edge with the smallest weight, we have a minimum spanning tree.

Algorithm 4.2.3 (The Prim-Jarník Algorithm for Finding an MST).   Given a connected weighted graph $G = (V, E)$ with n vertices and m edges.

STEP 0: (initialize)
>   Select an arbitrary initial vertex $s \in V$ to start from;
>   Let $E_T \leftarrow \varnothing$;
>   Let $C \leftarrow s$;

STEP 1: (iterate)
>   **while** E_T has fewer than $n - 1$ edges, **do**
>   select the edge of minimum weight between a tree and
>   a nontree vertex;
>   add the selected edge to E_T;
>   add the corresponding vertex to C;
>   **end**;

Output $T_{\min} = (V, E_T)$, a minimum spanning tree of G.

The Prim-Jarník algorithm can be implemented to run in $O(n^2)$ time.

A natural starting point in topological design is the minimum spanning tree. The problem, however, with using the MST as a network topology is that a flow may have to traverse as many as $n - 1$ links to reach its destination. The relative weighted path length in a subgraph $G' = (V, E')$ of a complete graph $G = (V, E)$, where $E' \subseteq E$, is defined by its stretch factor.

Definition 4.2.3.   Let $G = (V, E)$ be a complete graph with weight function w and let $G' = (V, E')$, $E' \subseteq E$, be a subgraph of G. Let $d_{G'}(i, j)$ denote the shortest weighted path between i and j in G', that is, the minimum of $w(p)$ over all paths p from i to j in G'. Then the stretch factor (or dilation) of G' is

$$\text{stretch}(G') = \max_{i,j \in V} d_{G'}(i, j)/d_G(i, j).$$

Thus, the stretch factor of a graph is a measure of how much longer a shortest path is in any subgraph G' of G as compared G itself. This measures the "quality" of a subgraph when the size and weight are decreased by removing some edges from G. It seems reasonable to assume that a good network topology candidate should have low size, weight and stretch. Another related measure is the graph diameter, the maximum number of edges any path connecting two arbitrary nodes i, j has.

Table 4.1   Some characteristics of the sample topologies in Figure 4.2

Property	MST	Mesh	Ring	Star
Size (links)	5	15	6	5
Weight (cost)	88	424	133	106
Min. degree	1	5	2	1
Max. degree	2	5	2	5
Diameter	4	1	3	2
Stretch	2.13	1	4.3	2.35

4.2.2 Example topologies

Some simple generic topologies are often used in practice. Such as the minimum spanning tree, topologies made up by the complete graph (often referred to as a full mesh topology), the maximum leaf spanning tree (also known as the star topology), and the 2-circulant (the ring topology). The physical distance is used as weight in Figure 4.2, which is proportional to the edge distances depicted.

We can use these examples for comparison of quality measures for topologies. The properties of these topologies are summarized in Table 4.2. The example topologies have the same order, but different size and weight, which usually determine the cost of the network. They also have different degrees, diameter, and stretch, which are related to network performance. The degree is the (minimum or maximum) number of neighbor vertices any vertex in a graph has.

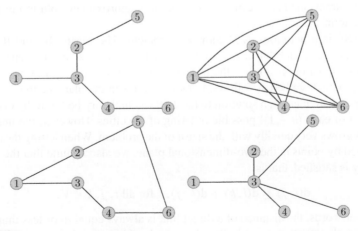

FIGURE 4.2

Some common network topologies: the minimum spanning tree (top left); the complete graph (top right); the shortest ring (bottom left); the minimum weight maximum leaf spanning tree (bottom right).

Table 4.2   Example of cost in discrete capacity steps

Variable	x_1	x_2	x_3	x_4
Capacity	2	4	8	17
Cost	2	3	5	9
Density	1	0.75	0.63	0.53

4.3 Ring topologies

A common network topology, especially for high-speed networks, are ring topologies. In Table 4.2 we can see that rings possess some desirable properties, such as low size, low stretch and provide path diversity with their minimum degree of 2. The problem of constructing such a ring while minimizing the cost is referred to as the traveling salesman problem (TSP). As the name suggests, the objective is to optimize the route (distance) of a "traveling salesman" that needs to visit n locations and return to the starting point so that the total distance of the trip is minimized.

Being one of the classics of combinatorial optimization, there are many methods for the TSP available. Some simple heuristics are easy to implement, but the result may be rather far from the optimum. Better results can be obtained by employing more sophisticated algorithms or combining different methods to improve on an initial approximation. Three heuristics that can be used as first approximations are the nearest neighbor, incremental insertion, and k-optimal methods (such as the Kernighan-Lin heuristic) (see, for example, Walukiewicz, 1991), which combine reasonable accuracy and fast execution. A solution or approximate solution to the TSP is called a tour.

The TSP is a notorious $\mathcal{NP}$-complete problem. The reason behind this is its general usefulness, and that it is easy to understand. Since a ring is a special type of network, its hardness implies the hardness of topological design in general. The problem can be described as follows. Let some points in the plane be the location of cities. If there are n cities, a solution to the TSP can in theory be found by evaluating the weight of each $(n-1)!$ possible ordering of the cities. However, the number of orderings grows very quickly with the size n of the problem. When letting the cities be represented by points in the two-dimensional plane, we also assume that the triangle inequality is satisfied, that is

$$d(i, j) \leq d(i, k) + d(k, j), \quad \text{for all } i, j, k \in V.$$

In other words, the distance of a direct link is always equal to or less than a path going through any other point in the plane. It should be noted that the TSP can be formulated as an integer program, but the methods described are more efficient (faster and/or simpler) than solving the integer program.

4.3.1 **The nearest neighbor algorithm**

One of the most straightforward methods to solve the traveling salesman problem is the nearest neighbor algorithm. Starting from an arbitrary vertex s, we connect to its nearest neighboring vertex and continue until vertex s is reached again. The algorithm, being a heuristic, does not guarantee any error bound on the solution, though. It may be necessary to try different starting vertices to find the shortest of n paths. It may also be beneficial to try the different solutions whenever there is a tie.

4.3.2 **Incremental insertion**

Incremental insertion represents another type of heuristic where vertices are inserted into a tour. Starting from a single vertex, it selects the farthest vertex not yet connected and forms a shortest path to determine the edges to connect this vertex with other vertices on this path. The farthest vertices are selected so that edges are added by decreasing cost. If s, i, and j are vertices and V_T is the set of vertices in the tour T, insert vertex j, such that

$$\max_{j \in V} \left\{ \min_{i \in V_T} \{ \mathrm{d}(s, i) + d_{ij} \} \right\},$$

where s is the starting vertex.

The minimization ensures that we insert the vertex in the position that adds the smallest amount of distance to the tour, while the maximum ensures that we pick the worst such vertex first. This works rather well, because by adding distant vertices first, paths to closer vertices are optimized better.

4.3.3 *k*-Optimal methods

Whereas the methods described so far provide a more or less accurate approximation, the k-optimal (also known as the Kernighan-Lin heuristic) provides a way of improvement by local search. The method randomly selects $k \geq 2$ edges from an initial tour, and the resulting components are reconnected by k new edges. If the cost decreases, an improvement has been found. Otherwise, the solution is discarded. The procedure is repeated until no further improvement can be made or until a pre-defined number of iterations have been performed. Usually, 2-optimal (2-OPT) or 3-optimal (3-OPT) methods are used. For $k > 3$, it is likely that the computation time becomes too great to be of practical use. The 2-optimal method in the context of topological design is also known as the branch exchange algorithm (see Figure 4.3).

When searching for edges to transform, we usually use a random selection method. It is important, however, that the edges are disjoint, i.e. do not have any vertices in common. If we search for vertices, then we also must assure that the edges actually exist in the input graph, and that the transformed edges do not exist in the input graph. If they do, the algorithm produces a degenerate solution. There are other practical considerations such as need for testing the graph for connectivity.

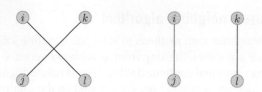

FIGURE 4.3

The 2-optimal method is applied on the traveling salesman problem where the initial solution is given by the nearest neighbor heuristic.

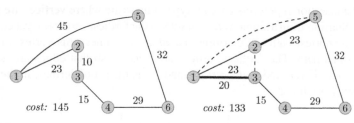

FIGURE 4.4

The best solution given by the nearest neighbor algorithm can be improved by the 2-optimal method.

Suppose that we have found an initial (heuristic) solution to the ring topology shown on the left Figure 4.4. Then, the 2-optimal method can be used to find an improvement starting from the initial topology, as shown on the right in Figure 4.4.

4.4　Spanning trees and spanners

Many heuristic network design algorithms rely on a generalization of the minimum spanning tree. One such generalization are spanners. This is a greedy approach, because the minimum spanning tree is guaranteed to connect the network at minimum cost.

Suppose then that we start with a minimum spanning tree to connect the network at minimum cost. As we have seen in Table 4.2, the minimum spanning tree does not have good properties in all respects (for example, the properties stretch and minimum degree). Therefore we would like to build upon the minimum spanning tree by adding edges in some way. This is the approach of the design method of Mansour and Peleg (1998).

4.4.1　Properties of spanners

We can see that the example topologies in Table 4.2 have different properties. For example, the mesh topology is the best possible in terms of minimum degree and stretch, but the worst in terms of size and weight. The other extreme is the minimum spanning tree, which has the lowest possible size and weight, but low minimum degree and large stretch. We ideally want to find a network with low size and weight, corresponding to the two components of the linear cost function (4.1). We would also like to maximize the performance of the network. Two simple performance measures are the minimum degree (which relates to reliability), and the stretch, which indicates the relative distance between two nodes.

Consider a graph $G = (V, E)$ with a weight function $w \to \mathbb{R}^+$ assigning a positive weight to each of the edges. Let $G' = (V, E')$ be a subgraph of $G = (V, E)$ with $E' \subseteq E$. Let $w(G')$ denote the total weight of G', that is, $w(G') = \sum_{e \in E'} w(e)$. The minimum-weight spanning tree T_{min} of G is then the spanning tree minimizing $w(T_{min})$.

Definition 4.4.1.   A subgraph $G' = (V, E')$ of a complete graph $G = (V, E)$ is called a t-*spanner* of G if for every $i, j \in V$, the distance $d_{G'}$ between i and j is less than a constant t times the shortest path in the complete graph, that is,

$$d_{G'}(i, j) \leq t \cdot d_G(i, j).$$

The value of t is the stretch factor of G'.

We can make a simple modification of Kruskal's algorithm to obtain an approximate algorithm for finding t-spanners. We take as input a weighted graph $G = (V, E)$ and a real number $t > 1$. The algorithm produces as a sparse subgraph $G' = (V, E')$ with stretch factor t. In general, the problem of finding a spanner for a graph with n nodes and at most m edges with minimal stretch factor is known to be $\mathcal{NP}$-hard.

4.4.2 **A greedy algorithm for spanner construction**

Consider the minimum cost network design problem for which a vertex set V, a flow demand matrix A, and a cost function $c(\cdot)$, are given.

The link capacity is assumed to be available in whole units of capacity. Let the cost associated with adding a link between two nodes i and j consist of two components: a fixed set-up cost α, and a dynamic cost component proportional to the physical distance between d_{ij} and link capacity C_{ij}. Note that the distance component can be measured in the number of hops as well as physical distance. In this case we can simply aggregate flows to determine C_{ij} and omit the factor d_{ij} altogether. Thus, we have the cost function for a link (i, j)

$$c_{ij} = \alpha + \beta \cdot \lceil F_{ij} \rceil \cdot d_{ij}, \tag{4.1}$$

where β is a constant, F_{ij} is the aggregate flow on link (i, j) which determines the capacity C_{ij}, and d_{ij} is the physical distance between the nodes i and j. We need not consider β, since the cost function can be rescaled so that $\beta = 1$.

Note that with the cost function (4.1), it is always cheaper to route one unit of commodity over a direct link than over an alternative path, consisting of several links. Non-existing edges can formally be assigned zero capacity.

By assumption, each demand between nodes i and j, a_{ij}, is sent along a single route r_{ij}. By specifying a set of routes R for all demands, the total cost incurred by R, denoted $c(R)$, is given by Equation (4.1). Let E' denote the set of links utilized by the routes $r_{ij} \in R$, that is, all links e which have an aggregate flow $F_e = \sum_{(i,j):e \in r_{ij}} f_{ij} \neq \emptyset$. Then,

$$c(R) = \alpha |E'| + \sum_{e \in E'} \lceil F_e \rceil \cdot d_e. \tag{4.2}$$

The goal is to find such a set of routes R which minimizes this cost.

The solution to this problem is completely characterized by specifying the routes r_{ij} for each pair of nodes (i, j). Since we have a single route r_{ij} per node pair by assumption, these routes unambiguously determine the amount of traffic carried by each link. Consequently, each link is assigned enough capacity to carry the corresponding traffic.

The minimum cost network design problem is $\mathcal{NP}$-hard to approximate within a factor less than $c \log(n)$ for some constant $c > 0$ (see Mansour and Peleg (1998)).

The present algorithm for constructing t-spanners is a generalization of Kruskal's algorithm for minimum spanning trees. It takes as input a weighted graph $G = (V, E, w)$, and a parameter $t > 1$ and constructs a t-spanner $G'(V, E')$ as follows. Create a set of edges E' and sort the edges in non-decreasing order of weights as $E = \{e_1, \dots, e_m\}$. Examine the edges one by one in that order. For each edge $e_k = (i, j)$, check whether $d_{G'}(i, j) \leq t \cdot w(e_k)$, where $G' = (V, E')$. If so, discard e_k. Otherwise, add e_k to E'.

Algorithm 4.4.1 (Greedy Algorithm for Spanner Construction).

STEP 1:   Given the cost matrix $C = (c_{ij})$ and a parameter t, construct a t-spanner G' for the graph $(V, V \times V, w)$ based on the weight function $w_{ij} = c_{ij}$.

STEP 2:   For every vertex pair i, j, set f_{ij} to follow some shortest path connecting i and j in the spanner G'.

The t-spanner so constructed has the following properties. Let $T_{\min}$ be the minimum spanning tree (MST) for G constructed by Kruskal's algorithm, considering edges in the same order as in Algorithm 4.4.1. Then the graph resulting from the algorithm G' has the following properties

Lemma 4.4.2. G' is a t-spanner of G.

Lemma 4.4.3. T is a subgraph of G'.

As discussed above, specifying the routes completely characterizes the solution, that is, the required link capacities and their total cost. It is also quite straightforward to verify that the algorithm runs in time polynomial in n. It remains to analyze the quality of the constructed solution, and compare it with the optimal one. For doing so, it is useful to first obtain some bounds on the cost of the optimal solution itself. We will rely on the following two straightforward bounds.

Theorem 4.4.4.   *For every weighted graph* $G = (V, E)$, *Algorithm 4.4.1 with parameter* $t = \log(n)$ *constructs a spanner* G' *for* G *with the following properties:*

(1) $Stretch(G') \leq \log(n)$,
(2) $|E(G')| = O(n)$, *and*
(3) $w(G') = O(\log(n)) \cdot w(MST)$.

In case we are using physical distances as weights, the graph is said to be Euclidean. In this case, stronger results follow.

Theorem 4.4.5.   *For every n-vertex Euclidean graph* G, *there is a polynomial algorithm for constructing a spanner* G' *for* G *with the following properties:*

(1) $Stretch(G') \leq O(1)$,
(2) $|E(G')| = O(n)$, *and*
(3) $w(G') = O(1) \cdot w(T_{\min})$.

We can also find some useful bounds for the network costs.

Lemma 4.4.6.   *The cost* c^* *of the optimal solution for the problem is bounded below as follows:*

(1) $c^* \geq w(T_{\min}) + \alpha \cdot (n - 1)$, *where* $w(T_{\min})$ *denotes the weight of the minimum-weight spanning tree for the graph,*
(2) $c^* \geq \sum_{1 \leq i, j \leq n} a_{ij} \cdot d_{ij} + \alpha \cdot (n - 1)$.

Proof.   The first claim follows from the fact that the network must be connected, since by our assumption G is connected. Consequently, it must contain (at least) a spanning tree of the graph, the minimum of which is $T_{\min}$. The second claim follows from the fact that even if we manage to satisfy the requirements exactly and not waste any capacity, we still must invest, for each requirement a_{ij}, at least the minimal price factor $a_{ij} d_{ij}$, by the triangle inequality.

Example 4.4.1.   Assume that we have six nodes that we want to connect in an efficient fashion for a backbone network. We also assume that the link costs are given, according to expression (4.1). For the construction of the t-spanner, we deploy Algorithm 4.4.1. By successively decreasing t, an incremental topology will form where edge after edge is included until the complete graphs have been formed. Computing the cost for each such topology gives a sequence of topologies, indexed by t, such that one has minimum cost, which we select as the optimal one.

Let the flow requirement be given by the symmetric matrix

$$A = \begin{pmatrix} 0\,2\,3\,1\,2\,1 \\ 2\,0\,1\,2\,3\,3 \\ 3\,1\,0\,2\,2\,1 \\ 1\,2\,2\,0\,3\,1 \\ 2\,3\,2\,3\,0\,2 \\ 1\,3\,1\,1\,2\,0 \end{pmatrix},$$

and the cost function be proportional to the flow and physical distances in the network

$$D = \begin{pmatrix} 0 & 23 & 20 & 32 & 45 & 51 \\ 23 & 0 & 10 & 23 & 23 & 37 \\ 20 & 10 & 0 & 15 & 29 & 32 \\ 32 & 23 & 15 & 0 & 32 & 20 \\ 45 & 23 & 29 & 32 & 0 & 32 \\ 51 & 37 & 32 & 20 & 32 & 0 \end{pmatrix}. \tag{4.3}$$

Let the set-up cost of a link be $\alpha = 21$ and the cost factor $\beta = 1$. Applying shortest path routing, we calculate the aggregated flow for each network candidate, that is, for each t-value which produces a new topology. Intuitively, it seems that the most cost efficient edges to include in the topology are the ones whose inclusion shortens the routes where we have substantial traffic. Starting from a value of the stretch factor of $t = 2.2$ and gradually decreasing it, the Algorithm 4.4.1 adds edges to the graph starting from the minimum spanning tree. By recording the cost and the corresponding network configuration, the algorithm produces the approximately cost optimal topology, shown in Figure 4.5, which has size $s(G') = 9$ and stretch $t = 1.2$. The total network cost based on Equation (4.3) and the assumptions made above is

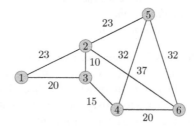

FIGURE 4.5

The 1.2-spanner for the example problem.

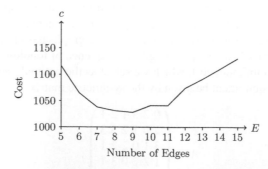

FIGURE 4.6

Costs of t-spanner designs depending on the number of edges E.

1028. Figure 4.6 shows how the cost of the solution varies with the number of edges given by the stretch factor.

4.5 Gomory-Hu design

We can formulate a topological design method based on the Gomory-Hu tree. The method is based on approximating a multi-commodity flow problem by a multi-terminal flow problem. It can be shown that the method is indeed exact for dominant single-commodity flows. Thus, the approximation lies partly in how to aggregate traffic from different origin-destination pairs. The method builds on the superposition of uniform capacity rings, suggesting that it may be suitable for design of high capacity packet networks. The network is transformed into equivalent representations, which can then be decomposed into capacity rings.

In Section 3.1, it is shown that for multi-terminal flows, every feasible network is flow-equivalent to a tree—the Gomory-Hu tree. The question here is: if we are able to construct a Gomory-Hu tree from the flow demands, can we use that to design an optimal network topology? The answer is indeed yes, under some constraints.

The Gomory-Hu tree is easily constructed from the demands. Given vertex demands a_{ij} between all vertex pairs (i, j), form the complete graph, denoted $K_n = (V, E_n)$, where there is an edge between all possible edge pairs (i, j). Let the weights on these edges be the flow requirements a_{ij}. This graph is called the complete requirement graph.

By constructing the maximum spanning tree of the complete requirement graph, we have the Gomory-Hu tree of any feasible topology. (This tree is also known as the dominant requirement tree.) The flows represented in the tree are the dominant flows and if the values are interpreted as capacities, these can accommodate any other flow requirement, considered as single-commodity flows. This follows from Theorem 3.1.1.

At this point, it is instructive to formulate the design problem as a linear program (using a similar notation as in Chapter 3). Let K_n be the complete graph with edge capacities u_{ij} and costs c_{ij}. Then the design problem can be stated as

$$\min \sum_{i,j \in V} c_{ij} u_{ij},$$

$$\sum_{i \in S; j \in \bar{S}} u_{ij} \geq \max_{i \in S; j \in \bar{S}} a_{ij} \quad \text{for all } S \subset V, \tag{4.4}$$

$$u_{ij} \geq 0 \quad \text{for all } i, j \in V,$$

using the convention that $u_{ij} = 0$ if the edge (i, j) is not present in the solution topology. Note that the constraints are for cuts, not for single edges. In words, the problem reads: given flow requirements a_{ij} between all vertex pairs, find edge capacities u_{ij} such that the maximum flow (minimum cut) is at least a_{ij} for each vertex pair (i, j), and among all such solutions, find the one which minimizes

$$\sum_{i,j \in V} c_{ij} u_{ij}.$$

The linear program has an exponential number of constraints and is therefore impractical to use in topology design.

If we have a Gomory-Hu tree, then this tree is feasible for the problem. Under the assumption of constant edge costs, it can be used for designing a topology which is optimal with respect to the sum of the edge capacities u_{ij}.

For general multi-commodity flow networks, the design principle can be extended to yield a heuristic procedure for more general edge costs subject to multi-commodity flows.

4.5.1 Constant edge costs

Consider a symmetric flow requirement a_{ij} between all vertex pairs (i, j). If we consider multi-terminal flows, that is, one flow at a time, and let all edge costs $c_{ij} = c$ be constant, it is possible to formulate a design method for constructing a topology that is optimal for single-commodity flows. Since the topology is optimal for the dominant flow, it may be used as an approximation to the optimal topology for multi-commodity flows. This procedure minimizes the sum of edge capacities only.

As shown above, the Gomory-Hu tree is given by the maximum spanning tree of the complete requirement graph. Starting from this, the second step is to decompose this tree into what are called uniform trees, having the same value u on all edges. For example, the minimum value of u on the tree will be the first uniform tree that can be removed from, or 'peeled off,' the tree. The first such tree contains all vertices, since it is the minimum u, and deducting this amount from the edge weights yields subtrees, on which the process is then repeated.

Each uniform tree is made into a ring by arbitrarily ordering the vertices, connecting them one after another, and finally connecting the last one connected with the first vertex. Then, the capacity of the peeled-off uniform tree can be divided by 2, since we have a ring and there are two paths for each demand. If there are only two vertices, the capacity remains unchanged on the edge between them. Note that since all edge costs are equal by assumption, the order of the vertices in the rings does not matter. The final topology is constructed by adding together all rings obtained in such a manner, adding the capacity requirements of the edges whenever two edges coincide.

We claim that such a solution to the design problem satisfies the flow demands and is optimal in the sense of minimizing the sum of the edge capacities. As a result of the design, the capacities are multiples of $\frac{1}{2}$ if the demands a_{ij} are integer values (see Figure 4.7).

Theorem 4.5.1. *Suppose a multi-terminal flow requirement is given and all edge costs $c_{ij} = c$ are constant. Then the Gomory-Hu design is optimal for the dominant requirement, considered as a single-commodity flow.*

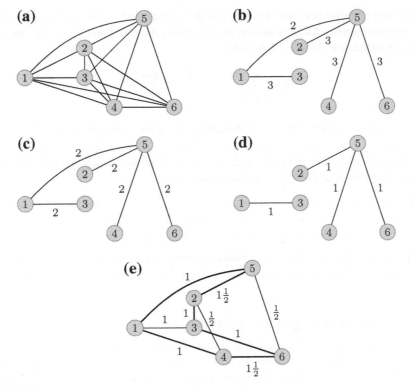

FIGURE 4.7

Example of topological design based on the Gomory-Hu tree, where edge costs are assumed to be constant.

Proof.   To show that the design is optimal for the problem, we formulate a lower bound on the value of the objective function and show that the corresponding solution is feasible, so it must indeed be optimal. Consider the complete requirement graph $K_n = (V, E_n)$. Define

$$\pi_i = \max_{j \in V} a_{ij} \quad \text{for all } i \in V. \tag{4.5}$$

From (4.4) we observe that for any feasible solution $G = (V, E_G)$ with capacities u_{ij}, we must have

$$\sum_{j:(i,j) \in E_G} u_{ij} \geq \pi_i \quad \text{for all } i \in V.$$

Summing over i, we have $\sum_{i \in V} \sum_{j \in V} u_{ij} \geq \sum_i \pi_i$. The sum of capacities is twice the sum of

$$\sum_{(i,j) \in E_G} u_{ij},$$

since each capacity is counted twice. For the constructed graph, we have equality, which must then be a lower bound on the objective function. Since $c_{ij} = c$ is constant by assumption, the objective function is

$$\min \sum_{(i,j) \in E_G} u_{ij} = \frac{1}{2} \min \sum_{i \in V} \pi_i.$$

Now, let $a'_{ij} = \min\{\pi_i, \pi_j\} = \min\{\max_k a_{ik}, \max_l a_{jl}\}$. Since $\max_k a_{ik} \geq a_{ij}$ and $\max_l a_{jl} \geq a_{ji} = a_{ij}$, clearly $a'_{ij} \geq a_{ij}$. This is the largest flow requirement such that the capacities u are not violated. Since the π values remain unchanged, the value of the objective function for a_{ij} and a'_{ij} is the same. $\square$

Example 4.5.1. Using (4.5) gives $\pi_1 = \pi_2 = \cdots = \pi_6 = 3$ and thus $\sum_i \pi_i = 18$. The sum of the capacities u_{ij} in Figure 4.9 is $\sum_{(i,j) \in E'} u_{ij} = 9 = \frac{1}{2} \sum_i \pi_i$. In contrast, the sum of the weights in the maximum spanning tree is 14.

The design procedure allows for many optimal topologies. It is more the strategy of constructing rings that is the main idea. This is extended to more general situations next.

4.5.2 Extension to general networks

We have seen that the Gomory-Hu design is optimal for the dominant single-commodity flow under the assumption of constant edge weights. This suggests that it might be a good topology for multi-commodity flows as well. In order to use the Gomory-Hu design for general networks, we need to cope with general edge costs as well as handle the effect of using multi-commodity flows rather than single-commodity flows. This extension of the Gomory-Hu design to general networks is described in Corne et al. (2000) where it is called the Union of Rings.

The immediate effect of using a more sophisticated cost function is that it adds constraints to the topological design phase. In the Gomory-Hu design, the first uniform tree is decisive for the network design. When a cost function, quite realistically, also depends on the edge distances, the order in which the edges are connected becomes important.

The cost is often assumed to be linear in link distance d_{ij} and nonlinear - possibly discrete - in capacity u_{ij}. We may therefore assume that each edge has a total cost c_{ij} which can be written

$$c_{ij} = g(u_{ij})d_{ij}, \tag{4.6}$$

where the nonlinear dependence on capacity u_{ij} is described by the function $g(u_{ij})$. Note that a fixed set-up cost per link can be incorporated by dividing the set-up cost by the distance and adding it to the total unit cost $g(u_{ij})$. As a consequence of such a cost function, we now have a TSP (discussed in Section 4.3) as a subproblem. This is in contrast to the Gomory-Hu design, where any ordering of the vertices would do.

Since all the u_{ij} are the same in the first uniform tree, we can simply write the cost $c_{ij} = \alpha_{ij} d_{ij}$ with α_{ij} is a constant defined for each edge, which may include a fixed set-up cost. These costs determine the topology, being the edge weights in the TSP problem.

The maximum spanning tree can be transformed into a linear flow-equivalent graph, which makes the decomposition easier. The linear representation is constructed by taking any vertex and the edge with the largest capacity and from this build a linear tree. The linearization is repeated as long as there are any vertices left. The procedure is summarized in Algorithm 4.5.2.

Algorithm 4.5.2 (Construction of the Linear Flow-Equivalent Graph).

STEP 1: Initially, let all vertices be marked as UNVISITED. Arbitrarily choose a starting vertex in T and mark it as VISITED. This vertex is the root of the linear graph L.

STEP 2: Select the edge with largest capacity from one of the VISITED vertices to one of the UNVISITED vertices and append that edge to L. If there is a tie, any vertex in the tie will do.

STEP 3: Mark the new vertex appended as VISITED.

STEP 4: If all vertices are VISITED, stop; otherwise go to STEP 2.

From the linear flow-equivalent graph, a minimum cost ring can be determined by solving the corresponding TSP. The capacity on each edge can then be reduced by half, still preserving the flow-equivalence. This is rather obvious since, in a ring, we can split traffic so that half going is in one direction and the other half in the other direction.

For the first uniform tree—the first tree to be peeled off the linear tree, any ordering of the vertices can be made without changing the flow-equivalence. This is used to form a minimum cost tour, by solving the corresponding TSP.

It is important to note that the Gomory-Hu tree is defined for multi-terminal flows. We arrive at a topology based on this assumption. When considering multi-commodity flows, however, these flows have to be assigned to the routes available in this topology. After assigning flows onto the topology, edge capacities need to be increased accordingly. To simplify matters, flows are routed along their shortest paths.

Note that the cost function enters twice in the procedure. First, when determining the topology, only the distance dependence is taken into account. After aggregating flows on this topology, capacities are assigned and link costs calculated using the cost function again. In this step, the cost no longer affects the primary topology.

Local adjustments may still be performed, in which the edge costs (4.6) are taken into account. Any edge which is not in the first ring can be tested by calculating the network cost with and without this edge. The traffic carried by the analyzed edge has to be rerouted onto remaining edges in the test scenario. Should the solution without this edge be cheaper, we delete the edge from the solution and reroute the traffic. A greedy approach is common, whereby the most expensive edge is tested first. The generalized Gomory-Hu design procedure can be summarized as follows.

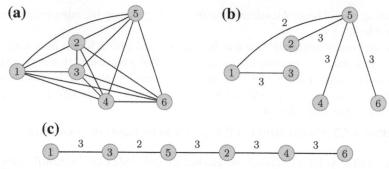

FIGURE 4.8

(a) The complete requirement graph; (b) the maximum spanning tree (dominant requirement tree); (c) the linear flow-equivalent graph.

Algorithm 4.5.3 (Extended Gomory-Hu Design Method).

STEP 1: Use the flow requirement matrix, $A = (a_{ij})$, to form the complete requirement graph K_n, labeling each edge (i, j) with the requirement a_{ij}.

STEP 2: Construct the maximum spanning tree $T_{\max}$ of K_n.

STEP 3: Convert $T_{\max}$ into a linear flow-equivalent graph L (see Algorithm 4.5.2).

STEP 4: Decompose L into a set of uniform capacity trees, connecting the first and last vertices to form a ring. The first tree is used as input to a TSP, which decides the order of the vertices. For subtrees, order the vertices as in the first ring and connect as necessary.

STEP 5: Superpose the rings formed in STEP 4 to a network topology G.

STEP 6: Aggregate traffic onto shortest paths, allocate capacity, and calculate new edge costs.

STEP 7: Analyze and remove any 'short cut' edges in G which are not cost efficient, by comparing a topology with and without this edge. When an edge is removed, reroute the flow onto other edges in G.

Steps 1 and 2 are identical to the original Gomory-Hu design method. Steps 3, 4, and 5 create the topology subject to a distance-dependent cost function. Step 6 defines the actual edge costs. In step 7, the cost efficiency may be tested as described.

Example 4.5.2. Using the same problem as in Section 4.4 and using the algorithms described, we arrive at the topology shown in Figures 4.8 and 4.9. From the linear flow-equivalent graph, we can extract the minimum capacity meeting which is 2. The ring is constructed by solving the traveling salesman problem, and the capacity is divided by 2. Next, two chains of the graph remain, the chain 2-5-6-4 and a link between nodes 1 and 3. The first forms a smaller ring of capacity $\frac{1}{2}$ and the latter just the link $(1,3)$ which needs to carry the full capacity of 1 (when wrapped around it can be seen as a double link, each with capacity $\frac{1}{2}$).

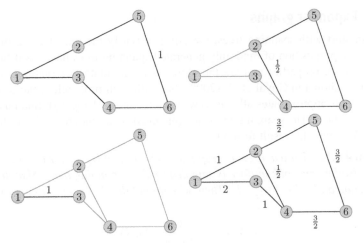

FIGURE 4.9

Gomory-Hu design. The first three panes show 4 rings that are merged to form the network in the bottom-right pane.

After the topology has been determined, flows are allocated according to their shortest paths, by careful bookkeeping. Using the cost function (4.1) for the capacities, we arrive at a solution cost $c = 1126$ higher than that for the t-spanner (1028). The t-spanner design is based on a linear cost, whereas a more general cost function can be used in the extended Gomory-Hu design. The Gomory-Hu design is in this case more efficient than the ring topology, subject to the same cost function.

Comparing this design with the t-spanner in Section 4.4, we can notice that the latter may be cheaper subject to the cost function. Since the Gomory-Hu design is not restricted to a linear cost function, the method more general. This follows from the fact that the topology design is less coupled to the cost function, due to that the flow assignment and capacity allocation are performed at a later stage. This is so, because the link costs are used twice, first when determining the optimal ring configuration, which only involves the link distances, and then again when the capacity assignment is performed. In the latter stage of the algorithm, any cost function can be used, in principle. Thus, we may arrive at a cheaper design compared to using the t-spanner if a different cost function is used.

4.6 Randomized topological design

Randomized methods are important in combinatorial optimization, because usually these problems are highly nonlinear and therefore more systematic approaches would not work. We here consider a way to randomly generate so-called splicers which can be used in Monte Carlo methods and genetic algorithms.

4.6.1 Expander graphs

We again start with spanning trees, but not necessarily the ones having minimum weight. A superposition of randomly generated spanning trees have, as it turns out, good resilience properties and are therefore good candidates for randomized algorithms. It is shown in Goyal et al. (2008) that with high probability, the union of k spanning trees approximates all cuts to within a factor of $O(\log(n))$. Such a union is called a k-splicer. The spanning trees are generated by a random walk on the graph vertices. We state this result formally.

Theorem 4.6.1. *For a d-regular graph $G = (V, E)$, let U_G^k be a random k-splicer, obtained by the union of k uniformly sampled random spanning trees. Also, let $\alpha > 0$ be a constant and $\alpha(k-1) \geq 9d^2$. Then, with probability $1 - o(1)$, for every $V' \subset V$, we have*

$$|\delta_{U_G^k}(V')| \geq \frac{1}{\alpha \log(n)} \cdot |\delta_G(V')|,$$

where $\delta_G(V')$ denotes the set of edges in G having exactly one end point in the set V'.

The proof can be found in Goyal et al. (2008). From Theorem 4.6.1 we note that the number of edges of the splicer divided by the number of edges of the complete graph is proportional to $\log(n)$. We define edge expansion as follows:

Definition 4.6.1. Let $G = (V, E)$ be an undirected graph, and for any set $V' \subseteq V$, let $\delta_G(V') = \{(u, v) \in E : u \in V', v \notin V'\}$. Then the edge expansion is

$$\min_{V' \subseteq V, 1 \leq |V'| \leq |V|/2} \frac{|\delta_G(V')|}{|V'|}.$$

We call a graph an edge expander if the edge expansion of the family is bounded below by a positive constant. Figure 4.10 shows the expansion of the t-spanner design in Example 4.4.1. The following algorithm due to Aldous and Broder (see Goyal et al., 2008) is a simple and easy to use algorithm for generation of k-splicers.

Algorithm 4.6.2 (Random Generation of k-Splicers).
    Given a graph $G = (V, E)$
STEP 1: Start a random walk at some arbitrary vertex of the complete graph corresponding to G.
STEP 2: When the walk visits a vertex for the first time, include into the tree the edge used to reach that vertex.
STEP 3: When all the vertices have been visited, stop. We have a spanning tree which is uniformly random regardless of the initial vertex.

Example 4.6.1. Consider the six nodes in Figure 4.11. Figure 4.11 shows two random walks on these nodes and the union of edges generated by the random walks. Note that the random walks have 5 edges each, and that the resulting graph has 7 edges, since the random walks have some edges in common.

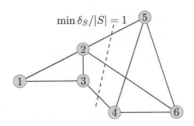

FIGURE 4.10

Expansion of the *t*-spanner design.

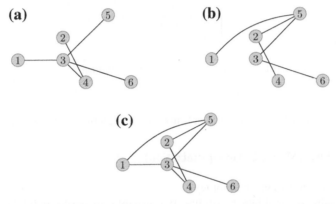

FIGURE 4.11

A 2-splicer constructed by forming the union of two randomly sampled spanning trees.

4.6.2 Monte Carlo optimization

A random algorithm is an algorithm which in some step makes a random choice. When a random algorithm may produce a result which is incorrect with a bounded probability, the algorithm is called a Monte Carlo method. Such an algorithm can be run multiple times with independent random choices in each run to decrease the probability of an erroneous result; the error in the solution can therefore be made arbitrarily small (see Figure 4.12).

The randomized generation of spanning trees can be used in Monte Carlo optimization. The principle is relying on sampling from a pre-defined class of candidates, each of which is then evaluated with respect to their feasibility and cost (see flow diagram in Figure 4.12).

In formulating a Monte Carlo optimization procedure, we need a way to generate candidate network topologies based on some general criteria. We assume that the search space is so large that we cannot generate all possible configurations and evaluate each of them. Therefore, the strategy is to generate topologies with some random sampling method, or using some heuristic or approximation as the ones mentioned in Sections 4.4 and 4.5.

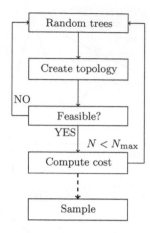

FIGURE 4.12

Flow diagram of the Monte Carlo topological design procedure.

The main steps of a Monte Carlo optimization design procedure may be described as follows:

Algorithm 4.6.3 (Monte Carlo Optimization).

STEP 1: Generate a random topology.
STEP 2: Evaluate candidate feasibility. If infeasible, discard candidate and go to step 1.
STEP 3: Calculate cost; if the cost is the lowest found so far, store this configuration and cost.
STEP 4: Repeat until a number of N iterations has been performed.

Output the optimal cost and network configuration.

We can use randomly generated splicers, and let the union of k uniformly sampled spanning trees form the candidate topology. Feasibility conditions can be reliability or performance requirements. (These topics are discussed in subsequent chapters). At present we use a simplified reliability condition: we want to find a topology where the node degree is 2 or greater. Note that this condition is not automatically fulfilled by Algorithm 4.6.2. This is clear from the fact that since the k spanning trees are sampled with replacement, they may happen to turn out to be the same tree. In this case, the resulting candidate graph will be the tree itself and at least two nodes will have degree 1.

Testing node degree is relatively easy. It is sufficient to count the number of neighbors of each node. Due to the flexibility of Monte Carlo methods, more advanced reliability conditions can easily be accommodated in the procedure. Any infeasible topology is discarded, and we generate a new one for feasibility evaluation. For any feasible topology, we next compute the cost. The Monte Carlo procedure should be repeated a large number of times, retaining the feasible topology with the lowest cost together with the cost value after each iteration. In the following example, we use the cost function (4.1). We therefore have a Monte Carlo optimization procedure for the same problem as in Section 4.4.

The most difficult task in a Monte Carlo optimization procedure may either be the generation of new candidates or the evaluation of feasibility or cost of the candidates. In this case, the generation of candidates is relatively straightforward, and so is the feasibility evaluation. The most computationally expensive part is the cost calculation. In order to compute the cost, all flows have to be determined and assigned to the edges in the network. We therefore need to solve a multi-commodity flow problem (see Section 3.3) for each network candidate before we can calculate the corresponding cost.

Example 4.6.2.   Consider the solution obtained in Example 4.4.1. Note that this topology can be obtained by a 2-splicer, that is, with two suitable spanning trees and that the complete graph cannot be formed by the union of two spanning trees. In the example, we have only nine edges which can be constructed by a 2-splicer.

The size of the search space for the Monte Carlo procedure based on 2-splicers is

$$\binom{15}{5} + \binom{15}{6} + \binom{15}{7} + \binom{15}{8} + \binom{15}{9} + \binom{15}{10} = 28,886,$$

where 15 is the total number of possible links in the graph. The number of possible cases that have to be discarded due to infeasiblity is

$$2^{(n-1)(n-2)/2} \cdot n = 6144.$$

This can be seen by considering the $n - 1$ subnetwork, that is, a network with node i, say, removed, which has a total of $2^{(n-1)(n-2)/2}$ configurations multiplied by $n - 1 + 1$ possible single links and the last one representing the case where node n is completely disconnected. The total number of feasible configurations is therefore 22,742 and the ratio of infeasible to total configurations is 21%.

The procedure 4.6.3 was run 1,00,000 times. The optimal cost was found to be 1028, corresponding to exactly the same topology as in Section 4.4 obtained by the greedy heuristic. The number of feasible topologies was counted to 75,519, giving an infeasible to total ratio of 24%. By counting the number of optimal cost configurations that was found, 9 out of 75,519, we can conclude (assuming that the candidate topologies are generated randomly) that the probability of the existence of a cheaper feasible solution is less than $1.19 \cdot 10^{-4}$.

Two important strengths of Monte Carlo design are the relative ease of implementation and the flexibility—any cost function may be used. Furthermore, the method gives an estimate of the probability of the solution being the optimal one.

Example 4.6.3.   By using a power law cost function with $\beta = 0.6$ and a fixed link set-up cost of 50, the Monte Carlo design method yielded a topology as depicted in Figure 4.13. The algorithm was run 20,000 times and the minimum cost topology was

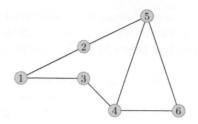

FIGURE 4.13

Monte Carlo design under a power law cost function.

found four times. Assuming randomness in the generation of topologies, the solution would be incorrect with probability $2.0 \cdot 10^{-4}$, as an upper bound.

4.7 Genetic algorithms

This section will not treat genetic algorithms in detail, but rather illustrate the principles by a relatively simple example. The literature on genetic algorithms is vast and this section is intended to serve only as an introduction of its application to network design. Genetic algorithms belong to a class of numerical methods called meta-heuristics. It mimics an evolutionary process where the "survival of the fittest" has a higher probability to reproduce, just like favorable genes promote better biological specimens. The algorithm has lent much of its vocabulary from biological evolution, such as 'population,' 'generation,' and 'reproduction.'

A genetic algorithm is based on a number of subroutines which require some care in its programming. Since it is a randomized algorithm, the quality of random number generators should be considered. Based on an initial population coded as binary strings, the procedure performs three steps in its search for an optimal solution. The steps are reproduction, recombination, and mutation, which represent the selection processes, but also allow random changes to the population in order to cover as large a search space as possible.

The first task is to find a proper binary coding that can represent any network topology. The main reason for this is that the recombination and mutation processes are harder to implement for non-binary representations. Given a number of binary represented strings, we need to evaluate their "fitness," representing the quality of a particular configuration, and which typically is a number reciprocal to its cost.

The reproduction (or crossover) operator combines randomly selected parts of two parent chromosomes to form an offspring chromosome. In addition, a random mutation is applied to the resulting configuration. The idea is that having two good solutions, combining them may lead to an even better one. Mutations are applied to allow for cases that may not have been represented in earlier generations.

We describe a genetic algorithm for network design by considering the implementation of the operators.

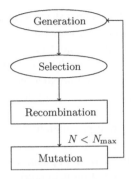

FIGURE 4.14

Flow diagram over a genetic algorithm for topological design.

Genetic algorithms are adaptive methods which may be used to solve search and optimization problems. They are based on the genetic process of biological organisms. Over many generations, natural populations evolve according to the principles of natural selection and survival of the fittest. By mimicking this process, genetic algorithms are able to 'evolve' solutions to real-world problems, if they have been suitably encoded.

Genetic algorithms work with a population of individuals, each representing a possible solution to a given problem. Each individual is assigned a fitness score according to how good a solution to the problem it is. The highly fit individuals are given opportunities to reproduce by cross breeding with other individuals in the population. This produces new individuals as 'offspring,' which share some features taken from each 'parent.' The least fit members are less likely to get selected for reproduction and 'die out.' A simplified flow diagram is shown in Figure 4.14.

A whole new population of possible solutions is thus produced by selecting the best individuals from the current 'generation,' and mating them to produce a new set of individuals. Thus new generation contains a higher proportion of the characteristics possessed by the good members of the previous generation. In this way, over many generations, good characteristics are spread throughout the population. By favoring the mating of the more fit individuals, the most promising areas of the search space are explored. If the genetic algorithm has been designed well, the population will converge to an optimal solution to the problem.

The evaluation function, or objective function, provides a measure of performance with respect to a particular set of parameters. The fitness function transforms that measure of performance into an allocation of reproductive opportunities. The evaluation of a string representing a set of parameters is independent of the evaluation of any other string. The fitness of that string, however, is always defined with respect to the other members of the current population. In the genetic algorithm, fitness is defined by f_i/f_A, where f_i is the evaluation associated with string i and f_A is the average evaluation of all the strings in the population.

Fitness can also be assigned based on a string's rank in the population or by sampling methods, such as tournament selection. The execution of the genetic algorithm is a two-stage process. It starts with the current population. Selection is applied to the current population to create an intermediate population. Then recombination and mutation are applied to the intermediate population. The process of going from the current population to the next population constitutes one generation in the execution of a genetic algorithm.

In the first generation the current population is also the initial population. After calculating f_i/f_A for all the strings in the current population, selection is carried out. The probability that strings in the current population are copied (that is, duplicated) and placed in the intermediate generation is in proportion to their fitness.

Individuals are chosen using stochastic sampling with replacement to fill the intermediate population. A selection process that will more closely match the expected fitness values is remainder stochastic sampling. For each string i where f_i/f_A is greater than 1.0, the integer portion of this number indicates how many copies of that string are directly placed in the intermediate population. All strings (including those with $f_i/f_A < 1.0$) then place additional copies in the intermediate population with a probability corresponding to the fractional portion of f_i/f_A. For example, a string with $f_i/f_A = 1.36$ places 1 copy in the intermediate population, and then receives a 0.36 chance of placing a second copy.

Remainder stochastic sampling is most efficiently implemented using a method known as stochastic universal sampling. Assume that the population is laid out in random order as in a pie graph, where each individual is assigned space on the pie graph in proportion to fitness. An outer roulette wheel is placed around the pie with N equally spaced pointers. A single spin of the roulette wheel will now simultaneously pick all N members of the intermediate population.

The first step in the implementation of any genetic algorithm is to generate an initial population. The splicers from Section 4.6 can be used for this purpose. It is necessary to decide the size of the initial population and the number of generations (that is, the number of iterations) N that the algorithm will be run. Usually a trial-and-error approach is the only way to determine these algorithm parameters.

Binary representation.   In order to implement the functions of a genetic algorithm, we need to have a binary representation of each candidate. Since the edges are the most important part of a network topology, it is advantageous to find a mapping from each node pair (i, j) to a number identifying the corresponding edge e_{ij}. Suppose that we have an enumeration of the vertices in graph. For any node pair (i, j), let l denote the largest number, and s the smallest number, so that $l = \max\{i, j\}$ and $s = \min\{i, j\}$. Then let the string position be indexed by $l \cdot (l - 1)/2 + s$. If an edge is present, we put a 1 in that position, if not a 0.

Fitness function.   The fitness function may include both feasibility constraints and cost evaluation. We may require, for example, that each candidate topology is connected.

The fitness function determines the probability assigned to each candidate that it will be chosen during the reproduction phase. Since we are interested in the lowest cost, we can use a number reciprocal to the cost, so that a lower cost yields higher probability. For example, we can assign the fitness value

$$C_{max}/C_i, \quad \text{for each candidate } i,$$

where C_{max} is the cost of the most expensive candidate in the population. To obtain probabilities, we then normalize the numbers so that they sum to one (see Figure 4.14)

Reproduction. Based on the fitness (or rather the normalized fitness), a selection of all candidates in the current population is made, proportional to their respective fitness. A common method is the so-called roulette wheel selection. An easy implementation of this is to generate a random number r uniformly in the interval [0, 1]. Next we sum the individual fitness values for the candidates, where edges are selected in the order of the binary representation of the topology, until the sum just but exceeds r. This is repeated as many times as there are candidates in the population. For each new generation, the fitness function for its members needs to recomputed.

Recombination (Crossover). The next step is to select two candidates, each with equal probability, and a random number c between zero and the length b of the binary representation. A new candidate is formed by combining the c first binary strings from the first candidate and the $b - c$ last binaries from the second candidate to form an 'offspring.' This is repeated for as many times as the population size.

Mutation. For each new candidate created by the recombination step, we 'scan' the binary string and generate a uniform random number for each allele of the chromosome. If the number is less than a small probability p, we change the current allele from its current value to its complement (that is, from 1 to 0 or vice versa). This step increases the search space. This is the final step in the creation of a new generation, after which the algorithm returns to the reproduction phase.

Example 4.7.1. A genetic algorithm can be formulated using the above principles. As feasibility condition we require that the graph corresponding to the topology is connected, otherwise would the routing of flow fail. This is therefore a minimal assumption. Thus, we have a larger search space than in Example 4.6.2.

Actually, the size of the search space is

$$\binom{15}{5} + \binom{15}{6} + \binom{15}{7} + \binom{15}{8} + \binom{15}{9}$$
$$+ \binom{15}{10} + \binom{15}{11} + \binom{15}{12} + \binom{15}{13} + \binom{15}{14}$$
$$+ \binom{15}{15} = 30,827,$$

where 15 is the total number of possible links in the graph. The number of possible cases that have to be discarded due to infeasiblity is

$$2^{(n-1)(n-2)/2} = 1024.$$

The number of infeasible configurations is therefore 3%. Using a population size of 40 and going through 500 generations, we find the same result as in Example 4.4.1 and in Example Example 4.6.3 with cost 1028. Thus, after evaluating and sampling 20,000 configurations, we arrive at the same result as previously. Note, however, that we have not imposed any other condition on the network than that it must be connected. This makes the search space here 31% larger than that of Example 4.6.2.

4.8 Resource allocation

So far, we have made the simplifying assumption that the capacity units available in the design process have a single discrete capacity unit. In reality, equipment usually comes in different (usually integer) capacities and costs. This then forms a subproblem whose solution will impact the cost of the final solution. The subproblem is referred to as capacity allocation and the mathematical equivalent is called the knapsack problem. Thus, after having assigned all flows to the edges in the network, taking overhead into account depending on network technology, such as requirements for minimum blocking or maximum delay, we want to select equipment of available capacities so that the sum of capacities per edge covers the required capacity demand at minimum cost. Since it may not be possible to match the required capacity demand exactly, an iterative procedure of capacity allocation and flow rerouting might be required to solve this design phase.

4.8.1 The Knapsack problem

In this section, we will consider solution methods of the integer knapsack problem:

$$v(K) = \min \sum_{j=1}^{n} c_j x_j,$$

$$\sum_{j=1}^{n} a_j x_j \geq b, \tag{4.7}$$

$$x_j \quad \geq 0 \text{ and integer}, \quad j = 1, \ldots, n.$$

Assuming that there are n link types available, the numbers c_j are the costs for a single link of link type j, x_j the selected quantity of link type j, a_j the capacity of link type j, and b_j the required capacity as given by solving a multi-commodity flow problem. Without loss of generality, we assume that all data in K are positive integers.

4.8.2 Dynamic programming

Dynamic programming is a powerful and easily implemented method for solving the integer knapsack problem. For dynamic programming to work, the flows and capacities must be integers. However, the costs may be real numbers. This is not much of a limitation, since rational numbers can be used by multiplying by an appropriate factor to yield integer values.

Dynamic programming solves the problem (4.7) by using the recursive relation

$$F_j(y) = \min\{F_{j-1}(y), F_j(y - a_j) + c_j\},$$
$$F_0(y) = \infty, \qquad \qquad (4.8)$$
$$F_j(y) = 0, \quad \text{for } y \le 0.$$

The equation gives $F_j(y)$, which is the minimum cost using only the first j link types on edges with flow y, that is

$$F_j(y) = \min \sum_{i=1}^{j} c_j x_j, \quad j < n,$$

with the condition that

$$\sum_{i=1}^{j} a_i x_i \ge y, \quad y < b.$$

Equation (4.8) works by deciding first how to best cover all flow values using only one line type. Then when a second line type is considered, it looks at all possible ways of dividing the flow between the two line types. When a third line types is added, Equation (4.8) is simply choosing the best amount of flow to cover with the third line type, leaving the rest of the flow to be covered optimally among the first two line types (it had decided those questions optimally after the first two iterations of the recursion). The term $F_{j-1}(y)$ means 'don't take any more of the ith line type,' while the term $F_j(y - a_j) + c_j$ means 'take at least one more of the ith line type' in the final decision. Since all the previous decisions have been made optimally, the only decision to make in Equation (4.8) is whether one more instance of line type i is necessary to cover the flow optimally.

If the unit costs are the same for all edges in the network, only one instance of the capacity assignment problem needs to be solved, with b being the maximum flow value of any edge in the network. If the local tariffs add a fixed charge per line (based on line type) in addition to the distance cost, then that fixed cost must be divided by the length of the line before being added to the cost per unit distance. The new unit cost would be computed as follows:

$$c_k(i, j) = c_k^1 + \frac{c_k^2}{d_{ij}}. \qquad \qquad (4.9)$$

In Equation (4.9), $c_k(i, j)$ is the unit cost of line type k on edge (i, j); c_k^1 is the cost per unit distance of line type k; c_k^2 is the fixed cost for line type k; and d_{ij} is the distance from node i to node j. When both cost per unit distance and fixed costs appear in our cost function, the capacity assignment problem must be recalculated for every edge in the topology (since unit cost is now a function of the distance of each edge). Still, the solution given by the dynamic programming method outlined above would be optimum.

Example 4.8.1.   A manufacturer of microwave transmission equipment provides, say, microwave equipment in capacities 2, 4, 8, and 17 Mbps, corresponding to 1, 2, 4, and 8 E1 links. Choosing equipment in a cost effective manner constitutes an integer knapsack problem. We assume that all links in the network have a distance such that one hop is required per link. Thus, the unit cost is the cost of the equipment and installation, which is assumed to be the same for all links.

Suppose the costs (in some arbitrary unit) are given as in Table 4.2. We can, by forming the ratio between cost and capacity, determine the cost density. This reflects the "economy of scale" as it decreases with capacity.

Let $v_k(y)$ be the value of the knapsack subproblem defined for the first k variables and for the right-hand side $b = y$,

$$v_k(y) = \max \left\{ \sum_{j=1}^{k} c_j x_j \Big| \sum_{j=1}^{k} a_j x \leq y, x_j \geq 0, x_j \in \mathbb{Z}, j = 1, \ldots, k \right\}. \quad (4.10)$$

If $k \geq 2$, then for $y = 0, 1, \ldots, b$, we may write (4.10) in the form

$$v_k(y) = \max_{x_k=0,1,\ldots,\lfloor y/a_k \rfloor} c_k x_k +$$

$$+ \max \left\{ \sum_{j=1}^{k-1} c_j x_j \Big| \sum_{j=1}^{k-1} a_j x_j \leq y - a_k x_k, x_j \geq 0, x_j \in \mathbb{Z}, j = 1, \ldots, k-1 \right\}.$$

The expression in the brackets equals $v_{k-1}(y - a_k x_k)$, so we may write (4.10) as

$$v_k(y) = \max_{x_k=0,1,\ldots,\lfloor y/a_k \rfloor} \{c_k x_k + v_{k-1}(y - a_k x_k)\}.$$

Putting $v_0(y) = 0$ for $y = 0, 1, \ldots, b$ we extend (4.10) for the case $k = 1$. The relation (4.10) expresses the so-called dynamic programming principle of optimality, which says that regardless of the number of the kth item chosen, the remaining space, $y - a_k x_k$, must be allocated optimally over the first $k - 1$ items. In other words, looking for an optimal decision at the n-stage process, we have to take an optimal decision at each stage of the process.

If, for a given y and k, there is an optimal solution to (4.10) with $x_k = 0$, then $v_k(y) = v_{k-1}(y)$. On the other hand, if $x_k > 0$, then in an optimal solution to (4.10) one item of the kth type is combined with an optimal knapsack of size $y - a_k$ over the first k items. Thus, we have

$$v_k(y) = \max\{v_{k-1}(y), c_k + v_k(y - a_k)\}, \qquad (4.11)$$

for $k = 1, \ldots, n$ and $y = 0, 1, \ldots, b$. Then, obviously $v(\mathbb{Z}) = v_n(b)$.

The computation of $v_k(y)$ requires by (4.11) comparison of two numbers. Thus the computational complexity of the dynamic programming is $O(nb)$. Dynamic programming is not a polynomial algorithm for solving K, since the length of data is K is $O(n \log(n))$.

4.8.3 Branch-and-bound

The branch-and-bound approach can easily be described by the example above. In essence, the problem is solved as a linear program, probably giving a non-integer number as the optimum values. We then choose the integer upper and lower bounds of one of these variables and solve two new linear programs with this restriction. In Example 4.8.1, the optimal value is choosing a variable with the lowest density. We then choose the upper bounding integer (the ceiling) and the lower bounding integer (the floor), (which then must be feasible) and solve the problem again for these two different cases. By continuing fixing variables, we obtain a tree of solutions, which in one of its bottom leaves gives the optimal solution.

4.9 Summary

The topic of the chapter is network topology optimization with respect to capacity. This assumes that the most expensive components of a network are the links in the network. The network cost obviously also depends on the modeling of the cost of the links. The simplest cost model is a linear and continuous function of distance and capacity, but a more realistic model is both discrete and upwards convex, which reflects the scale of economy. In other words, the cost per unit decreases as the capacity increases.

A number of methods are presented. The Mansour-Peleg (Mansour and Peleg, 1998) and the Gomory-Hu (Corne et al. 2000) methods can be described as combinatorial approximations. Applied to the same problem, the methods give different results. This illustrates the complexity of topology design: changing some assumptions may yield a fundamentally different solution. A Monte Carlo optimization method and a genetic algorithm are used for the same problem as the one approximated by spanners. These methods give the same result as the approximation and, in addition, a probabilistic error bound. It should be noted, however, that the Gomory-Hu approximation is exact for constant link costs and can therefore be used for constructing an upper bound of the cost. The traveling salesman problem appears as a subproblem in the Gomory-Hu design. Some methods are described in the chapter (see for example Walukiewicz, 1991). Examples of references discussing Monte Carlo are Thompson (2000), Gilks et al. (1996), and genetic algorithms in Beasley et al. (1993), Whitley (1993), and Ladd (2000).

The topological design has tremendous impact on other design activities. Should a link be added or removed, the network flow and performance metrics need to be recalculated. On the other hand, a topology may have to be revised, should, for example, traffic volume or distribution change. By including a fixed set-up cost for links, the cost model is made somewhat insensible to changes due to small changes in traffic pattern. This cost can reflect not only actual charges, but also the cost of manpower in reconfiguration work. It is therefore a parameter of great importance.

Stochastic Processes and Queues

Traffic is inherently random. Calls arrive at random points in time and last for random intervals of time. Nevertheless, we would like to say something about the system and the service it provides. The flows discussed in Chapters 2, 3, and 4, have all been assumed to be constant. On a macroscopic level, this may be justified, but not on a link level, where random variations in traffic demand may lead to deterioration of the performance of the network.

In this chapter, some principles of traffic modeling and queueing theory are discussed. Traffic is modeled by a *stochastic process* (random process), and a node in a network is modeled by a queue, a model which can be used to analyze the characteristics of the traffic and the system. We refer to this as *performance analysis* of the queue or network.

Often, we have a system given as well as a traffic pattern—measured or assumed— and seek some measures of the performance of the system and its interaction with the traffic flow. This activity is called performance analysis of the queue, system of queues or network. Being able to calculate system performance measures enables design of a system which meets expected target performance limits. Being able to predict system performance, it is possible to devise a trial and error design methodology (at a minimum), but many much more systematic and efficient procedures exist.

For the most part, in the sequel, we assume that the network topology is given, as well as some estimation of traffic characteristics - measured or estimated in some way.

Some fundamentals on traffic arrival processes and inter-arrival time distributions are given in Section 5.1. Modeling with queues and results about state probabilities and performance measures are discussed in Section 5.2. Some general theories of Markov chains applied to queues and systems of queues follow in Section 5.3. Next, the Erlang B-formula and some of its extensions are derived in Section 5.4, which constitute the basis for classical performance analysis. This also forms the basis of the overflow theory of Section 5.5.

5.1 Traffic and blocking

The purpose of a communication network is to carry traffic. Traffic can be divided into offered traffic, the total amount of traffic actually generated by users; carried traffic, the proportion of offered traffic that is carried by the network; and blocked traffic. The carried traffic is the difference between the offered traffic and the blocked traffic. Blocking is the result of an instantaneous insufficiency of network resources.

The rate, in particular if Poisson-distributed, at which calls arrive from external users to the network is denoted by λ in the text unless otherwise stated. Offered traffic within a network is usually denoted by ν or γ, depending on the context.

We will refer to a unit of traffic as a call or a *job*, regardless whether it is traditional speech telephony or some other type of service. An entity generating calls or jobs is referred to as a user. In this chapter, we introduce the classical theory of loss networks - networks where a call is connected only if there are available resources end-to-end at call set-up.

A user will experience the likelihood of being denied service by the network at its service request as a quality measure of the service, called the Grade of Service (GoS). It is commonly defined as the number of blocked calls divided by the number of offered calls. As such, it can also be seen as a measure of customer satisfaction with the service provider. Other quality measures of a single connection such as delay and packet loss, are related to packet-switched networks. Such measures are referred to as Quality of Service (QoS) parameters, and are discussed in Chapters 7 and 8. Thus, GoS refers to the call being rejected before connection is set-up and QoS the quality of the connection once the connection has been established. In this text, we will sometimes consider the GoS as a QoS parameter on system level, since call blocking and packet loss are closely related.

Blocking (or congestion) can be defined in different ways. Call blocking is the proportion of call attempts that are denied service to the total number of calls generated. In contrast, time blocking is the probability that network resources are in a state where they are fully occupied and therefore cannot accept any more traffic. A third measure is traffic blocking which is the total offered traffic minus carried traffic, divided by the offered traffic. If the arrival intensity λ is independent of system state, which is a property of the Poisson arrival process, these three measures are identical. This property is often referred to as the PASTA property—Poisson Arrivals See Time Averages. Traffic blocking is the most representative measure as it expresses the actual proportion of traffic lost. Time blocking, on the other hand, is often easier to compute and is sometimes used in place of traffic blocking. Blocking can be defined per link, node, path, or the entire network.

When a call is serviced by the network, it is assumed to occupy network resources for a period of random duration. This period of time is called the holding time of the call and its probability distribution the holding time distribution. The mean of this distribution is denoted by $1/\mu$. In many contexts, it is convenient to consider traffic intensity, the mean number of simultaneous calls in progress, which is denoted by $\rho = \lambda/\mu$. This chapter is an introduction to traffic theory, which is the probabilistic relations between traffic, network resources, and quality of service. These relations are shown schematically in Figure 5.1.

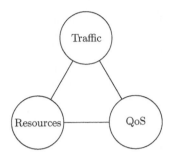

FIGURE 5.1

The traffic-resource-QoS triangle.

5.1.1 Point processes

Calls arrive to a resource (a node or a link) at random times. At each point in time, we usually assume that only a single call may arrive. This is a reasonable assumption if the calls arrive independently of each other and the time unit is short enough. The arrivals in time are described by a stochastic process (or random process). The special feature of arrivals being single at any random time instant is called a (simple) stochastic point process, or just point process. In a simple point process we can therefore distinguish any two arrivals from each other in time. Figure 5.2 shows an arrival process, which is a point process. This process only describes the pattern of arrivals, but nothing else about the call, such as call duration.

The most important point process is the Poisson process, shown in Figure 5.2. It has a very fundamental role due to its properties. It is defined in continuous time and takes on values $\{0, 1\}$ only (on sufficiently short time scales). It therefore has a discrete state space, consisting of the possible values it can assume.

We can construct a so-called counting process from a point process as follows. Let T_i be the arrival time of the i th call. The arrivals are ordered in time, so that

$$0 = T_0 < T_1 < T_2 \ldots$$

We can take the first observation at time $T_0 = 0$. The total number of arrivals up to time t, defining the half open interval $[0, t)$, is denoted N_t. The variable N_t is the sum of the arrivals up to t, or the arrival count. This process is also a random variable in continuous time and having discrete state space. The value of N_t is clearly

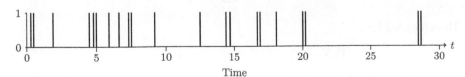

FIGURE 5.2

A Poisson call arrival process.

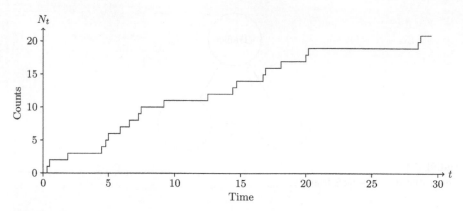

FIGURE 5.3
The counting process associated with a Poisson process.

non-decreasing. The counting process corresponding to the arrival point process in Figure 5.2 is shown in Figure 5.3.

The time difference between two successive arrivals is

$$X_i = T_i - T_{i-1}, \quad i = 1, 2, \ldots \tag{5.1}$$

The quantity X_i is called the inter-arrival time, which again is a random variable, and the distribution of this interval is called the inter-arrival time distribution. The ith arrival time can be expressed $T_i = \inf\{t : N(t) = i\}$, that is, the shortest time it takes to record i arrivals.

Corresponding to the two random variables N_t and T_i, a point process can be characterized either in number representation N_t, where the time interval t is kept constant, and the accumulated number of calls arriving during any such time interval is the random variable N_t. Note that for such a fixed t, the number of arrivals can be greater than a single arrival. In interval representation, on the other hand, the variable T_i is observed until a constant number i of arrivals has occurred. The distinction in representation is important in constructing statistical estimators and simulation methods.

The fundamental relationship between the two representations is given by the relation

$$N_t < i \text{ if and only if } T_i = \sum_{j=1}^{i} X_j \geq t, \quad i = 1, 2, \ldots \tag{5.2}$$

Therefore, we have

$$\mathbf{P}(N_t < i) = \mathbf{P}(T_i \geq t), \quad i = 1, 2, \ldots,$$

known as the Feller-Jensen identity.

Analysis of point processes can be based on both these representations, as they are equivalent. When $i = 1$ in T_i, we obtain call averages by dividing by the total number of calls. The number representation N_t is averaged over time and result in time averages. Mostly, the number representation is used.

For the period of analysis, we will usually require that an arrival process does not change its characteristics with time, so that it is stationary. This is defined by:

Definition 5.1.1. The process X_t is called strictly stationary, if the joint distributions of $(X_{t_1}, \ldots, X_{t_k})$ and $(X_{t_1+h}, \ldots, X_{t_k+h})$ are the same for all positive integers k and all integers $t_1, \ldots, t_k, h$.

This means intuitively that the graphical representation over two time intervals of equal length should have the same statistical properties, or look the same. We usually tacitly assume that any data provided has been collected so that it is approximately stationary. Another important property is independence. This can be expressed in the future evolution of a process only depending upon its current state.

Definition 5.1.2. The probability that i number of events ($i, j \geq 0$ integer) take place in $[t_1, t_1 + t_2)$ is independent of events before time t_1, if, for a process X_t at time points t_0, t_1, and t_2, we have

$$\mathbf{P}(X_{t_2} - X_{t_1} = i | X_{t_1} - X_{t_0} = j) = \mathbf{P}(X_{t_2} - X_{t_1} = i)$$

If this is true for all t, then the process is called a Markov process, so that the future evolution only depends on the present state, but is independent of how this state was reached. This is also referred to as the lack of memory property. If this property only holds for certain time points (such as arrival times), these points are called equilibrium points or regeneration points. The process then can be described as having limited memory, and the details of its evolution need only be recorded back to the latest regeneration point.

5.1.2 The Poisson process

Now, if we fix a point in time t and look ahead in a short time interval $t + h$, then a call may or may not arrive in the interval of length h. If h is small and the arrivals independent, then we may assume that the probability of a call arriving in the interval of length h is approximately proportional to h. Since h is small, the probability of two or more arrivals in this interval can be considered negligible.

We formally define a Poisson process as follows. We change notation from N_t to $N(t)$ to highlight that the Poisson is a discrete process in continuous time.

Definition 5.1.3. A Poisson process with intensity λ is a stochastic process $X = \{N(t) : t \geq 0\}$ taking values in $S = \{0, 1, 2, \ldots\}$ such that

(a)
$$N(0) = 0, \text{ and if } s < t \text{ then } N(s) \leq N(t), \tag{5.3}$$

(b)
$$\mathbf{P}(N(t + h) = n + m | N(t) = n) = \begin{cases} \lambda h + o(h) & \text{if } m = 1 \\ o(h) & \text{if } m > 1 \\ 1 - \lambda h + o(h) & \text{if } m = 0 \end{cases}, \tag{5.4}$$

(c) if $s < t$ then the number $N(t) - N(s)$ of arrivals in the interval $(s, t]$ is independent of the times of arrivals during $[0, s]$.

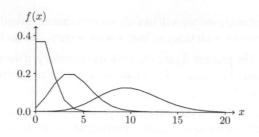

FIGURE 5.4

The Poisson distribution $Po(\lambda)$ for $\lambda = 1, 4, 10$.

The process $N(t)$ represents the number of arrivals of the process up to time t, where $N(t)$ is the counting process. The Poisson process is one of the simplest examples of continuous-time Markov processes. (A Markov process with discrete state space is usually referred to as a Markov chain).

Theorem 5.1.1. *The counting process $N(t)$ has the Poisson distribution with parameter λt, that is*

$$\mathbf{P}(N(t) = j) = \frac{(\lambda t)^j}{j!}e^{\lambda t}, \quad j = 0, 1, \dots. \tag{5.5}$$

The probability densities for three Poisson distributed variables with different intensities are shown in Figure 5.4.

Proof.   Conditioning of $N(t + h)$ on $N(t)$ gives

$$\mathbf{P}(N(t + h) = j) = \sum_i \mathbf{P}(N(t) = i)\mathbf{P}(N(t + h) = j)|N(t) = i)$$

$$= \sum_i \mathbf{P}(N(t) = i)\mathbf{P}((j - 1) \text{ arrivals in } (t, t + h])$$

$$= \mathbf{P}(N(t) = j - 1)\mathbf{P}(\text{one arrival})$$

$$+ \mathbf{P}(N(t) = j)\mathbf{P}(\text{no arrival}) + o(h).$$

Thus, $p_j(t) = \mathbf{P}(N(t) = j)$ satisfies

$$p_j(t + h) = \lambda h p_{j-1} - \lambda p_j, \quad \text{if } j \neq 0,$$

$$p_0(t + h) = (1 - \lambda h)p_0(t) + o(h).$$

Subtracting $p_j(t)$ from each side of these equations, dividing by h, and letting $h \to 0$ gives

$$p'_j(t) = \lambda p_{j-1}(t) - \lambda p_j(t), \quad \text{if } j \neq 0, \tag{5.6}$$

$$p'_0(t) = -\lambda p_0(t). \tag{5.7}$$

On the boundary,

$$p_j(0) = \begin{cases} 1 \text{ if } j = 0 \\ 0 \text{ otherwise} \end{cases}.$$

These expressions form a system of differential-difference equations for the probabilities $p_j(t)$. Solving (5.7) subject to the condition $p_0(0) = 1$ gives

$$p_0(t) = e^{-\lambda t}. \tag{5.8}$$

Substitution into (5.6) with $j = 1$ yields

$$p_1(t) = \lambda t e^{-\lambda t}, \tag{5.9}$$

and by induction, we have

$$p_j(t) = \frac{(\lambda t)^j}{j!} e^{\lambda t}, \tag{5.10}$$

$\square$

Theorem 5.1.2. $X_1, X_2, \ldots$ *are independent exponential random variables with parameter* λ, *that is,* X_i *has the cumulative distribution function*

$$\mathbf{P}(X_i \le t) = 1 - e^{-\lambda t}. \tag{5.11}$$

Proof. For the variable X_1 we have

$$\mathbf{P}(X_1 > t) = \mathbf{P}(N(t) = 0) = e^{\lambda t},$$

and so X_1 is exponential. Now, conditional on X_1,

$$\mathbf{P}(X_2 > t | X_1 = t_1) = \mathbf{P}(\text{no arrival in } (t_1, t_1 + t] | X_1 = t_1).$$

The event $\{X_1 = t_1\}$ relates to arrivals during the time interval $[0, t_1]$, whereas the event {no arrival in $(t_1, t_1+t]$} relates to arrivals after t_1. These events are independent, by Definition 5.1.3, and therefore:

$$\mathbf{P}(X_2 > t | X_1 = t_1) = \mathbf{P}(\text{no arrival in } (t_1, t_1 + t]) = e^{\lambda t}.$$

Thus, X_2 is independent of X_1, and has the same distribution. Similarly,

$$\mathbf{P}(X_{i+1} > t | X_1 = t_1, \ldots, X_i = t_i) = \mathbf{P}(\text{no arrival in } (t_1, t_1 + t]),$$

where $T = t_1 + t_2 + \cdots + t_i$, and the proof is completed by induction on i. $\square$

It can be seen that the process $N(t)$, constructed by (5.2) from a sequence $X_1, X_2, \ldots$, is a Poisson process, if and only if, the X_i are independent identically distributed exponential variables (using the lack of memory property). It turns out that the exponential distribution is the only continuous-time distribution with this property. The probability densities for two exponentially distributed variables with different intensities are shown in Figure 5.5.

5.1.3 Characterization of traffic

We already mentioned that the traffic intensity is defined as

$$\rho = \lambda / \mu, \tag{5.12}$$

where λ is the number of arrivals per time unit and μ^{-1} is the mean holding time of the call. The unit Erlang is commonly used, although the quantity is dimensionless.

The traffic intensity describes the mean number of calls simultaneously in progress. It is used for offered traffic, carried traffic, and blocked traffic.

Example 5.1.1.   The traffic offered to a switch is $\lambda = 1200$ per hour, and the mean holding time is 4 minutes. Then the traffic intensity is $1200 \cdot 4/60 = 80$ Erlang. It is dimensionless, because we have connections per hour times hour.

The typical intensity per user is in the range 0.01–0.06 Erlang, and possibly as much as ten times higher for switch board operators. Blocked traffic is the difference between offered traffic and carried traffic, $\rho_b = \rho_o - \rho_c$. The traffic measured over a certain time interval is called the traffic volume and is measured in Erlang-hour or call-minutes. By dividing the traffic volume by the time interval of the record, we have the average traffic intensity.

It is not economically feasible to design a network for rare, extreme traffic peaks. There may also be other reasons for an unsuccessful call, such as a number error or that the receiving party is unavailable. Rather, a representative, typical high traffic load is used as design parameter, allowing for a small amount of blocked (lost) traffic. When measuring traffic volume, the time interval should be chosen long enough to get statistically reliable measurements of typical high load, while smoothing out extreme peaks, but not so long that the long-term time variations impact the reading. Such time intervals, giving busy hour traffic, can be defined in many ways (ITU-T, 1993). The blocking probability is often chosen in the range 0.5–2%.

5.2   Modeling with queues

In networks, we are mostly interested in modeling links, since we usually assume that nodes have sufficient capacity of handling the traffic demand on each site, and links therefore are the only source of blocking. A link can be understood as a sender/receiver device on each side of a transmission line with a channel for each device pair. In traditional digital networks, time division duplex (TDD) is often used, where a channel is a time slot used for sending and receiving data by a dedicated sender/receiver pair.

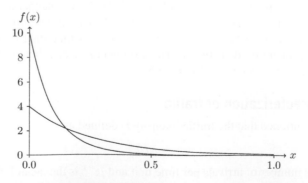

FIGURE 5.5

The exponential distribution $Exp(\lambda)$ for $\lambda = 4, 10$.

In loss systems, a link is traditionally modeled by an $M/M/n/n$ queue. In this model, calls are arriving according to a Poisson process having exponential holding times to a link equipped with n servers (channels) where no call is allowed to wait for service. When all servers are busy, any additional arriving traffic is lost.

In contrast, a data link is usually modeled by an $M/M/1/b$ queue, with a single server and a buffer b which can be finite or infinite. Chapters 7 and 8 discuss modeling of packet-switched networks, where this link model will be deployed.

In this chapter we will focus on the former system, the $M/M/n/n$ queue, starting with some general remarks about queueing systems. We use the terms system and queue rather synonymously in the text. Strictly speaking, a queue has a waiting facility which allows the formation of a queue, but a system may not.

5.2.1 Characteristics of queueing processes

In general terms, a queueing system consists of a number of facilities (or servers) which are serving arriving jobs (or customers). We use the term 'job' to describe a call or service request of some kind. A job may thus enter a system and either be served immediately and leave after completion of service or, if a server is not immediately available, the job has to wait for service or be turned away and be lost.

In most cases, the following six characteristics can be used to describe a queueing system:

(1) The arrival pattern of jobs—the probability distribution of inter-arrival times and the probability of jobs arriving in batches;
(2) The service pattern of servers—the probability distribution of service times;
(3) The queueing discipline, where the two most important are first come, first served (FCFS) and processor sharing (PS) disciplines;
(4) The system capacity;
(5) The number of servers or channels;
(6) The number of service stages.

In most queueing problems in telecommunications, the arrival process of jobs is stochastic, and it is necessary to know the probability distribution of the times between successive arrivals of jobs. In some cases jobs may arrive simultaneously (batch or bulk arrivals), but this situation is not considered here.

An inter-arrival time distribution may change over time, be nonstationary, but for our purposes it is assumed to be stationary, describing the peak traffic, measured during busy hour. A special situation of state-dependent arrivals appears in situations where we have different priority classes of jobs, for example. A similar argument may apply to the characteristics of service times as well. The two most common assumptions on service times are that they are deterministic or exponentially distributed. Some results for general service times can also be derived.

The most likely situation is having random arrivals and service times, and the state of the queue—the number of jobs in service or waiting for service—will therefore also follow a stochastic pattern. It is usually assumed that the arrival process and departure process are mutually independent.

The queue discipline refers to how jobs are selected for service in the queue. The most common in the context of telecommunication design are first-come, first-served (FCFS), random selection for service (RSS), or (generalized) processor sharing (PS) disciplines, as well as various job priority schemes.

In most practical situations, there is a physical limitation on the number of jobs that can be allowed to be waiting for service at any instant. We will call the waiting space in a queue for the *buffer* and its size for the buffer size. A limited buffer means that some jobs may have to be turned away if the buffer is full, and until some jobs have been processed, to leave space for new jobs to enter the system. Such queueing systems are called finite systems.

When there are multiple channels, we assume that the system is fed by a single line, so that it will not happen that some channels are idle while jobs are waiting for service. The service processes belonging to each of the parallel channels are assumed to be independent of each other.

A queueing system may have multiple service stages. These will be referred to as queueing networks even if the structure is as simple as two queues in series.

To describe a queueing system succinctly, we adopt the standard notation originating from Kendall (see Table 5.1). A queueing process is described by symbols with slashes between them, such as $A/B/X/Y/Z$, where A is the distribution of inter-arrival times, B the distribution of the service times, X the number of parallel servers or channels, Y the restriction on system capacity or buffer size, and Z the queueing discipline. Usually, the symbol Y is omitted if the queue is infinite and the symbol Z is omitted if the queue discipline is first come, first served (FCFS).

The symbol G represents a general probability distribution, with no assumption on the distribution, but possibly with mean and variance of the distribution known. The

Table 5.1 Some common symbols used in the Kendall notation to describe a queueing system

Characteristic	Symbol	Name
A/B	M	Exponential
A/B	D	Deterministic
A/B	E_k	Erlang type k
A/B	G	General
A	MAP	Markovian additive process
X/Y	$1, 2, \ldots, \infty$	No channels/system capacity
Z	FCFS	First come, first served
Z	LCFS	Last come, first served
Z	RSS	Random selection for service
Z	(G)PS	(Generalized) processor sharing
Z	PR	Priority
Z	GD	General discipline

general distribution is required to represent independent and identically distributed random variables, however.

5.2.2 Measuring system performance

The goal of queueing analysis is to determine the performance and effectiveness of a queueing system. These are related to the waiting time a job might experience, the length of a queue (system occupancy), and a measure of the idle time (or the reciprocal, utilization) of the servers.

The most detailed information is given by the equilibrium probabilities of these measures. In many cases, however, the mean values can suffice. We can distinguish between jobs being served and jobs waiting for service. Thus, the waiting time can be described for all jobs in the system or only for jobs waiting to be served, and the queue length can show all jobs in the system or only jobs waiting. The idle time of servers may refer to the fraction of time any particular server is idle or the time when the entire system is empty.

5.2.3 Some general results

We present some general results and relationships for $G/G/1$ and $G/G/n$ queues. These results will prove useful in the following chapters, as well as providing insight into some principles of queueing analysis. Note that these are infinite systems, so that the amount of carried traffic equals the amount of offered traffic.

Denoting the average rate of jobs entering the system by λ and the average rate of jobs being serviced by μ^{-1}, a general measure of traffic congestion for an n-server system is $\rho \equiv \lambda/n\mu$. When $\rho \geq 1$, the average number of arrivals into the system exceeds the maximum average service rate of the system. The queue length therefore increases indefinitely and no steady state exists, unless, at some point, customers are blocked from entering, in which case we have a finite system. In order for a steady state to exist generally we must therefore have $\rho < 1$.

We may consider the counting process $N(t)$ for the number of jobs in the system at time t. This number can be divided into jobs waiting in the queue $N_q(t)$ and those being serviced $N_s(t)$. We consider the state probabilities $p_i(t) = \mathbf{P}(N(t) = i)$ and the equilibrium probabilities $p_i = \mathbf{P}(N = i)$. For an n-channel system we are interested in the two expected values

$$L = \mathbf{E}(N) = \sum_{i=0}^{\infty} i p_i, \tag{5.13}$$

and the expected number in queue,

$$L = \mathbf{E}(N_q) = \sum_{i=n+1}^{\infty} (i - n) p_i. \tag{5.14}$$

Similarly, we let T_q be the time a job spends waiting in the queue before being serviced and let T be the total time a job spends in the system. Then

$$T = T_q + S,$$

where S is the time the system spends servicing the job. The expectations are denoted

$$W_q = \mathbf{E}(T_q),$$

$$W = \mathbf{E}(T).$$

Then the following relations hold, known as Little's formulas

$$L = \lambda W,$$

$$L_q = \lambda W_q.$$

By this very general result, it is sufficient to find only one of the four expected values to calculate the expectation of any other measure, observing that

$$W = \mathbf{E}(T) = \mathbf{E}(T_q) + \mathbf{E}(S) = W_q + 1/\mu.$$

The formula can be proven by looking at Figure 5.9, showing the state evolution of a queue with respect to time. Let the total time of observation be T. There are four calls arriving at times t_1, t_2, t_3 and t_6, and four departures at t_4, t_5, t_7 and t_8. The states are represented by the columns with bases $(t_1, t_2), (t_2, t_3), \ldots, (t_7, t_8)$. The mean queue length L is the mean height of the enclosed area, so that:

$$L = \frac{1}{T}((t_2 - t_1) + 2(t_3 - t_2) + 3(t_4 - t_3) + 2(t_5 - t_4) + (t_6 - t_5) + 2(t_7 - t_6) + (t_8 - t_7)).$$

On the other hand, viewing each call as a bar with its length representing the holding time of the call, the same area can be expressed as a collection of bars stapled on each other. Suppose call 1 has duration $(t_4 - t_1)$, call 2 $(t_5 - t_2)$, call 3 $(t_7 - t_3)$, and call 4 $(t_8 - t_6)$. Then the average holding time is

$$W = \frac{1}{N}((t_4 - t_1) + (t_5 - t_2) + (t_7 - t_3) + (t_8 - t_6)).$$

Since $T \cdot L$ and $N \cdot W$ contain the same area, we have, with $N/T = \lambda$,

$$L = \frac{N}{T} W = \lambda W$$

on average. Note that the individual call arrival and departure times do not matter. We can equally take the durations to be $(t_4 - t_3), (t_5 - t_2), (t_7 - t_6), (t_8 - t_1)$ and arrive at the same result. Also, $L - L_q = \lambda(W - W_q) = \lambda/\mu$. But $L - L_q = \mathbf{E}(N) - \mathbf{E}(N_q) = \mathbf{E}(N - N_q) = \mathbf{E}(N_s)$, that is, the expected number of customers being serviced in the steady state is $\lambda/\mu = r$. The quantity r is the offered load. For a single server ($n = 1$) system, $r = \rho$.

Then,

$$L - L_q = \sum_{i=1}^{\infty} i p_i - \sum_{i=1}^{\infty} (i - 1) p_i = \sum_{i=1}^{\infty} p_i = 1 - p_0.$$

The expected number of jobs per server is r/n and the probability $p_{\bar{0}}$ that any given server is busy in a multi-server system in steady state is

$$r/n = \rho = 0 \cdot (1 - p_{\bar{0}}) + 1 \cdot p_{\bar{0}}.$$

For the $G/G/1$ queue, this is simply

$$p_0 = 1 - p_{\bar{0}} = 1 - \rho.$$

5.2.4 Simple Markovian queues

Many queues are birth-death processes and can be analyzed with a straightforward method that consider the transition rates out from and into neighboring states. A birth-death process can only move from a state to a neighboring state. Let λ_i be transition rate from state i to state $i + 1$, and μ_i the transition rate from state $i + 1$ to state i. It can be shown that with some conditions on λ_i and μ_i, a steady-state solution exists for the birth-death process and can be found by solving the stochastic balance conditions, that is, flow into and out from a state. The flow is the state probabilities of neighboring states times the transition probabilities into and out from the states, which should sum to zero (see Figure 5.10 for an example of a birth-death process). The stochastic balance conditions generate in case of a birth-death process a set of equations which are referred to as the detailed balance equations,

$$0 = -(\lambda_j + \mu_j) p_j + \lambda_{j-1} p_{j-1} + \mu_{j+1} p_{j+1}, \quad j \geq 1$$

$$0 = -\lambda p_0 + \mu_1 p_1.$$

These equations can be rewritten as

$$p_n = \frac{\lambda_{n-1} \lambda_{n-2} \cdots \lambda_0}{\mu_n \mu_{n-1} \cdots \mu_1} p_0$$

$$= p_0 \prod_{i=1}^{n} \frac{\lambda_{i-1}}{\mu_i}, \quad n \geq 1, \tag{5.15}$$

which can be shown by induction. In general, n may be infinite. and since the probabilities must sum to unity, we have

$$p_0 = \left(a + \sum_{n=1}^{\infty} \prod_{i=1}^{n} \frac{\lambda_{i-1}}{\mu_i} \right)^{-1}. \tag{5.16}$$

From this, we see that a necessary and sufficient condition for the existence of a steady-state solution is the convergence of the infinite series in (5.16).

Equations (5.15) and (5.16) are very useful for analyzing queues, and by changing λ_i and μ_i accordingly, closed-form expressions for many specific systems can be derived.

The $M/M/1$ Queue. By setting $\lambda_i = \lambda$ and $\mu_i = \mu$, we get equations for the $M/M/1$ queue. It is rather straightforward to find the steady-state probabilities (Gross and Harris, 1998) as

$$p_i = (1 - \rho)\rho^i, \quad \rho = \lambda/\mu < 1. \tag{5.17}$$

The effective measures using the notation introduced above are

$$L = \frac{\rho}{1 - \rho} = \frac{\lambda}{\mu - \lambda}$$

$$L_q = \frac{\rho^2}{1 - \rho} = \frac{\lambda^2}{\mu(\mu - \lambda)}$$

$$W = \frac{L}{\lambda} = \frac{1}{\mu - \lambda}$$

$$W_q = \frac{L_q}{\lambda} = \frac{\rho}{\mu - \lambda}.$$

Queues with Multiple Channels. Consider a multi-server system with independent and identically distributed exponential service times, $M/M/n$ and arrival process Poisson, where n is the number of servers (or channels). By the knowledge of the transition rates between the queue states, we can formulate a system of equations. Since the arrival process is Poisson, $\lambda_i = \lambda$ for all states i. Even if there are more than n jobs in the system, the transition rate from such a state cannot exceed the number of servers. Thus,

$$\mu_i = \begin{cases} i\mu, & 1 \leq < i < n \\ n\mu & i \geq n \end{cases}. \tag{5.18}$$

From (5.15) we obtain

$$p_i = \begin{cases} \frac{\lambda^i}{i!\mu^i} p_0 & 1 \leq i \leq n \\ \frac{\lambda^i}{n^{i-n} n!\mu^i} p_0 & i \geq n \end{cases}, \tag{5.19}$$

$$p_0 = \left(\sum_{i=0}^{n-1} \frac{r^i}{i!} + \frac{r^n}{n!(1 - \rho)} \right)^{-1} \quad r/n = \rho < 1. \tag{5.20}$$

The equation reduces to the one for the $M/M/1/\infty$ queue when $n = 1$. The performance measures are

$$L_q = \left(\frac{r^n \rho}{n!(1-\rho)^2} \right) p_0,$$

$$L = r + L_q,$$

$$W_q = \left(\frac{r^n}{n!n\mu(1-\rho)^2} \right) p_0,$$

$$W = \frac{1}{\mu} + W_q.$$

Finite Multiple Channel Queues.   Finally, we consider an $M/M/n/b$ queue, with a state space which is truncated at system size b which is approximately the same as a system buffer. Care should be taken to how this system size is defined. It usually includes the jobs being serviced and therefore, strictly speaking, may not be equivalent to a physical buffer, and then $b \geq n$. This slight abuse of terminology is overlooked in this text, but whether b represents the number of jobs in the system or only the jobs waiting for service should be clear from the context. Figure 5.10 shows the transition diagram for this queue. Similarly to, the systems discussed previously we have:

$$p_n = \begin{cases} \frac{\lambda^i}{i!\mu^i} p_0 & \text{for } 1 \leq i < n \\ \frac{\lambda^i}{n^{i-n}i!\mu^i} p_0 & \text{for } n \leq i \leq b \end{cases}, \tag{5.21}$$

$$p_0 = \begin{cases} \left(\sum_{i=0}^{n-1} \frac{r^i}{i!} + \frac{r^n}{n!} \frac{1-\rho^{b-n+1}}{1-\rho} \right)^{-1} & \text{for } \rho \neq 1 \\ \left(\sum_{i=0}^{n-1} \frac{r^i}{i!} + \frac{r^n}{n!}(b-n+1) \right)^{-1} & \text{for } \rho = 1 \end{cases}. \tag{5.22}$$

The performance measures are found to be

$$L_q = \frac{p_0 r^n \rho}{n!(1-\rho)^2} \left(1 - \rho^{b-n+1} - (1-\rho)(b-n+1)\rho^{b-n} \right), \quad \rho \neq 1,$$

$$L = L_q + \lambda_{eff}/\mu = L_q + \lambda(1-p_b)/\mu,$$

$$W = \frac{L}{\lambda_{eff}} = \frac{L}{\lambda(1-p_b)}, \text{ and}$$

$$W_q = W - \frac{1}{\mu} = \frac{L_q}{\lambda_{eff}}.$$

Note that the arrival rate is modified to an effective arrival rate by multiplication of λ by the factor $(1 - p_b)$.

In the case of $M/M/1/b$, we have the state probabilities

$$p_n = \begin{cases} \frac{(1-\rho)\rho^n}{1-\rho^{b+1}} & \rho \neq 1 \\ \frac{1}{b+1} & \rho = 1 \end{cases}, \tag{5.23}$$

$$p_0 = \begin{cases} \frac{(1-\rho)}{1-\rho^{b+1}} & \rho \neq 1 \\ \frac{1}{b+1} & \rho = 1 \end{cases}. \tag{5.24}$$

We also have

$$L_q = \begin{cases} \frac{\rho}{1-\rho} - \frac{\rho(b\rho^b+1)}{1-\rho^{b+1}} & \rho \neq 1 \\ \frac{b(b-1)}{2(b+1)} & \rho = 1 \end{cases},$$

with $L = L_q + (1 - p_0)$ with $1 - p_0 = \lambda(1 - p_b)/\mu$, with compensation for the effective output. Figures 5.6, 5.7 and 5.8 show the occupancy and, overflowing and residual traffic in an $M/M/n/n$ system, respectively.

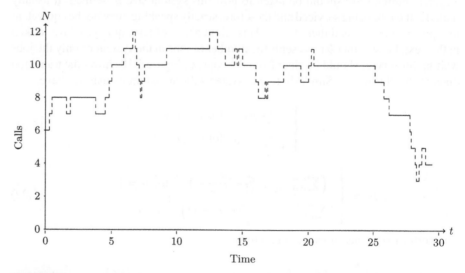

FIGURE 5.6

The evolution in time of the states of an $M/M/n/n$ queue with large n; the number of calls in progress N at time t.

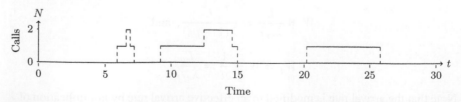

FIGURE 5.7

Overflow traffic from an $M/M/n/n$ queue, the number of calls in progress N at time t.

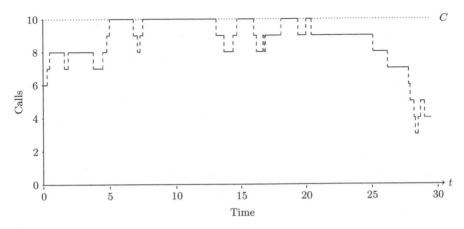

FIGURE 5.8

The evolution in time of the states of an *M/M/n/n* queue with congestion; the number of calls in progress *N* at time *t*.

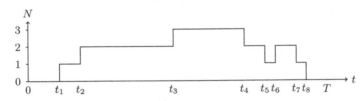

FIGURE 5.9

Illustrative proof of Little's formula.

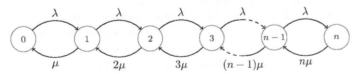

FIGURE 5.10

State transition probabilities for an *M/M/n/n* system.

5.3 Markov chain analysis

The power of using Markov chains in analysis of queues lies in that the states are discrete variables but the transition rates are random (either discrete or continuous), which make them suitable for numerical methods. The states are typically represented by a matrix where all the states of the system are enumerated, and represented by the rows and columns, and the transition rates are represented by the entries in this matrix. A Markov chain is characterized by the fact that it is memoryless. In other words, the process does not possess any memory of its past states, and any transition to a new state is only described by the current state and the transition rates to other states.

The advantage of this representation lies in that it is possible to find a steady-state probability vector by solving a linear system of equations. The drawback is that the number of equations increases very fast with the number of queues and the number of capacity units per queue, which make up the state space. Markov chain analysis can also in some situations be used in order to formulate approximation methods.

5.3.1 Discrete-time Markov chains

One of the most important models of random arrivals are *Markov chains*. These can be defined in discrete time, which can be though of as "clock ticks", when a change of the state or value of the process only can occur at discrete time values. They can also be defined in continuous time, when a change may happen anytime. Formally, we define a stochastic process as a collection of random variables. In both cases, we assume that the states in the state space S can be enumerated and therefore be described by integers.

Definition 5.3.1 (Markov property in discrete time).   The process X_n is a *discrete-time Markov chain* if it satisfies the *Markov property*

$$\mathbf{P}(X(t_n) = j | X(t_1) = i_1, X(t_2) = i_2, \ldots, X(t_{n-1}) = i_{n-1})$$
$$= \mathbf{P}(X(t_n) = j | X(t_{n-1}) = i_{n-1}) \tag{5.25}$$

for all $n \geq 1$ of discrete times and states $j, i_1, \ldots, i_{n-1} \in S$.

This defines the important property of the Markov chain being memoryless. The theory of Markov chains is fundamental to analysis of queueing systems, but the computational complexity increases rapidly with the size of the state space, so for large systems, approximations often have to be used.

The evolution of a discrete-time Markov chain at any clock tick n can be expressed in its n-step transition probabilities. The transition probabilities are conveniently represented in matrix form and is called the *transition matrix P* of the Markov chain.

Let the transition probabilities from state i to state j be

$$p_{ij} = \mathbf{P}(X_n = j | X_{n-1} = i),$$

where we assume that p_{ij} only depends on i and j, but not on n. The chain is then called *homogeneous*. The transition matrix is the matrix formed by all p_{ij},

$$P = (p_{ij}),$$

which is of order $|S| \times |S|$, since $i, j \in S$. The transition from state i to state j is then the (i, j) entry in the transition matrix. The transition matrix $P = (p_{ij})$ is a so-called *stochastic matrix*, meaning that is has the properties

$$p_{ij} \geq 0 \quad \text{(The entries in } P \text{ are non-negative)}$$

$$\sum_{j \in S} p_{ij} = 1 \quad \text{(The rows in } P \text{ sum to unity)}.$$

The transition matrix P describes the evolution of the chain for a single time step. The evolution after n steps can be shown to descibed by the matrix P^n, the nth power of single step transition matrix.

By defining an initial state probability vector of the chain as $\pi^{(0)}$ where $\pi_i^{(0)} = \mathbf{P}(X_0 = i)$, the state vector after n steps is given by the expression

$$\pi^{(n)} = \pi^{(0)} P^n. \tag{5.26}$$

Thus, knowing the transition matrix and an initial state distribution, the state probabilities of the chain at any later time n can easily be calculated by multiplying π by P, n times. If the chain reaches state probabilities which do not change further in time, it is said to be in an *equilibrium* or *steady state*. This is usually the situation of a system that has been run for a long time. Before a steady state has been reached, it may be in *transient* states. We will only analyze equilibrium state probabilities. Since a steady state does not change from one step to the next, we must have $\pi^{(n)} = \pi^{(n)} P$ from some value of n. Dropping the superscript, we can find this state probabilities by solving the equation

$$\pi = \pi P \tag{5.27}$$

for π. A straightforward method to solve this equation is to take *any* initial state distribution $\pi^{(0)}$, for example $(1, 0, 0, \ldots, 0)$ and iterate (5.27) until π does not change anymore. There are more efficient methods of solving (5.27) which are discussed when this problem is encountered in later chapters.

Now turning to continuous-time Markov chains, let $X = \{X(t) : t \geq 0\}$ be a process, that is, a collection of random variables taking values in some countable state space S and indexed by $t \in [0, \infty)$. We denote the chain by $X(t)$ to stress that it is a function of a continuous time variable t. Again, we shall assume that S is a subset of the integers, so that states can be represented by discrete, but possibly infinite, integer values. The process X is called a (continuous-time) *Markov chain* if it satisfies.

Definition 5.3.2 (Markov property in continuous time). The process $X(t)$ is called a (continuous-time) *Markov chain* if it satisfies the *Markov property*

$$\mathbf{P}(X(t_n) = j | X(t_1) = i_1, X(t_2) = i_2, \ldots, X(t_{n-1}) = i_{n-1})$$
$$= \mathbf{P}(X(t_n) = j | X(t_{n-1}) = i_{n-1})$$

for all $j, i_1, \ldots, i_{n-1} \in S$ and any sequence $t_1 < t_2 < \cdots < t_n$ of times.

The Markov property shows the important property of Markov chains being *memoryless*. Being in the current state, the next state depend only on the current state and the transition probabilities, but not on how the current state was reached.

In continuous time, there is no exact analogue for P since the time is a continuous variable. We may define another matrix, the so-called *infinitesimal generator Q*, which plays a similar role to P. This will be defined in a few steps, starting with the transition probabilities.

Definition 5.3.3 (Transition probability). The *transition probability* $p_{ij}(s, t)$ is defined as

$$p_{ij}(s, t) = \mathbf{P}(X(t) = j | X(s) = i) \quad \text{for } s \leq t.$$

Similarly to the discrete-time case, the Markov chain is homogeneous if the transition probabilities do not depend on the times s and t individually but only on their difference $t - s$. (In a discrete-time chain, this difference is always unity, and hence independent of n.)

Definition 5.3.4 (Homogeneous Markov chain).   *A Markov chain is called homogeneous if*

$$p_{ij}(s, t) = p_{ij}(0, t - s) \quad \text{for all } i, j, s, t,$$

and we can then write $p_{ij}(t - s)$ for $p_{ij}(s, t)$.

In the sequel, we suppose that $X(t)$ is a homogeneous chain, and we write P_t for the $|S| \times |S|$ transition matrix with entries $p_{ij}(t)$. These transition probabilities are always assumed to be continuous.

Theorem 5.3.1.   *The family $\{P_t : t \geq 0\}$ is a* stochastic semigroup, *that is, it satisfies the following:*

(i) $P_0 = I$ *(the identity matrix);*
(ii) P_t *is stochastic, that is P_t has non-negative entries and row sums 1;*
(iii)   *the* Chapman-Kolmogorov *equations,* $P_{s+t} = P_s P_t$ *if $s, t \geq 0$.*

Proof.
(i)   Obvious. For $t = 0$ no transitions can take place, and so the result follows.
(ii)   Let **1** denote the column vector of ones.

$$(P_t \mathbf{1})_i = \sum_j p_{ij}(t) = \mathbf{P}(\cup_j \{X(t) = j\} | X(0) = i) = 1. \tag{5.28}$$

(iii)

$$p_{ij}(t + s) = \mathbf{P}(X(t + s) = j | X(0) = i) \tag{5.29}$$

$$= \sum_k \mathbf{P}(X(t + s) = j | X(s) = k)\mathbf{P}(X(s) = k | X(0) = i) \tag{5.30}$$

$$= \sum_k p_{ij}(s) p_{kj}(t). \tag{5.31}$$

$\square$

The evolution of $X(t)$ is specified by the stochastic semigroup $\{P_t\}$ and the distribution of $X(0)$. Most questions about X can be rephrased in terms of these matrices and their properties.

Now, suppose that the chain is in state $X(t) = i$ at time t. Two things may happen in a small time interval $(t, t + h)$:

(a)   the chain remains in state i with probability $p_{ii}(h) + o(h)$. The small error term takes into account the possibility that the chain moves out of i and back to i again in the interval;
(b)   the chain moves to a new state j with probability

$$p_{ij}(h) + o(h).$$

The term $o(h)$ represents the probability that two or more transitions occur in the interval $(t, t + h)$. This probability is small for small h. The transition probabilities $p_{ij}(h)$ are approximately linear in h for small h. Thus, there exist constants $\{q_{ij}, i, j \in S\}$ such that

$$p_{ij}(h) \simeq q_{ij}h \quad \text{if } i \neq j, \tag{5.32}$$

$$p_{ii}(h) \simeq 1 + q_{ii}h. \tag{5.33}$$

We must have $q_{ij} \geq 0$ for $i \neq j$ and $q_{ii} \leq 0$ for all i, since the probabilities are positive. The matrix $Q = (q_{ij})$ is called the *infinitesimal generator* of the chain and takes over the role of the transition matrix P for the discrete-time Markov chains.

Combining (5.32) with the two cases (a) and (b) above for the small time interval $(t, t + h)$, starting from $X(t) = i$, we have

(a) the chain remains in state i with probability $1 + q_{ii}h + o(h)$,
(b) the chain moves to state $j \neq i$ with probability $q_{ij}h + o(h)$.

We expect that the transition probabilities in each row sum to unity for every t, $\sum_j p_{ij}(t) = 1$, and in particular

$$1 = \sum_j p_{ij}(h) \simeq 1 + h \sum_j q_{ij}$$

so that

$$\sum_j q_{ij} = 0 \quad \text{for all } i, \text{ or } Q\mathbf{1} = \mathbf{0}, \tag{5.34}$$

where $\mathbf{1}$ and $\mathbf{0}$ are column vectors of ones and zeros.

Formally, we can write (5.32) as

$$\lim_{h \to 0} \frac{1}{h}(P_h - I) = Q,$$

which is equivalent to the statement that P_t is differentiable at $t = 0$. Therefore Q can be found from knowledge of P_t. The converse is also usually true. Suppose that $X(0) = i$, and conditioning $X(t + h)$ on $X(t)$ gives that

$$p_{ij}(t + h) = \sum_k p_{ik}(t)p_{kj}(h)$$

$$\simeq p_{ij}(t)(1 + q_{jj}h) + \sum_{k \neq j} p_{ik}(t)q_{kj}h$$

$$= p_{ij}(t) + h \sum_k p_{ik}(t)q_{kj},$$

or

$$\frac{1}{h}(p_{ij}(t + h) - p_{ij}(t)) \simeq \sum_k p_{ik}(t)q_{kj} = (P_t Q)_{ij}.$$

Now, letting $h \to 0$ we obtain the so-called *forward equations*,

$$p'_{ij}(t) = \sum_k p_{ik}(t)q_{kj},$$

or

$$P_t' = P_t Q$$

in matrix form, where P_t' denotes the matrix with entries $p_{ij}'(t)$. By a similar argument, conditioning $X(t + h)$ on $X(h)$, we have the *backward equations*,

$$p_{ij}'(t) = \sum_k q_{ik} p_{kj}(t),$$

or

$$P_t' = Q P_t.$$

These two equations show how P_t are related to Q for a continuous-time chain. Using the boundary condition $P_0 = I$, a unique solution can usually be found by

$$P_t = \sum_{n=1}^{\infty} \frac{t^n}{n!} Q^n \tag{5.35}$$

of powers of matrices (with $Q^0 = I$). Equation (5.35) is deducible from the forward equations by solving the system of differential equations in a similar way as we can show that a function of the single variable $p(t) = e^{qt}$ solves the differential equation $p'(t) = q p(t)$. Thus, the representation (5.35) for P_t can be written

$$P_t = e^{Qt}, \tag{5.36}$$

where e^A is interpreted as $\sum_{n=0}^{\infty} (1/n!) A^n$ which is defined whenever A is a square matrix.

We have indicated that, subject to some technical conditions, a continuous-time chain has an infinitesimal generator Q which specifies its transition probabilities.

A continuous-time Markov chain can further be described in terms of its *interarrival times*. Suppose that the chain is in state $X(t) = i$ at time t. The development of $X(t + s)$ for $s \geq 0$ can be described as follows. Let

$$T = \inf\{s \geq 0 : X(t + s) \neq i\}$$

be the time until the chain changes its state. The variable T is called a *holding time*. We have

Proposition 5.3.1. The variable T is exponentially distributed with parameter $-q_{ii}$.

For the transition probabilities, we also have

Proposition 5.3.2. The probability that the chain jumps to $j (\neq i)$ is $-q_{ij}/q_{ii}$.

A more detailed discussion including proofs can be found in, for example Grimmett and Stirzaker, 1992.

We will also distinguish between *reducible* and *irreducible* chains. An irreducible chain is such that each state j can be reached (eventually) from any other state i, and therefore the state space cannot be decomposed into subspaces. We have

Definition 5.3.5. The chain is called *irreducible* if for any pair i, j of states we have that $p_{ij}(t) > 0$ for some t.

Furthermore, a state can be *periodic*, in which a return to a state i only can happen in $k > 1$ transitions. This is equivalent to say that the n-step transition probabilities $p_{ii}(n) = 0$ unless n is a multiple of k. If $p_{ii}(n) > 0$ for all $n \geq 1$, the chain is called *aperiodic*. This is usually the case in the applications we encounter.

Just as for discrete-time chains, we are usually interested in the asymptotic behavior of the chain $X(t)$ for large t. A equilibrium or steady state distribution (also known as a stationary distribution) is defined as follows.

Definition 5.3.6. The vector π is an *equilibrium distribution* of the chain if $\pi_j \geq 0$, $\sum_j \pi_j = 1$ and $\pi = \pi P_t$ for all $t \geq 0$.

For any Markov chain, we have

Proposition 5.3.3.

$$\pi = \pi P_t \text{ for all } t \text{ if and only if } \pi Q = \mathbf{0},$$

where $\mathbf{0}$ is a vector consisting of zeros only.

The statement in Proposition 5.3.3 is of great practical importance. Whenever we need to find the equilibrium distribution of an irreducible continuous-time Markov chain with infinitesimal generator Q, we only need solve the equation $\pi Q = \mathbf{0}$ for π.

We will sometimes find it very useful to define the *reverse* of a Markov chain X_t with equilibrium distribution π. Let $Y_t = X_{-t}$ be the chain X evolving backwards in time. Then, Y_t is also a Markov chain with equilibrium distribution π, and we have

Theorem 5.3.2. The Markov chain X with equilibrium distribution π and transition matrix P is time reversible if and only if

$$\pi_i p_{ij} = \pi_j p_{ji}, \quad \text{for all } i, j \in S. \tag{5.37}$$

The theory of continuous-time Markov chains is fundamental to analysis of queueing systems, but the computational complexity increases rapidly with the size of the state space, so for large systems, it is often necessary to make approximations.

5.3.2 Numerical solution of Markov chains

There is a large number of methods available to solve the equations in Proposition 5.3.3. Two approaches are solution by iteration and direct methods. The simplest iterative method is the *power method*. Since for an equilibrium distribution, $\pi = \pi P_t$, we have after k iterations and with initial distribution vector $\pi^{(0)}$

$$\pi^{(k)} = \pi^{(0)} P_t^k.$$

When the chain is finite, aperiodic and irreducible, the vectors $\pi^{(k)}$ converge to the equilibrium distribution π for any choice of initial distribution vector $\pi^{(0)}$. That is

$$\lim_{k \to \infty} \pi^{(k)} = \pi.$$

The matrix P_t can be obtained from the infinitesimal generator by

$$P_t = I + Q\Delta t,$$

which is the first two terms in the series (MacLaurin) expansion of (5.36). Here Δt is chosen so that $\Delta t \leq 1/\max_i |q_{ii}|$. With $\Delta t < 1/\max_i |q_{ii}|$, the transition matrix has diagonal elements $p_{ii} > 0$ and the Markov chain is aperiodic. In Stewart, 1994, the value $\Delta t = 0.99/\max_i |q_{ii}|$ is suggested. Under these conditions, the power method can be shown to converge. The method may however converge slowly and is computationally expensive, but the intermediary steps give an idea of how the equilibrium distribution is reached from an initial distribution.

Alternatively, the equations in Proposition 5.3.3 can be solved using linear algebra. Such methods are referred to as *direct methods*. These are based on an LU-decomposition of the transpose Q^T of the infinitesimal generator Q, so that $Q^T = LU$ where L is a lower triangular matrix and U is an upper triangular matrix. The equation $Q^T\boldsymbol{\pi} = \mathbf{0}$ is then equivalent to $(LU)\boldsymbol{\pi}$ and is solved in two steps: $U\boldsymbol{\pi} = \mathbf{z}$ and $U\mathbf{z} = \mathbf{0}$. The system of equations is singular, so some values can be set arbitrarily, and the vector $\boldsymbol{\pi}$ can then be found by back-substitution and normalization.

One way to handle the singularity is by following method, referred to as *inverse iteration*. Suppose that we have an LU-factorization with $U = (u_{ij})$ of an n-state system infinitesimal generator Q. Due to the singularity of Q, the matrix U has a zero pivot, that is, a diagonal element equal to zero. If the zero pivot is in the last row n of U, this entry u_{nn} is modified and set to some small number $u_{nn} = \epsilon$. (Depending on how the LU-decomposition is performed, the zero pivot may or may not be in the last row n of U.) Call this modified matrix $\bar{U}$. By choosing the right-hand side to $\mathbf{e}_n$, where $\mathbf{e}_n^T = (0, 0, \ldots, 0, 1)$, we then need to solve $\bar{U}\mathbf{x} = \mathbf{e}_n$. By normalizing $\mathbf{x}$, the solution $\boldsymbol{\pi}$ is obtained, that is, $\boldsymbol{\pi} = \mathbf{x}/||\mathbf{x}||$. The practical use of direct methods is discussed in greater detail in Section 8.2. A third approach, called *randomization* is a fast approximate method, which is discussed next.

5.3.3 Randomization

Markov chains are the core of most queueing analyses, and are the preferable method when applicable, meaning when computational power allows. However, using Markov chains for analysis becomes quickly cumbersome as the state space grows. In particular, infinite queues cannot be analyzed directly as the transition matrix is of infinite dimension. Large or infinite systems can be analyzed approximately, however, by using randomization and truncation.

The randomization method relies on approximating a continuous-time Markov chain by a discrete-time Markov chain. It also relies on the memorylessness of the Poisson process.

It can be shown that if the total transition rate form every state in a homogeneous continuous time Markov chain (where no parameters is changing in time), the transitions can be described by a Poisson process. The transition rates may not be, in general, the same for all system states, but can be scaled (or uniformed) by the introduction of fictitious "self-transitions".

In a homogeneous continuous time Markov chain, the transition rates out from states n are $\lambda + \min\{n, s\}\mu$, which are represented by the diagonal entries in the matrix Q.

When a transition occurs, the next state is chosen according to the transition probability matrix $P = I + Q/\Lambda$, which has the following nonzero entries

$$p_{i,i+1} = \lambda/\Lambda$$
$$p_{i,i} = 1 - (\lambda + \min\{i, n\}\mu)/\Lambda$$
$$p_{i+1,i} = \min\{i, n\}\mu/\Lambda$$

for all $i = 0, 1, \ldots$.

The matrix P can be interpreted as a transition probability matrix for a discrete time Markov chain, with the transition times being randomized according to a Poisson process with rate L.

In effect, the continuous-time Markov chain is decomposed into a discrete-time Markov chain (often referred to as the uniformed or the imbedded Markov chain) and a Poisson process.

With $P = I + Q/\Lambda$, we obtain the equilibrium distribution from

$$\pi(t) = \pi(0) \sum_{k=0}^{\infty} e^{-\Lambda t} \frac{(\Lambda t)^k}{k!} P^k. \tag{5.38}$$

It should be noted that the variable t has to be chosen so that the mean Λt of the Poisson distribution is not too large. A possible choice Stewart, 1994 is to let $t = 0.99/\Lambda$. The upper bound $t < \ln(1/\epsilon_t)/\Lambda$ with $\epsilon_t = 10^{-30}$ is suggested in Gross and Harris, 1998.

In order to implement the randomization method, the infinite series in (5.38) has to be truncated after the first K terms. The constant K can be determined as follows. Fix a tolerance $\epsilon > 0$ such that at the truncation constant K,

$$\sum_{K+1}^{\infty} e^{-\Lambda t} \frac{(\Lambda t)^k}{k!} < \epsilon.$$

In other words, we search for K such that

$$\sum_{k=0}^{K} \frac{(\Lambda t)^k}{k!} > (1 - \epsilon)/e^{-\Lambda t},$$

which is easily performed by iteration. The constant ϵ is then an error bound on π. In Ingolfsson et al., 2002, the value $K = \lceil \Lambda t + 5\sqrt{\Lambda t} + 4.9 \rceil$ is suggested.

Calculation of the sum in (5.38) is typically done recursively to avoid numerical overflow. We may for example start with $\pi^{(0)} = (1, 0, 0, \ldots, 0)$ and set $\mathbf{y} := \pi^{(0)}$

and $\pi := \pi^{(0)}$ and let, for $k = 1, \ldots, K$,

$$\mathbf{y} := \mathbf{y}P \cdot \frac{\Delta t}{k},$$
$$\pi := \pi + \mathbf{y}.$$

Finally, the equilibrium distribution is given by $\pi \cdot e^{-\Lambda t}$.

The randomization method works, just as exact method, on any problem which can be modeled by a continuous-time Markov chain. The computation time is dependent on model structure and on the truncation parameter K for the infinite series. It is, however, in general much faster than any exact method. By Ingolfsson et al., 2002 of the order of 25% of the computational requirements of an exact method.

Example 5.3.1. Suppose that we want to analyze an $M/M/7/7$ system but we only have the transition probabilities of an $M/M/8/8$ system. These transition probabilities are given by the infinitesimal generator matrix:

$$Q = \begin{pmatrix} -20 & 20 & 0 & 0 & 0 & 0 & 0 & 0 \\ 3 & -23 & 20 & 0 & 0 & 0 & 0 & 0 \\ 0 & 6 & -26 & 20 & 0 & 0 & 0 & 0 \\ 0 & 0 & 9 & -29 & 20 & 0 & 0 & 0 \\ 0 & 0 & 0 & 12 & -32 & 20 & 0 & 0 \\ 0 & 0 & 0 & 0 & 15 & -35 & 20 & 0 \\ 0 & 0 & 0 & 0 & 0 & 21 & -41 & 20 \\ 0 & 0 & 0 & 0 & 0 & 0 & 24 & -24 \end{pmatrix}$$

The value of Γ is $\Gamma = \max_i |p_{ii}|$, that is the largest absolute value on the matrix diagonal. We find $\Gamma = 41$. Next, we construct the transition matrix of the embedded discrete time Markov chain as $\bar{P} = I - Q/\Gamma$, where I is the identity matrix, so that

$$\bar{P} = \begin{pmatrix} 0.512 & 0.488 & 0 & 0 & 0 & 0 & 0 & 0 \\ 0.073 & 0.439 & 0.488 & 0 & 0 & 0 & 0 & 0 \\ 0 & 0.146 & 0.366 & 0.488 & 0 & 0 & 0 & 0 \\ 0 & 0 & 0.220 & 0.239 & 0.488 & 0 & 0 & 0 \\ 0 & 0 & 0 & 0.293 & 0.220 & 0.488 & 0 & 0 \\ 0 & 0 & 0 & 0 & 0.366 & 0.146 & 0.488 & 0 \\ 0 & 0 & 0 & 0 & 0 & 0.512 & 0 & 0.488 \\ 0 & 0 & 0 & 0 & 0 & 0 & 0.585 & 0.415 \end{pmatrix}$$

Notice the reverse transformation to the one in (5.34). It is also necessary to determine values for t and ϵ. Let us say we set $\epsilon = 0.001$. The parameter t can be chosen as $t < 1/\Lambda$, for example, $t = (1 - \epsilon)/\Lambda$. It is also necessary to determine the number N after which the calculations may be truncated. Either $\Phi^{(n)}$ can be iterated until no significant change is observed from two consecutive iterations, or a number N can be determined as the N, such that the sum

$$\sum_{n=0}^{N} \frac{(\Gamma t)^n}{n!} < (1 - \epsilon)e^{\Gamma t}.$$

Table 5.2 Enumeration of states for two queues in tandem	
State	**(n_1, n_2)**
1	$(0, 0)$
2	$(1, 0)$
3	$(0, 1)$
4	$(2, 0)$
5	$(1, 1)$
6	$(0, 2)$
7	$(3, 0)$
8	$(2, 1)$
9	$(1, 2)$
10	$(0, 3)$
$\vdots$	$\vdots$

Starting from any initial state vector π_0, each iteration n is giving an adjustment to π_n, calculated from

$$\pi_n = \pi_{n-1} + \pi_{n-1}\bar{P}\frac{\Gamma t}{n}, \quad n \geq 1.$$

This is just the Taylor expansion of the exponential of Λt. Iterating to $n = N$, the final result is given by $\pi = \pi_N e^{-\Gamma t}$. The numerical results are shown in Table 5.3. The approximation is very good compared to the exact values.

5.3.4 Truncation of state space

When the state space of a system is infinite or very large, the methods discussed cannot be applied. It is, however, possible to approximate an infinite (or very large) Markov chain by a finite (smaller) one. This is achieved by *state-space truncation*. This amounts to constructing an infinite generator $\hat{Q}$ representing a Markov chain on the truncated state space, and solving for $\hat{\pi}$ as an approximation to the sought equilibrium distribution π. We expect that $\hat{\pi}$ will approach π as the state space of the approximating chain increases towards the state space of the original chain.

Let Q be the infinitesimal generator of a Markov chain with a large (possibly infinite) state space having a unique equilibrium distribution. We now form the matrix $\hat{Q}$ from the "north-west corner" of Q, that is, the restriction of Q to the states below some limit k. The matrix $\hat{Q}$ has to be *augmented* in order to have the necessary properties of an infinitesimal generator of a Markov chain. This augmentation is often made by modifying the last column of the approximating generator $\hat{Q}$. In this case, the augmentation simply amounts to modifying the last column so that each row sum is zero.

The approximate equilibrium distribution can now be computed from

$$\hat{Q}\hat{\pi} = 0$$

using any solution method for Markov chains. The choice of the state space restriction k and the resulting approximation error are difficult to asses. In general, the errors are largest in the lowest and the highest states of the approximation.

Table 5.3 Approximation of an $M/M/7/7$ system by randomization and state-space truncation of an $M/M/7$ system with $\lambda = 20$ and $\mu = 3$ per server. The error parameter is set to $\epsilon = 0.001$

State	Exact	Randomization
0	0.002	0.002
1	0.011	0.011
2	0.037	0.037
3	0.081	0.081
4	0.136	0.136
5	0.201	0.201
6	0.192	0.191
7	0.160	0.159

5.3.5 Two Stations in Tandem

Modeling a queue or system of queues involves a number of steps that have to completed:

(1) Choice of a representation of the state space;

(2) Enumeration of all possible transitions between the states;

(3) Construction of the transition matrix or the infinitesimal generator from transition probabilities; and,

(4) Finding equilibrium probability vectors of the Markov chain, which can be used to find performance measures of the system.

It is particularly convenient to model external job arrivals to the system by a Poisson process and letting the service times be exponentially distributed. State transitions may or may not be state dependent. In the latter case, the number of jobs in the system is typically determining transition probabilities between states. Finite queues are modeled so that the nature of departure are taken into account. A job may for example have to wait in one queue for free resources in another queue before any state transition can take place. This is known as a *delay system*. Alternatively, a job may be released for service in subsequent systems and is lost, should no resources be available in a later service stage at the time of transition. This scenario corresponds to a *loss system*.

Choice of state space representation and enumeration of the states are usually performed at the same time. Most practical systems allow only transition between nearby states. The transition matrix (infinitesimal generator) is therefore usually very large but *sparse*, where most of the matrix entries are zero. This property should be taken advantage of in numerical solutions of Markov chains.

To conclude this section, we consider the analysis of two systems in tandem. For this system, the arrival process is Poisson with intensity λ, each station consists of a single server with mean service time $1/\mu_s$ at server $s = 1, 2$. The scheduling of jobs is first come first served (FCFS), and there is only one class of customer. Each of the two queues has infinite capacity.

In this example we may denote the state of the system by the pair (i, j), where i represents the instantaneous number of customers in station 1 and j denotes the number in station 2 (including the customers in service). For any nonnegative integer values of i and j, the pair of integers (i, j) represents a feasible state of this queueing system. The state enumeration is such that every pair of occupaid states in the individual systems (i, j) is mapped onto a single state number, say k, for the whole network. In generating the Markov chain we need to consider the single-step transitions that can occur between any two states, say k and l. The $(k, l)^{\text{th}}$ element of the infinitesimal generator matrix Q denotes the rate of transition from the kth state to the lth state.

Depending on the state enumeration, transitions to higher states are usually by transitions due to externally arriving jobs and transitions from the first system to the second. In the first case, jobs arrive at rate λ, and in the second at rates $i\mu_1$, where i again is the number of jobs in system one. These transition are transitions in the upper right-hand corner of the transition matrix. Transitions to lower states are represented by the rates $j\mu_2$. These values are in the lower left-hand part of the matrix. For two systems in tandem, we have the following transitions

(1) $(i, j) \rightarrow (i + 1, j)$ at rate λ,
(2) $(i, j) \rightarrow (i - 1, j + 1)$ at rate μ_1,
(3) $(i, j) \rightarrow (i, j - 1)$ at rate μ_2.

For two systems in tandem with $n_1 = 3$ and $n_2 = 3$ we have $(n_1 + 1) \cdot (n_2 + 1) = 16$ states, and each state has to be represented. This illustrates the main problem with this appoach. The number of equations is growing very fast. Although being a very important method of analyzing single queues, it is not a very practical way of analyzing queueing networks, depite the accuracy of the method.

The systems operate as follows. A job arrives at the first system and is immediately serviced if there are free resources, otherwise it is lost. After completion of service in system one, it is transferred to the second system. If there are free resources in the second system, the job is processed in system two. If there are no free resources in system two, we can distinguish between two scenarios; a job may leave system one and be lost, should system two not have resources free at the transition time, or, alternatively, the job may have to wait in system one for a free resource in system two. The two scenarios lead to different solutions.

Suppose we have two $M/M/3/3$ systems in series, with $\lambda = 5$ at the first system and $\mu_1 = \mu_2 = 2$. Note that for a single system, we would have $r = \rho/n = 5/3 \cdot 2$. This is also the case for the first system in scenario one. For the second system, the offered input rate is the output rate of system one, which is no longer Poisson due to blocking. Furthermore, in scenario two, the resources in system one are occupied longer than $\mu_1 = 2$, so the state probabilities are shifted towards higher states.

We can solve for the equilibrium distribution of the tandem system exactly, and compare with randomization of the exact system and of a truncated state space of an infinite system. Starting with scenario one, where a job leaving system one is lost if there are no free resources at system two, we have the generator

$$
\begin{pmatrix}
-5 & 5 & 0 & 0 & 0 & 0 & 0 & 0 & 0 & 0 & 0 & 0 & 0 & 0 & 0 & 0 \\
0 & -7 & 2 & 5 & 0 & 0 & 0 & 0 & 0 & 0 & 0 & 0 & 0 & 0 & 0 & 0 \\
2 & 0 & -7 & 0 & 5 & 0 & 0 & 0 & 0 & 0 & 0 & 0 & 0 & 0 & 0 & 0 \\
0 & 0 & 0 & -9 & 4 & 0 & 5 & 0 & 0 & 0 & 0 & 0 & 0 & 0 & 0 & 0 \\
0 & 2 & 0 & 0 & -9 & 2 & 0 & 5 & 0 & 0 & 0 & 0 & 0 & 0 & 0 & 0 \\
0 & 0 & 4 & 0 & 0 & -9 & 0 & 0 & 5 & 0 & 0 & 0 & 0 & 0 & 0 & 0 \\
0 & 0 & 0 & 0 & 0 & 0 & -6 & 6 & 0 & 0 & 0 & 0 & 0 & 0 & 0 & 0 \\
0 & 0 & 0 & 2 & 0 & 0 & 0 & -11 & 4 & 0 & 5 & 0 & 0 & 0 & 0 & 0 \\
0 & 0 & 0 & 0 & 4 & 0 & 0 & 0 & -11 & 2 & 0 & 5 & 0 & 0 & 0 & 0 \\
0 & 0 & 0 & 0 & 0 & 6 & 0 & 0 & 0 & -11 & 0 & 0 & 5 & 0 & 0 & 0 \\
0 & 0 & 0 & 0 & 0 & 0 & 2 & 0 & 0 & 0 & -8 & 6 & 0 & 0 & 0 & 0 \\
0 & 0 & 0 & 0 & 0 & 0 & 0 & 4 & 0 & 0 & 0 & -13 & 4 & 5 & 0 & 0 \\
0 & 0 & 0 & 0 & 0 & 0 & 0 & 0 & 6 & 2 & 0 & 0 & -13 & 0 & 5 & 0 \\
0 & 0 & 0 & 0 & 0 & 0 & 0 & 0 & 0 & 0 & 4 & 0 & 0 & -10 & 6 & 0 \\
0 & 0 & 0 & 0 & 0 & 0 & 0 & 0 & 0 & 0 & 0 & 6 & 4 & 0 & -15 & 5 \\
0 & 0 & 0 & 0 & 0 & 0 & 0 & 0 & 0 & 0 & 0 & 0 & 0 & 6 & 6 & -12
\end{pmatrix}.
$$

The states are enumerated using the principle in Table 5.2 and rows corresponding to infeasible states have been removed.

A comparisom between the equlibrium probability distributions for the exact model, the randomization method and the truncation of the infinite state space is shown in Table 5.4.

Table 5.4 Example of two systems in tandem; a job departing from system one is lost whenever there are no free resources in system two			
State	**Exact**	**Randomization**	**Truncation**
(0, 0)	0.0120	0.0150	0.0150
(1, 0)	0.0310	0.0375	0.0374
(0, 1)	0.0300	0.0328	0.0328
(2, 0)	0.0433	0.0508	0.0504
(1, 1)	0.0784	0.0838	0.0836
(0, 2)	0.0370	0.0365	0.0365
(3, 0)	0.0728	0.0831	0.0808
(2, 1)	0.1176	0.1245	0.1228
(1, 2)	0.0956	0.0931	0.0928
(0, 3)	0.0294	0.0270	0.0270
(3, 1)	0.1102	0.1098	0.1027
(2, 2)	0.1161	0.1090	0.1075
(1, 3)	0.0660	0.0585	0.0591
(3, 2)	0.0734	0.0648	0.0518
(2, 3)	0.0616	0.0522	0.0559
(3, 3)	0.0257	0.0208	0.0430

Table 5.5 Example of two systems in tandem: marginal probabilities for each queue (loss system)

State	System 1	State	System 2
$(0, -)$	0.108	$(-, 0)$	0.159
$(1, -)$	0.271	$(-, 1)$	0.336
$(2, -)$	0.339	$(-, 2)$	0.322
$(3, -)$	0.282	$(-, 3)$	0.183

Table 5.6 Example of two systems in tandem: marginal probabilities for each queue (waiting system)

State	System 1	State	System 2
$(0, -)$	0.067	$(-, 0)$	0.144
$(1, -)$	0.220	$(-, 1)$	0.320
$(2, -)$	0.353	$(-, 2)$	0.327
$(3, -)$	0.359	$(-, 3)$	0.209

By collecting the entries (i, j) for $j = 0, 1, 2, 3$ of joint probabilities in Table 5.4, we obtain the marginal state probabilities for each queue in the loss system, shown in Table 5.5.

It is easily seen that system one is independent of system two, and its equilibrium state probabilities can be found using analysis on this system in isolation. The state probabilities of system two, however, depends on the states of both systems.

If a job exiting system one has to wait for free resources in system two, the transition matrix (5.3.5) has to be modified so that, for example, transition probabilities from state $(1, 3)$ to $(0, 3)$ is zero. The diagonal elements in the matrix is adjusted accordingly. The marginal probabilities for each queue in a waiting system is shown in Table 5.6. The state probabilities of system one is shifted towards higher states due to jobs waiting, and so are the state probabilities for system two, due to that no jobs are lost.

5.4 The Erlang B-formula and generalizations

The Erlang B-formula is aimed at determining the probability of loss based on the offered traffic and the number of circuits. It is often necessary to evaluate it with a prescribed accuracy. As it is used repeatedly in some applications, the speed of the evaluation might be of significant importance.

The basic form of the Erlang B-formula for an integer number of circuits is

$$E_B(\rho, n) = \frac{\frac{\rho^n}{n!}}{\sum_{i=0}^{n} \frac{\rho^i}{i!}}, \tag{5.39}$$

where $B_n = E_B(\rho, n)$ is the loss probability, ρ is the offered traffic, and n is the number of circuits.

A generalized version of the formula is available for a non-integral number of circuits,

$$E_B(\rho, x) = \frac{\rho^x e^{-\rho}}{\int_\rho^\infty t^x e^{-t} dt}. \tag{5.40}$$

5.4.1   Integer number of circuits

Direct evaluation of $E_B(\rho, n)$ according to the definition would be time consuming. Therefore, a recurrent form of the formula is used widely,

$$E_B(\rho, n+1) = B_{n+1} = \frac{\rho \cdot B_n}{n + 1 + \rho \cdot B_n}, \tag{5.41}$$

$$B_0 = 1. \tag{5.42}$$

Concerning a real implementation, it is better to evaluate the reciprocal value of $E_{1,N}(A)$,

$$I_n(\rho) = \frac{1}{E_B(\rho, n)} = 1 + \frac{n}{\rho} I_{n-1}, \tag{5.43}$$

$$I_0 = 1. \tag{5.44}$$

5.4.2   Non-Integer Number of Circuits

In some circumstances, it is required to evaluate blocking probabilities for non-integral number of circuits, such as in analysis of overflow systems and non-Poissonian traffic approximations. In such cases, a generalized formula can be used,

$$E_B(\rho, x) = \frac{\rho^x e^{-\rho}}{\int_\rho^\infty t^x e^{-t} dt} = \frac{\rho^x e^{-\rho}}{\Gamma(x, \rho)}, \tag{5.45}$$

where the number of circuits x can be any real number. The expression requires numerical evaluation of the (upper) *incomplete gamma function*,

$$\Gamma(x, \rho) = \int_\rho^\infty t^{x-1} e^{-t} dt.$$

By partial integration it can be shown for integer values of x that

$$\Gamma(x+1, \rho) = \int_\rho^\infty t^x e^{-t} dt = x! e^{-\rho} \left(1 + \frac{\rho}{1!} + \cdots + \frac{\rho^x}{x!} \right),$$

so that (5.45) indeed reduces to (5.39) for integer arguments n.

A numerical method to evaluate the incomplete gamma function is described in Press et al.. The convergence of (5.45) for $x > \rho$ may be slow, which makes it time

consuming to obtain a result with an appropriate accuracy. In such cases, it more effi-
cient to evaluate $E_B(\rho - 1 + \theta, \rho)$, where θ is the fractional part of x. The final value of
the loss probability is then evaluated using Equation (5.41) or (5.43), with n (an integer
number) replaced by x (a real number) and using the initial value $E_B(\rho - 1 + \theta, \rho)$.

Instead of direct evaluation of $E_B(\rho, x)$ for $x \in \mathbb{R}$, $x \geq 0$, an approximation can
be used.

5.4.3 Approximations

The Rapp Approximation The approximation is based on the equality of $E_B(\rho, x)$
and its first derivative with the approximating function and its first derivative in the
points $x = 0$ and $x = 1$.

$$E_B(\rho, x) \approx C_0 + C_1 x + C_2 x^2, \tag{5.46}$$

$$C_0 = 1, \tag{5.47}$$

$$C_1 = -\frac{\rho + 2}{(1 + \rho)^2 + \rho}, \tag{5.48}$$

$$C_2 = \frac{1}{(1 + \rho)((1 + \rho)^2 + \rho)}. \tag{5.49}$$

The approximating function is directly used in the interval $x \in (0, 1)$. For $x > 1$ a
fractional part of the loss probability is determined. The final value of the approximation is evaluated using Equation (5.41), which is also valid for $N \in \mathbb{R}$.

The Szybicky Approximation. Another approximation intended to be used on interval $x \in (0, 2)$ and it equals $E_{1,x}(A)$ in points $x = 0, 1, 2$.

$$E_B(\rho, x) \approx \frac{(2 - x) \cdot A + A^2}{x + 2A + A^2}.$$

For $x > 2$, it is again possible to use the approximation in conjunction with the
recurrent formula (5.41).

The Hedberg Approximation. Another approximation, proposed by Hedberg
(Iversen, 2013) and again based on an approximation by a hyperbolic function, is

$$E_B(\rho, x) \approx \frac{1}{1 + \rho} \left(\frac{C(1 + C)}{x + C} + \rho - C \right),$$
$$C = \frac{1}{2} \left(\rho(3 + \rho) + \sqrt{\rho^2(3 + \rho)^2 + 4\rho} \right).$$

The approximation is valid for $0 \geq x < 1$.

5.4.4 The Erlang C-formula

The Erlang C-formula is used for *delay systems*, and expresses the probability that a job arriving according to a Poisson process would have to wait. Such systems are used to describe packet switched networks and some signaling networks. A delay system is described by a $M/M/n$ queueing system with Poisson arrivals process, exponential service times, n servers and an infinite number of waiting positions. The state of the system is defined by the total number of customers in the system (either being served or waiting in the queue). Letting $\rho = \lambda/\mu$, we have the steady state probabilities

$$p(i) = \begin{cases} p(0)\frac{\rho^i}{i!} & \text{if } 0 \leq i \leq n, \\ p(0)\frac{\rho^i}{n! \cdot n^{i-n}} & \text{if } i \geq n \end{cases} \qquad (5.50)$$

By normalization of the state probabilities, we obtain $p(0)$ from

$$\sum_{i=0}^{\infty} p(i) = 1,$$

or, with the series expanded,

$$1 = p(0)\left(1 + \frac{\rho}{1} + \frac{\rho^2}{2!} + \cdots + \frac{\rho^n}{n!}\left(1 + \frac{\rho}{n} + \frac{\rho^2}{n^2} + \cdots\right)\right).$$

The innermost brackets is a geometric series with quotient ρ/n. Statistical equilibrium is only obtained for $\rho < n$. Otherwise, the queue length would increase towards infinity with time. We obtain

$$p(0) = \frac{1}{\sum_{i=0}^{n-1} \frac{\rho^i}{i!} + \frac{\rho^n}{n!}\frac{n}{n-\rho}}, \quad \rho < n, \qquad (5.51)$$

and equations (5.50) and (5.51) yield the steady state probabilities $p(i)$, $i > 0$.

Letting W denote the random variable of waiting time, we have the Erlang C-formula

$$E_C(\rho, n) = \mathbf{P}(W > 0) = \sum_{i=n}^{\infty} p(i),$$

where $p(i)$ is the state probability of i jobs in the system, and n is the number of channels or circuits. Expanding this gives

$$E_C(\rho, n) = \frac{\rho^n}{n!}\frac{n}{n-\rho}\sum_{i=0}^{n-1}\frac{\rho^i}{i!} + \frac{\rho^n}{n!}\frac{n}{n-\rho}, \quad \rho < n. \qquad (5.52)$$

The similarities between Equations (5.39) and (5.52) can be used to evaluate the Erlang C-formula using a method for the B-formula. We have, in fact, the relationship

$$E_C(\rho, n) = \frac{n E_B(\rho, n)}{n - \rho(1 - e_B(\rho, n))}, \quad \rho < n. \qquad (5.53)$$

From (5.53) it is also clear that $E_B(\rho, n) < E_C(\rho, n)$, and for $\rho \geq n$, $E_C(\rho, n) = 1$, since all jobs are delayed.

Comparing (5.39) and (5.52), we note that a recurrence expression for the latter would be identical to the former apart from the last term

$$\frac{\rho}{n} \frac{n}{n - \rho}$$

A direct recurrence relation can therefore be contructed by using (5.41)–(5.42) up to state $n - 1$, and then finding the final term from

$$E_C(\rho, n) = \frac{\rho E_B(\rho, n - 1)}{n - \rho(1 - E_B(\rho, n - 1))}.$$

The carried traffic equals the offered traffic $Y = \rho$. The queue length is a random variable L, for which we have

$$\mathbf{P}(L > 0) = \sum_{i=n+1}^{\infty} p(i) = p(n)\frac{\rho}{n - \rho}.$$

We obtain the mean queue length $L_n = \mathbf{E}(L)$ from the state probabilities

$$L_n = 0 \cdot \sum_{i=0}^{n} p(i) + \sum_{i=n+1}^{\infty} (i - n)p(i) = \sum_{i=n+1}^{\infty} (i - n)p(n)\left(\frac{\rho}{n}\right)^{i-n}.$$

from which we get

$$L_n = E_C(\rho, n)\frac{\rho}{n - \rho}.$$

and, using Little's formula, the mean waiting time

$$W_n = \frac{L_n}{\lambda} = E_C(\rho, n)\frac{1}{\lambda}\frac{\rho}{n - \rho}.$$

5.4.5 Derivatives

There are situations where the derivative of the Erlang B-formula is required. This occurs sometimes when using a numerical method for solving a set of equations or in optimization problems.

Derivative with respect to traffic. Consider the Erlang B-formula given by (5.39), where $(n, \rho) > 0$ are non-negative real numbers. Denoting the denominator by G_n, the normalization constant, we find the derivative with respect to ρ as

$$\frac{\partial E_B(\rho, n)}{\partial \rho} = \frac{G_n \frac{\rho^{n-1}}{(n-1)!} - \frac{\rho^n}{n!}\frac{\partial G_n}{\partial \rho}}{G_n^2},$$

where

$$\frac{\partial G_n}{\partial \rho} = 1 + \frac{\rho}{1} + \cdots + \frac{\rho^{n-1}}{(n-1)!} = G_{n-1},$$

where G_{n-1} is the normalizing constant for a system with $n-1$ channels.

From the recursion formula, we have

$$E_B(\rho, n-1) = \frac{n E_B(\rho, n)}{\rho(1 - E_B(\rho, n))},$$

which gives

$$\frac{G_{n-1}}{G_n} = (1 - E_B(\rho, n)).$$

Putting this together we arrive at

$$\frac{\partial E_B(\rho, n)}{\partial \rho} = \frac{\frac{n}{\rho} \frac{\rho^n}{n!}}{G_n} - E_B(\rho, n) \frac{G_{n-1}}{G_n}$$

$$= \frac{n}{\rho} E_B(\rho, n) - E_B(\rho, n)(1 - E_B(\rho, n)) = \left(\frac{n}{\rho} - 1\right) E_B(\rho, n)$$

$$+ \left(E_B(\rho, n)\right)^2. \tag{5.54}$$

Higher order derivatives can be found in the same way. A discussion on the derivative with respect to n can be found in Iversen (2013).

5.4.6 Inverse

A common task in resource allocation is to evaluate an *inverse* of the Erlang B-formula, such as finding n as a function of (ρ, E) or ρ as a function of (n, E), where E is a pre-specified blocking rate. This can be accomplished by using *Newton-Raphson's method*, iterating

$$x_{k+1} = x_k + \frac{f(x_k)}{f'(x_k)},$$

where x_k is the sought quantity obtained after iteration k. The iteration is initiated with a starting value x_0, which is typically a first "best guess". If ρ is the quantity of interest, we can use $f(\rho) = \rho(E_B(\rho, n) - E)$ and $\rho_0 = n/(1 - E)$. For n, $f(n) = E_B(\rho, n) - E$, with initial number of channels x are chosen so that, $n_0 - 1 < x \leq n_0$ where n_0 is chosen so that $E_B(\rho, n_0) \leq E < E_B(\rho, n_0 - 1)$.

For the solution, one can also use approximation formulae, for example, that suggested by Rapp:

$$\rho^* = v + 3z(z - 1), \tag{5.55}$$

$$m^* = \frac{\rho^*(\rho + z)}{\rho + z - 1} - \rho - 1, \quad z = v/\rho. \tag{5.56}$$

The approximation is not accurate if ρ is small and z is large.

5.5 Overflow theory

The theory of overflow traffic describes how traffic characteristics are affected by overflow or blocking. It is instructive to study such systems for several reasons. One is to gain understanding of how blocking is affecting traffic characteristics, which happen, for example, at traffic rerouting triggered by overflow. Another is that the theory can be used to approximate non-Poissonian traffic which is offered a system by constructing an equivalent system offered Poissonian traffic.

An overflow model consists of two systems (or groups), called the primary and the secondary system. The primary system consists of n_1 channels and is offered Poissonian traffic of intensity $\rho_a = \lambda_a / \mu_a$, where λ_a is the arrival rate and μ_a the rate of the exponential service distribution. Incoming traffic is carried by the primary system as long as there are free channels. In case of congestion, overflowing traffic is routed to the secondary system, equipped with n_2 channels. This configuration is shown in Figure 5.11. If the primary system is congested and all the servers in the secondary system are occupied, any additional traffic is blocked and lost. Note that we consider unidirectional traffic in this section.

Clearly, the primary system by itself forms an $M/M/n_1/n_1$ system. (It does not see what happens to the traffic that overflows, whether it is blocked or rerouted. The primary and the secondary systems form together an $n = n_1 + n_2$ loss system, that is an $M/M/n/n$ system. This system is shown in Figure 5.12.

Since overflow traffic is generated only when the primary system is in a blocked state, the time periods in which the primary system is congested form intervals in which a traffic stream to the second is switched on. During other times, the overflow traffic stream is switched off. The traffic offered to the secondary system (the overflow traffic) is therefore called an interrupted Poisson process (IPP).

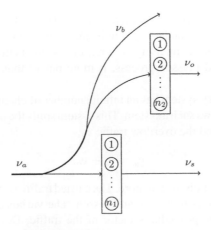

FIGURE 5.11

Overflow network with two systems, each of which has n_1 and n_2 channels, respectively.

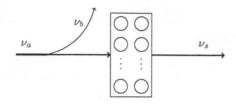

FIGURE 5.12
Overflow network where the two systems are considered as one; this is equivalent to the two systems in terms of blocked traffic.

In Figure 5.11, the intensity of the overflow traffic, v_o, is

$$v_o = v_a \cdot E_B(v_a, n_1),$$

where $E_B(\cdot)$ is the Erlang B-formula (5.39). We will usually denote a traffic intensity modified by offered a system by v, which does not need to be Poisson distributed, and use λ to denote the intensity of external Poisson arrivals. (We allow, however, exemptions to this rule in some cases. Here, we let $\lambda = v_a$ for notational consistency.)

When both systems are congested, the intensity of the blocked (and lost) traffic is

$$v_B = v_a \cdot E_B(v_a, n_1 + n_2),$$

where we let B denote blocking rates (in the time blocking sense). Since the overflowing traffic is not Poisson, different blocking measures are numerically different. The call blocking of the overflowing traffic can be found as the ratio of the overflowing traffic intensities for the primary system and the total system, which both are offered Poisson traffic. We have, denoting the blocking of the overflow traffic by B_o,

$$B_o = \frac{v_a E_B(v_a, n_1 + n_2)}{v_a E_B(v_a, n_1)} = \frac{E_B(v, n_1 + n_2)}{E_B(v, n_1)}.$$

It is possible to show that $B_o > E_B(v_o, n_2)$. Thus, the blocking that the overflowing traffic encounters in the secondary system is greater than had the overflowing traffic been Poisson with the same intensity. The reason for this is that the overflow traffic, which is an interrupted Poisson process, is more bursty than the ordinary Poisson process.

In case the secondary system has an infinite number of channels, congestion never occurs. Figure 5.12 shows such a system. This system splits the traffic into two streams, the carried traffic v_s, and the overflow traffic v_o,

$$v_a = v_s + v_o.$$

Neither traffic stream is Poisson anymore. The carried traffic is smoother than Poisson, and the overflow traffic is burstier than Poisson. The variance to mean ratio of the occupancy is called the peakedness factor of the traffic. Describing the degree of traffic burstiness by its peakedness, we see that the peakedness of the carried traffic is

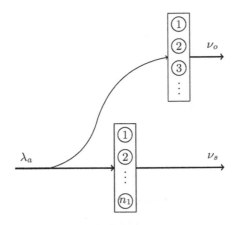

FIGURE 5.13

Overflow network where the second system has infinite capacity. In this case, no blocking occurs at the two systems.

$Z_c < 1$, whereas the peakedness of the overflow traffic $Z_o > 1$. It is possible to derive some properties of the resulting traffic streams, which are useful in many situations.

Consider two systems with n_1 channels in the first and $n_2 = \infty$ in the second (see Figure 5.13), with a traffic intensity of v being offered to the first system. Then, the expectation and the variance of the occupancy N of the second system, representing the overflow traffic, can be calculated.

Denoting the expectations by m_k and the variances v_k for systems $k = 1, 2$, we have, for the first system

$$m_1 = v(1 - E_B(v, n_1)),$$

$$Z_1 = \frac{v_1}{m_1} = 1 - v(E_B(v, n_1 - 1) - E_B(v, n_1)),$$

and for the second

$$m_2 = v E_B(v, n_1),$$

$$Z_2 = \frac{v_2}{m_2} = 1 - m_2 + \frac{v}{n_1 + 1 - v + m_2},$$

$$E(N) = v E_B(v, n_1) = \alpha.$$

Also, the variance of N can be calculated in this case. The result is known as the Riordan formula,

$$\text{Var}(N) = \alpha \left(1 - \alpha + \frac{v}{n_1 + 1 - v + \alpha} \right).$$

In the case of a Poisson arrival process the occupancy distribution is a $Po(v)$ distribution with peakedness factor 1. The peakedness of the overflowing traffic is

$$Z = \frac{\text{Var}(N)}{E(N)} = 1 - \alpha + \frac{v}{n_1 + 1 - v + \alpha},$$

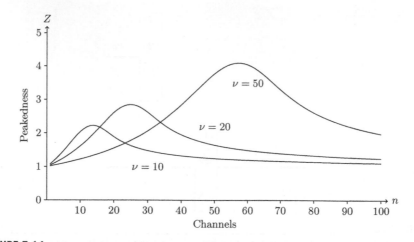

FIGURE 5.14

Peakedness for $\nu = 10$, $\nu = 20$ and $\nu = 50$.

where $\alpha = \nu E_B(\nu, n_1)$. The peakedness factor is in this case a function of n_1 and ν. We notice that when ν is held fixed and small n_1, $Z \approx 1$ (that is, all the traffic flows over). When n_1 is increased, Z attains a maximum (when $n_1 \approx \nu$) and then decreases for large n_1 toward unity (as the overflow events occur rarely). This is shown in Figure 5.14 for different values of ν.

5.5.1 The Hayward approximation

The Hayward approximation is an approximate method to calculate the blocking probability for non-Poissonian traffic, for example overflow traffic. It is based on the observation that for the occupancy N in an infinite server system,

$$\mathbf{E}(N) = \nu,$$

$$\text{Var}(N) = \nu, \quad N \sim Po(\nu).$$

For non-Poissonian traffic in general we have $\text{Var}(N) \neq \mathbf{E}(N)$. The Hayward approximation uses an equivalent Poisson traffic to describe the non-Poissonian traffic, and then uses the Erlang B-formula to determine the blocking rate.

We investigate the relations for occupied capacity R, rather than for the occupancy N—the number of connections—directly. Thus, we let c be the bandwidth (number of trunks) required by a single connection and $R = N \cdot c$ the bandwidth occupied in state N. The parameter c then works as a scaling parameter between N and R. For Poissonian traffic, we have

$$\mathbf{E}(R) = \mathbf{E}(c \cdot N) = c \cdot \nu$$

$$\text{Var}(R) = \text{Var}(c \cdot N) = c^2 \cdot \nu,$$

so that

$$\frac{\text{Var}(R)}{\text{E}(R)} = c.$$

In case of a non-Poissonian source with known mean m and variance v of occupation N, clearly

$$\text{E}(N) = m,$$

$$\text{Var}(N) = v.$$

and

$$\text{E}(R) = c \cdot m$$

$$\text{Var}(R) = c^2 \cdot v,$$

Denote with an apostrophe traffic related to the equivalent system with a Poissonian arrival process. This non-Poissonian traffic is now replaced by Poissonian traffic where v' is the traffic intensity and c' is the bandwidth requirement of a single connection. The mean and the variance of the occupied capacity should be the same for the original non-Poissonian traffic and the model Poisson traffic, that is

$$\text{E}(R) = \text{E}(R'), \quad \text{Var}(R) = \text{Var}(R'),$$

where $R' = N \cdot c'$ is the bandwidth occupied in state N by the equivalent Poissonian traffic. Note, that in the approximation by the equivalent Poissonian traffic, the bandwidth required by a single connection is taken as a free parameter, which may be a real value. Comparing the two first moments leads to the following conditions for v' and c',

$$v' \cdot v' = m \cdot c$$

$$v' \cdot (c')^2 = v \cdot c^2,$$

which gives

$$v' = \frac{m}{v}$$

$$c' = c \cdot \frac{v}{m},$$

which represent the equivalent intensity and the equivalent bandwidth, respectively. The size of the system is modified correspondingly. If the original system has n trunks with a capacity of $n \cdot c$ bandwidth units, then it can carry $n \cdot c/c'$ equivalent connections. Thus, the equivalent system has n' trunks, given by

$$n' = \frac{n \cdot c}{c'} = n \cdot \frac{m}{v}.$$

Now the blocking probability is approximated by the blocking rate of the equivalent Poissonian traffic, so that

$$B \approx E_B(v', n') = E_B(\frac{m^2}{v}, n \cdot \frac{m}{v}) = E_B(\frac{m}{Z}, \frac{n}{Z}), \tag{5.57}$$

where $Z = v/m$. The load per channel is the same as before, that is, $(m/Z)/(n/Z) = m/n$. When $Z > 1$, the system is scaled down to a smaller one, and the blocking therefore increases.

The non-Poissonian traffic may originate from several independent sources. If for each source the mean m_i and the variance v_i of the occupancy are known, then the corresponding parameters for the aggregate traffic stream are

$$m = \sum_i m_i$$

$$v = \sum_i v_i.$$

The Hayward approximation then gives the approximate total blocking rate for the aggregate traffic stream in a system with n trunks by (5.57).

Example 5.5.1.   Let a non-Poissonian traffic stream be characterized by mean $m = 10$ and variance $v = 30$ and offered to a system with $n = 20$ channels. The peakedness of the traffic is $Z = v/m = 3$ and by Hayward's approximation, the blocking is given by scaling the mean and the number of channels by the peakedness

$$E_B \left(\frac{m}{Z}, \frac{n}{Z} \right) = E_B(3.33, 6.67).$$

Here we need to evaluate the Erlang B-formula for non-integer number of channels by (5.45). The result is $B = 0.0434$.

5.5.2   The Wilkinson-Bretschneider method

The Wilkinson-Bretschneider method (also known as the Equivalent Random Theory (ERT) method) provides another approximate method to calculate the blocking probability for non-Poissonian traffic.

The Wilkinson-Bretschneider method is based on approximating the non-Poissonian traffic (m, v) by overflow traffic from a fictitious system where v^* is the offered traffic and n^* is the number of channels in the first system. The parameters v^* and n^* are determined so that the overflow traffic in the fictitious system has the intensity m and variance v. Matching the moments using Riordan's formulas, we have

$$m = v^* \cdot E_B(v^*, n^*)$$
$$v = m \left(1 - m + \frac{v^*}{n^* + 1 - v^* + m} \right).$$ \hfill (5.58)

If the system to which the non-Poissonian traffic is offered has n trunks, the intensity m_l of the ultimately overflowing traffic can be calculated as

$$m_l = v^* E_B(v^*, n^* + n).$$

An estimate of the traffic blocking given by

$$B = \frac{m_l}{m}.$$

Example 5.5.2. Suppose we wish to approximate the blocking experienced by the same non-Poissonian traffic as in Example 5.5.1 with mean $m = 10$ and variance $v = 30$, offered to a system with $n = 20$. In order to solve the first of the moment-matching Equations (5.58), we can use the Newton-Raphson method. Let

$$f(v^*, n^*) = m - v^* E_B(v^*, n^*) = 0.$$

The parameter m can be considered fixed and is omitted from the arguments. Letting $x^{(n)}$ be the iterative improvement of v^*, we can iterate $x_{n+1} = x_n + h_n$ with $h_n = -f(x_n)/f'(x_n)$ until it converges, which will then be the value of v^* we require. For this, however, we need the derivative of $f(v)$ which can be found using (5.54). The value of n^* is recalculated in each iteration which rearranged gives

$$n^* = x_n(m + Z)/(m + Z - 1) - m - 1.$$

After iterating until the values converge, we find that $v^* = 46.70$ and $n^* = 39.60$. The blocking is found to be $B = v^* E_B(v^*, n^* + n)/m = 0.0484$. An alternative is to use Rapp's approximation, which involves much less computation. Then,

$$v' = v + 3Z(Z - 1),$$

$$n' = v'(m + Z)/(m + Z - 1) - m - 1,$$

which give $v' = 48$ and $n' = 41$, and $B = E_B(v', n' + n) = 0.0499$.

5.6 Summary

The chapter serves as an introduction to some important stochastic processes in general, and stochastic processes in queues, in particular. First, the concepts of traffic, system, and blocking are discussed. Thorough discussions can be found in Iversen (2013) and Hajek (2006) in the context of telecommunications. Fundamentals on probability theory is presented in Grimmett and Stirzaker (1992), for example. The methods discussed are applied to various problems in subsequent chapters.

The components in a network subject to random traffic are modeled as queues. Two important queues are discussed; the $M/M/1$ queue, used mainly as a model in packet networks and $M/M/n/n$ systems which model trunk groups in loss networks. Some important performance measures of these systems are presented. Two useful references are Gross and Harris (1998), and more advanced topics on general queues, Cohen (1969).

Markov chain analysis is of fundamental importance to queueing theory. It is important to be familiar with some techniques of finding equilibrium distributions to

Markov chains. Many numerical methods can be found in Stewart (1994). However, difficulties arise when state spaces are infinite or large. Then, it is possible to make an approximation by truncating the state space (Zhao and Liu, 2004).

The fundamental Erlang formulas and their computational aspects and some generalizations and extensions presented in Section 5.4. These constitute some of the most often used formulas in teletraffic engineering. Some of the vast references on these equations are Iversen (2013) and Hodousek (2003).

The final section discusses overflow theory. This is important for many reasons. Overflow traffic results whenever alternative routing is allowed. When traffic overflows, and the original traffic is split onto two trunk groups, their characteristics change. Traffic in the primary group becomes smoother and in the secondary group more bursty. This can be used to approximate blocking of non-Poissonian traffic. A comprehensive account can be found in Iversen (2013).

Loss Networks

We now turn to the design of loss networks with focus on routing and capacity allocation. Even if traditional circuit-switched technology is becoming obsolete and networks are migrated to more resource efficient technologies, the study of such networks is still important.

Analysis of traditional loss networks provides the foundation of the engineering of networks for circuit-switched services such as voice telephony—the by-far dominating service provided by many network operators. Furthermore, the methods available for more complex networks are in most cases extensions of traditional methods.

A loss network (or circuit-switched network) is a network for telecommunication services where the system attempts to reserve necessary resources between the end nodes for establishing a communication channel before call set-up. If there are insufficient resources at the time of call, the call is lost. Once established, the communication channel remains occupied by the call for its entire duration. This functionality limits the number of calls that can be handled by the network at any time instant. We therefore distinguish between the carried traffic as opposed to the offered traffic which is the demand of service on the network, some of which is typically lost. A network operator, however, would not wish that the discrepancy between offered and carried network is too large, as this leads to a substandard service level for the users. Note that the communication channel reserves a constant bandwidth to an admitted call regardless of whether anything is transmitted or not.

In this chapter, we will generally assume that the network topology is given, together with a stochastic demand between the nodes in the network and a number representing the desired service level, that is, the maximum allowable end-to-end blocking in the network. This number is often represented by the Grade of Service.

We assume that we have a network of size n and an $n \times n$ traffic matrix A, which defines the traffic demand between each pair of nodes. We also assume that are given an $n \times n$ cost matrix C which defines cost structure of the links. Finally, we also expect the grade of service requirements that should be met to be known.

The design process is iterative in nature. The first step is to find a feasible network topology. We then create a routing strategy based on this topology, subject to the transmission technology of the network. The routing strategy is then used to assign traffic flows to the links, and capacities are assigned to each link so that (a) the grade-of-service target is met; and (b) the overall cost is minimized. The overall cost represents the objective function of the network. We may repeat the procedure by changing the topology slightly—searching the neighborhood of the topology—to see if we can find a better solution, that is, a network with similar or better performance (Grade of Service) at a lower cost.

First, fixed-point methods are introduced as an efficient iterative method of estimating blocking in networks.

Section 6.3 explains some elementary routing strategies and trunk reservation, a simple and robust mechanism for admission control in loss networks. An optimization method for networks with dynamic routing and trunk reservation is described in Section 6.4. Simulation of loss networks is discussed in Section 6.5. Finally, some qualitative results on network stability and efficiency are presented in Section 6.6.

6.1 Calculating blocking in a network

The Erlang B-formula (5.39) allows exact calculation of the blocking rate of a single link with known offered traffic ρ. In a network, however, links are typically shared by different traffic streams. When a route contains more than one link, blocking further down the route, as seen from the source, affects the amount of traffic carried. This effect is referred to as thinning, and the result is that less traffic is offered to a link on a route when there is blocking elsewhere on the route. The blocking rate in the first link is therefore lower than it had been without blocking in other links on a route. Note that the traffic streams and link blocking rates in a network are not independent in general. Calculating the exact blocking rates in large networks shared by many traffic streams is therefore difficult. Fortunately, we can use an iterative procedure known as the Erlang fixed-point approximation (or the reduced load approximation). Based on the (approximate) link blocking rates, we can then calculate the end-to-end blocking rate over a route, that is, the blocking as seen from the source through to the destination.

In the following, we denote a link by l and a route by r. A route may consist of one or more links, so that the source and the destination are connected. Using v to denote traffic intensities within the network, define the traffic v_l offered to a link l as the sum of traffic streams v_r following any route r that includes l. Formally, we write this as

$$v_l = \sum_{r:l \in r} v_r,$$

that is, we sum over all routes and add together the contributions to v_l from routes which use link l.

First, we state two bounds for the blocking rates in a network. An upper bound for the link blocking rate B_l on a link l is given by

$$B_l \leq E_B(v_l, n_l), \tag{6.1}$$

where v_l is the traffic offered to l and n_l its capacity. To see that this bound is valid, note that the right-hand side remains unchanged if the capacities of all other links except l are infinite. Since the traffic offered to link l is dependent on blocking in other links, B_l must be smaller than or equal to this quantity. If the blocking probabilities B_l are very small, the end-to-end blocking rates B_r can be approximated by

$$B_r \approx \sum_{l \in r} B_l.$$

Usually, a better approximation for end-to-end blocking rates is one derived from (6.1). It follows from the fact that the upper bound on link blocking implies that the links are mutually independent. Thus,

$$B_r \leq 1 - \prod_{r:l \in r} (1 - E_B(v_l, n_l)). \tag{6.2}$$

The two bounds (6.1) and (6.2) can be useful approximations when blocking rates are small. However, this may be an overly optimistic assumption.

When the link blocking rates cannot be assumed to be small, the thinning effect has to be taken into account. Then link blocking affects the offered traffic which again affects the blocking. The method builds on a nonlinear system of equations, one set of equations for the aggregate traffic offered to each link l and one set for the link blocking rates. The equations can be iterated by successive substitution until the equations converge to the fixed point, which is the final result. We denote by $\hat{B}_l$ and $\hat{v}_l$ the successive approximations. Then,

$$\hat{B}_l = E_B(\hat{v}_l, n_l), \tag{6.3}$$

where

$$\hat{v}_l = \sum_{r:l \in r} v_r \prod_{k \in r - \{l\}} (1 - \hat{B}_l). \tag{6.4}$$

Note that in (6.4), the product representing the thinning effect includes all links in a route r using link l, except for link l itself. The blocking on link l is given by (6.3). It can be shown (Kelly, 1986), that the equations have a unique solution, called the Erlang fixed point. The end-to-end blocking on route r is given by

$$B_r \approx 1 - \prod_{l \in r} (1 - \hat{B}_l). \tag{6.5}$$

The end-to-end blocking thus includes blocking from all links l along route r. Equation (6.4) can be solved iteratively. Initially, we may take $\hat{B}_l = 0$ for all l, and substitute into (6.4). This gives a value for $\hat{v}_l$ for all l, which is substituted into (6.3) to give new values of $\hat{B}_l$, and so on. During the iteration, the values are alternatingly higher and lower than the fixed-point value. When the values do not change from one iteration to the next, the fixed point has been reached. Finally, the end-to-end blocking (6.5) can be computed.

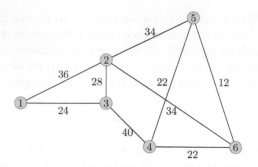

FIGURE 6.1

Estimation of end-to-end blocking by the Erlang fixed-point method; link capacities.

Example 6.1.1. Consider the small network depicted in Figure 6.1 showing the topology and the capacities in the number of channels for each link. For simplicity, we let $\nu = 12$ for each node pair, and move on to estimate the link end-to-end blockings.

We use shortest path routing, as follows. The shortest routes between any node pair are selected, but there is possibly more than one such route. It this route is unique (possible the direct route), we use this route. If there are two or more shortest routes, these are used by equally splitting the traffic load between them.

We get the paths for each node pair as listed in Table 6.1. Denote the blocking of a link by $\hat{B}_{ij}$, where (i, j) identifies the link. Then the system of equations is

Table 6.1 Routes and the corresponding end-to-end blocking

r	Stream	Route	B_r
1	(1,2)	1–2	0.091
2	(1,3)	1–3	0.129
3	(1,4)	1–3–4	0.190
4	(1,5)	1–2–5	0.173
5	(1,6)	1–3–4–6	0.173
6	(2,3)	2–3	0.112
7	(2,4)	2–3–4	0.200
8	(2,5)	2–5	0.091
9	(2,6)	2–6	0.091
10	(3,4)	3–4	0.070
11	(3,5)	3–2–5	0.194
12	(3,6)	3–4–6	0.194
13	(4,5)	4–5	0.134
14	(4,6)	4–6	0.134
15	(5,6)	5–6	0.199

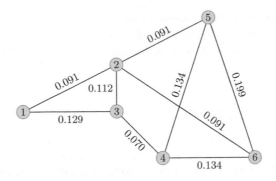

FIGURE 6.2

Estimation of end-to-end blocking by the Erlang fixed-point method; link blocking rates.

$$\hat{v}_{12} = 12 + 12(1 - \hat{B}_{25}) + 12(1 - \hat{B}_{26})$$
$$\hat{v}_{13} = 12 + 12(1 - \hat{B}_{34})$$
$$\hat{v}_{23} = 12 + 4(1 - \hat{B}_{34}) + 6(1 - \hat{B}_{25}) + 6(1 - \hat{B}_{26})$$
$$\hat{v}_{25} = 12 + 12(1 - \hat{B}_{12}) + 4(1 - \hat{B}_{45}) + 6(1 - \hat{B}_{23})$$
$$\hat{v}_{26} = 12 + 12(1 - \hat{B}_{12}) + 4(1 - \hat{B}_{46}) + 6(1 - \hat{B}_{23})$$
$$\hat{v}_{34} = 12 + 12(1 - \hat{B}_{13}) + 4(1 - \hat{B}_{23}) + 6(1 - \hat{B}_{45}) + 6(1 - \hat{B}_{46})$$
$$\hat{v}_{45} = 12 + 4(1 - \hat{B}_{25}) + 6(1 - \hat{B}_{34})$$
$$\hat{v}_{46} = 12 + 4(1 - \hat{B}_{26}) + 6(1 - \hat{B}_{34})$$
$$\hat{v}_{56} = 12$$
$$\hat{B}_{ij} = E_B(\hat{v}_{ij}, n_{ij}) \quad \text{for } (i, j) \text{ as above.}$$

Note that the equation for $\hat{v}_{56}$ can be omitted from the system as it is independent of the other equations. The link carries no traffic apart from between its end points, and the blocking is therefore trivially given by the Erlang B-formula. After a few iterations, the equations converge to the fixed point, with link blockings as shown in Figure 6.2.

The method assumes that the thinning in different links is independent (and therefore can be formulated as a product of those). It also assumes that the traffic streams are Poisson distributed along each route. These assumptions are idealizations of reality, and the fixed-point method is therefore an approximation. In fact, the blocking rates in different links are not independent, and the traffic is not Poisson distributed when blocking occurs; the blocking makes the traffic smoother by removing traffic peaks. Nevertheless, the approximation is good when many different traffic streams of the same magnitude are sharing the links. That is, the result is accurate for large networks with high capacity links, many traffic streams, and routing diversity.

Fixed-point methods have proven to be very useful in network performance analysis. Such methods can be formulated for loss networks with dynamic routing and trunk reservation, as well as in performance analysis of packet networks and multi-service networks.

6.2 Resource allocation

Being a fundamental part of network design, we revisit the topic of resource allocation (also known as dimensioning) in the context of loss networks and by using available information. Such information may be economical relationships, physical network characteristics, or performance metrics. A greedy heuristic due to K. Moe provides a simple, yet flexible, method of resource allocation. We start with an underprovisioned network, where the assigned capacities are too low for the blocking to meet required levels. In its original form, Moe's principle states that the cost of an additional line must be covered by the revenue generated by this line.

The method is iterated until no net gain can be achieved in the network, alternatively until some performance target is met. Moe's principle can be formulated in economical terms or in terms of some network performance metric. In general, both the cost and the gain or performance are represented by nonlinear functions of the capacities.

Instead of using the cost of a line, a performance metric such as end-to-end blocking can be used. In the latter case, a limit on the maximum end-to-end blocking is known, and Moe's principle is successively applied to the network until all end-to-end blocking is below this limit. A system will sometimes be referred to as a trunk group for traditional reasons, and this term also reflects that the static number of channels in the system is being modified.

Resource allocation of a network constitutes a nonlinear optimization problem. We assume that a topology is given, as well as a traffic matrix A definiting the offered traffic between all node pairs, an upper bound on the blocking rate B, and the link cost structure. The task is to allocate channels n_{ij}, whenever (i, j) is a link, so that for all traffic streams, the end-to-end blocking is below the bound B and the cost of the network is minimized. Section 6.1 considers the estimation of end-to-end blocking in a network by the means of the Erlang fixed-point method. We will combine this with Moe's principle to find a robust, albeit heuristic, procedure for the resource allocation problem.

6.2.1 Moe's principle

As a consequence of the Erlang B-formula, small trunk groups are less efficient than large ones for a given level of blocking. On the other hand, if the offered traffic v is increased proportionally for both types of groups, the loss increase is higher for large groups than for small groups. Based on these observations, Moe's principle

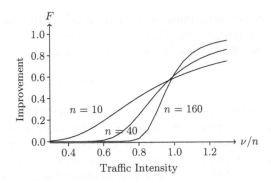

FIGURE 6.3

The Erlang Improvement Formula for capacities $n = 10$, $n = 40$, and $n = 160$.

formulates a strategy for allocation of resources both to small and large trunk groups in an incremental approach. This is shown in Figure 6.3. Smaller trunk groups improve more at lower ratios of v/n, whereas larger trunk groups improve more at higher v/n ratios.

Starting from an original underprovisioned network, where basically all end-to-end blockings are greater than the target value B, we denote the number of channels on a link by n. We may formulate a cost function $c(x, n)$ as

$$c(x, n) = \begin{cases} 0, & \text{if } x \leq n \\ c_0(x - n), & \text{if } x > n \end{cases},$$

where $x \geq 0$ is the expected value of occupied channels on a route and $c_0(\cdot)$ is the cost of adding additional channels to a route. The quantity x will be discrete if the distribution of occupied channels is discrete, and continuous if the distribution is continuous. The cost function is assumed to be non-decreasing. In case the cost function $c(\cdot)$ is linear, we have

$$c(x) = \begin{cases} 0, & \text{if } x \leq n \\ c_o + c_1(x - n), & \text{if } x > n \end{cases},$$

where c_0 and c_1 are constants. Other types of cost functions can be used, such as a piecewise linear or continuous convex functions. The classical Moe's principle can be expressed as:

Proposition 6.2.1. *The costs of new lines on a route must be covered by the revenues of those lines.*

This is a very general principle. Here, we will assume that offered traffic arrive as Poisson processes, so that x is discrete. We can formulate Moe's principle mathematically as follows. Since the offered traffic is Poisson with mean intensity v, for any trunk group the lost and carried traffic is given by the Erlang B-formula as $\nu E_B(\nu, n)$ and $\nu(1 - E_B(\nu, n))$, respectively. If the number of channels of the trunk group are increased by one, then the increase in traffic on this group is

$$F(\nu, n) = \nu(E_B(\nu, n) - E_B(\nu, n + 1)). \tag{6.6}$$

The function $F(\nu, n)$ is called the improvement function of Poisson traffic. Just like the Erlang B-formula, the improvement function $F(\nu, n)$ is a function of traffic intensity and the number of channels n. Figure 6.3 illustrates the improvement function for different sizes of trunk groups. It shows the improvement in traffic achieved for different ratios ν/n by adding a single additional channel to the existing number of channels.

If α denotes the revenue per traffic and time units, the additional revenues given by the trunk group are $\alpha F(\nu, n)$. Moe's principle can now be stated as

$$\alpha F(\nu, n) > c(\nu, n + 1) - c(\nu, n) = \Delta c(\nu, n). \tag{6.7}$$

It is instructive to see how this can be formulated as a general optimization problem. Let the carried traffic in a network with m links (trunk groups), each with n_l channels, be described by the function $f(n_1, n_2, \ldots, n_m)$. Since we consider loss systems, we will always have a positive improvement whenever the number of channels on some link is increased, and so

$$\frac{\partial f}{\partial n_i} > 0.$$

The function $f(\cdot)$ is just a 'continuous' version of (6.6). Let the target amount of carried traffic be described by $\bar{f}$. This value is, in principle, determined by the maximum allowed end-to-end blocking rate B. We denote the individual costs per link by c_i for $i = 1, 2, \ldots, m$.

Now, we wish to minimize the network costs $c = \sum_{i=1}^{m} c_i$ of the resource allocation. It is possible to use the standard optimization technique of Lagrange multipliers. Denoting the multiplier ϑ, the necessary condition for the minimum solution is

$$c_i - \vartheta \frac{\partial f}{\partial n_i} = 0, \quad \text{for } i = 1, 2, \ldots, m,$$

or,

$$\frac{1}{\vartheta} = \frac{\partial f / \partial n_1}{c_1} = \cdots = \frac{\partial f / \partial n_m}{c_m}. \tag{6.8}$$

This indicates that the optimum is achieved when resources are distributed in a way, such that the marginal increase in carried traffic per link divided by the cost, is constant. If there are different unit revenue factors α_i per link, we get

$$\frac{1}{\vartheta} = \alpha_1 \frac{\partial f / \partial n_1}{c_1} = \cdots = \alpha_m \frac{\partial f / \partial n_m}{c_m}.$$

6.2.2 An optimization procedure

Suppose that we are given an upper bound B to end-to-end blocking probabilities in a network, the offered traffic is Poisson and the initial number of channels n_l per link $l = 1, 2, \ldots, m$ are set so that the network is underprovisioned, that is, the end-to-end blockings are larger than B. We can initially choose n_l so that the blocking of each individual link is larger than B. We also have a routing scheme defined on the topology, which determines the fixed-point equations for end-to-end blocking rates.

We may formulate an iterated approach as follows. First, using the fixed-point equation for the network with underprovisioned resources, we get the end-to-end blocking rates. It is tricky to identify the best link on which to increment the capacity from this result directly, since that would require knowledge of the links used by each route and the aggregated traffic v_l on these links. It is much easier to determine the relative gain through improved blocking rates for each link with one channel added by the fixed-point method.

Thus, we can find the maximum $F_l(v, n)/c_l$, where $F_l(v, n)$ is the improvement function (6.6) and c_l the cost of increasing the capacity on link l by one. This link is then chosen as the one where to increase the channels by one. The procedure is repeated until all end-to-end blocking rates are below the bound B.

It should be noted that this is a heuristic based on the greedy principle, and although an optimum solution is not guaranteed, the result is usually quite good. It may require many iterations to reach a satisfactory result.

Example 6.2.1.   Consider the network in Figure 6.1, with the routing scheme described there, and suppose that we require the end-to-end blocking to be less than $B = 0.05$. We saw in Section 6.1, that with the capacities assigned, the end-to-end blockings do not fulfill this criterion. By using the procedure described, we can adjust the capacities of the links to improve the result.

Assuming a convex cost structure proportional to distance and capacity, we reach after 81 iterations the results link capacities shown in Figure 6.4 and link blocking rates in Figure 6.5. The total end-to-end blocking rates are listed in Table 6.2.

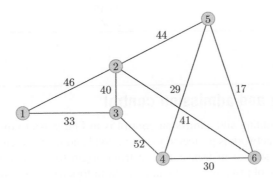

FIGURE 6.4

Resource allocation in a network based on Moe's principle, final capacities. (The end-to-end blocking is required not to exceed 5%.)

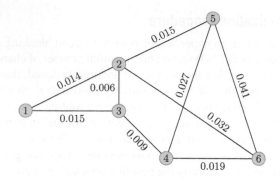

FIGURE 6.5

Resource allocation in a network based on Moe's principle, final link blocking rates. (The end-to-end blocking is required not to exceed 5%.)

Table 6.2 End-to-end blocking for all node pairs in Example 6.2.1	
Node Pair (i, j)	B_{ij}
(1,2)	0.014
(1,3)	0.015
(1,4)	0.024
(1,5)	0.029
(1,6)	0.047
(2,3)	0.006
(2,4)	0.036
(2,5)	0.015
(2,6)	0.033
(3,4)	0.009
(3,5)	0.029
(3,6)	0.033
(4,5)	0.027
(4,6)	0.019
(5,6)	0.041

6.3 Routing and admission control

The purpose of routing and admission control is to find ways to direct and control traffic so that the network is protected from overload and traffic streams are provided the best possible service of transportation from origin to destination. This section gives an overview of principles that can be used in loss networks. We will return to methods applicable in other network types in subsequent chapters. Admission controls considered here are of trunk reservation type.

6.3.1 **Routing**

There is a large number of principles for traffic routing. As a general rule, a good routing scheme improves the utilization of the network, and allows for load sharing of existing resources, also referred to as resource pooling. Such schemes can also compensate for inaccuracies in traffic estimation or network design. It is well known that a system with multiple channels is most efficient when these channels are fed by a single queue. This means that no channel is idle whenever there are jobs waiting to be served. This is the principle behind resource pooling. Figures 6.6 and 6.7 show two logical topologies. Figure 6.6 depicts the situation when traffic is routed only through dedicated links. For example, if the link between nodes 1 and 2 is blocked, then without any rerouting scheme, no traffic between the nodes may be transported though the network, even if the links between nodes 1 and 3, and nodes 3 and 2 have spare capacity.

By allowing rerouting, possibly subject to some conditions, the network can be made to operate more similar to the situation shown in Figure 6.7. Any route from, say node 1, may be used to reach its destination, say node 2. This principle usually increases the utilization and performance of the network, but may be detrimental under high load. This situation is discussed in Section 6.6.

The gain by resource pooling can be expressed in mathematical terms. Consider the minimal network in Figure 6.6. If we let the offered traffic intensities and link

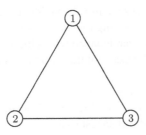

FIGURE 6.6

A network with static routing.

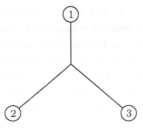

FIGURE 6.7

A network with dynamic routing.

capacities be equal, say v and n, the maximum carried traffic is $3v(1 - E_B(v, n))$. If, on the other hand, a complete resource pooling had been possible, the carried traffic would be $3v(1 - E_B(3v, 3n))$. To have a numerical example, we use $v = 15$ and $n = 15$, and obtain in the first case a carried traffic of 36.9 and in the second 40.0. The second case in an upper bound of the efficiency it is possible to achieve in a loss network where l is the index of the resource, that is

$$\sum_l v_l \left(1 - E_B \left(\sum_l v_l, \sum_l n_l\right)\right) \tag{6.9}$$

Routing decisions should be made locally, in a decentralized manner. A centralized control node may be subject to failure and delays. Routing decisions should rather be decentralized to achieve robustness. The logic should also be simple, so that a decision is made fast, even if based on a limited amount of data. It should still be efficient enough, making effective use of available resources and protecting the network from overload situations.

Common routing strategies. The structure of a network greatly influences the routing possibilities, of course. In a hierarchical network, there are dedicated transit nodes which some traffic is forced to go through. Traditional POTS (plain old telephone service) networks were often structured in this manner, with local switches, transit switches and international gateways. A long-distance call would have to go through the local switch to the transit switch, connecting to another transit switch at the other end.

In contrast, modern networks are mostly having a *flat* topology with no dedicated transit nodes, and any link connecting two nodes is feasible to be part of a route.

The levelled structure of hierarchical networks limits the flexibility and robustness of a flat network topology. The usage of transit nodes is not further considered in the text.

In static routing (fixed routing) there is only a single route for each connection. This route is typically the route on which a call occupies least network resources, which may be referred to as the cheapest route.

In loss networks, the common strategy is to define a number of possible routes, with a minimum of two, for each origin-destination pair. These are defined in routing tables. One route is defined as the primary route, and the others secondary routes. The primary route is usually a direct link, or the shortest path in a network. When there is no congestion on the primary route, this is used. In case of congestion on the primary route, the system tries to establish connection on one of the secondary routes. If there are no resources available on any secondary route either, the call is lost. This is called hierarchical routing (or sequential routing), since routes are tried in hierarchical order. This routing principle should not be confused with routing in a network with hierarchical structure mentioned earlier.

Many different routing strategies, can be devised, depending on what event should trigger rerouting and how to choose an alternative route. The "length" of a route is not necessarily its physical distance. It may be the number of links on a route—the number of hops, or some metric based on free resources along a path, or some other performance metric. In packet networks, transit delay is often used as metric. When the routing decision is based on some performance metric, the method is often called adaptive routing.

Rather than trying a set of routes hierarchically, an alternative route may be chosen at random from a given set. Such a scheme is called random alternative routing (RAR).

Routing algorithms can be formulated in terms of shortest paths where "distance" is based on some ficticious cost. The cost per route may be *additive*, equaling the sum of the cost of the links in a route, *multiplicative*, being the product of the metric values along a path, or *concave*. Examples of additive metrics are hop counts, geometric distance or delay. Multiplicative metrics are reliability measures and admissibility, the compliment of loss probability. More complex metrics depend dynamically on available bandwidth or load along a path. In theory, a metric for routing can be a function of many arguments, but it is vital that it remains stable when data is unavailable and that it is capable of making fast routing decisions. In Baumann et al. (2006), the authors suggest to use a combination of bandwidth, delay and reliability for a routing decision, three metrics which may be considered mutually independent.

A type of adaptive alternative routing is the following scheme. The system first tries to set up a call along a direct route, if there is one. If congested an alternative route with the largest number of end-to-end free resources is selected. In general, given a list of available routes, the scheme finds the link along each path which has the smallest number of free resources. It then selects the route where this is the maximum. This is referred to as a min-max scheme, and the method is called least loaded routing (LLR).

The problem with most routing principles is that it selects an alternative route, which is more expensive than the primary route, with higher probability at high network load. When more calls are routed on alternative paths, more network resources are needed per call. A call, which otherwise may have been routed onto a direct link, may find this congested due to calls routed through alternative routes using this link. The network then tries to reroute this call, increasing the proportion of traffic using alternative routes. From a network perspective, such a scenario leads to uneconomical use of resources.

Trunk reservation. In view of the possibility of suboptimal distribution of resources under heavy load by many routing principles, it is a good idea to reserve some resources for direct calls. Such a method is trunk reservation, which takes a parameter that can be defined per service type and link—the trunk reservation parameter—reserving a number of circuits on a link for traffic types that are to have priority (for example, direct traffic) on the link under heavy load. In order to set up a call of a type that is not prioritized, the number of free resources has to be at least the required number of channels plus the value of the trunk reservation parameter. The method can be viewed as a form of admission control, since it limits the amount of traffic of certain types when the load is high. Trunk reservation provides a very simple and effective way of traffic prioritization. The method is analyzed in some detail in Section 6.6.

The disadvantage of trunk reservation is that it is a local strategy which only considers a single trunk group at a time, not the total end-to-end connection. The advantage of using the number of free resources as a routing metric is therefore that the trunk reservation parameter can be accounted for at call set-up.

Repacking. The routing strategies mentioned so far all have the routes determined at call set-up and are the only class of routing schemes possible for circuit-switched networks. For packet-switched networks, on the other hand, each packet can follow a different route from the previous packet. This allows for *repacking* of the traffic.

When a call (or session) is broken up in packets and each packet can be sent on different routes at any time, a call can be rerouted onto a different path whenever the network state changes, such as when a call arrives that has higher priority on a link. This allows for a very efficient utilization of resources. In fact, in the limit as the packets size decrease to zero, such a scheme enables or come close to complete resource pooling, where all resources are shared among all demands at all times. The upper bound on the achievable efficiency in a loss network is given by Equation (6.9).

6.4 Network programming

A circuit-switched network can be modeled by a linear program, incorporating both dynamic alternative routing and trunk reservation, which reveals important qualitative properties of such networks (Gibbens and Kelly, 1995). The main difficulty in the analysis is that each source-destination pair may be allowed to have several possible routes, each available subject to network state and relative cost, making the number of flows increase manyfold. The model is an extension of the multi-commodity flow problem discussed in Section 3.3 to handle blocking effects due to stochastic traffic.

A simplified explanation of the method is as follows. Depending on the routing strategy, we can identify three classes of traffic: traffic on primary routes on a direct link, traffic on primary routes consisting of more than one link, and traffic on alternative routes. These classes define traffic "priorities," so that direct traffic—the most profitable—has the highest priority, and is limited by an extension of the Erlang bound for trunk reservation and competing traffic, the M-function. The second highest priority has traffic on primary routes along paths with no direct link. Traffic along such paths are routed onto the "complement" of the this Erlang bound, given by the M-function. Finally, traffic on alternative routes are routed on spare capacity determined by "saturated cuts" in the network.

6.4.1 Network bounds

Denote by K the number of call demands arriving at each link in a network, where each link $j \in E$ is offered traffic at rate v_j and having capacity C_j. Thus, each origin-destination pair k defines a call type uniquely. We consider a dynamic alternative routing strategy, where the performance metric is the hop count (for simplicity of the discussion only; any metric can be used to define a set of primary and alternative routes). A call of type k can use any route $r \in \mathcal{R}_k$, where $\mathcal{R}_k$ is the set of routes defined for call type k. The set $\mathcal{R}_k$ is defined according to the following rules. If a call can be routed onto a direct link, this link is considered primary and all of the traffic is routed onto this route. If there are more than one shortest path (with more than a single hop), these routes are considered primary and the traffic rate is equally split onto these routes. If there is a direct link present, any routes with the second largest hop count is considered an alternative route. These routes are not used when the network is moderately loaded.

When a call of type k is sent onto a route $r \in \mathcal{R}_k$, it uses one circuit from each link $j \in r$ for the duration of the call. The network checks that at least one circuit is available on all links along a routes, starting from the primary route, before the call is set up. Should no such route be found, the call is blocked and lost. No assumptions are made on the distribution of the holding time of the call, and the mean is assumed equal to unity, independently of the chosen route. There is no loss of generality here, since the model is dependent on the offered traffic load $\rho_j = \nu_j/\mu_j$, and scaling by an arbitrary value of μ_j does not affect the result.

Suppose that acceptance of a call of type k generates a reward of w_k. Then flow problem can be formulated as

$$\max \sum_k w_k f_k$$

$$f_k = x_k + \sum_{r \in \mathcal{R}_k} y_r \le \nu_k,$$

$$x_j + \sum_{r:j \in r} y_r \le C_j, \tag{6.10}$$

$$x_j \ge 0, \quad j \in E$$
$$y_r \ge 0, \quad r \in \mathcal{R}.$$

Here, $\mathcal{R}$ is the collection of all routes defined in the network. We will not concern ourselves with different rewards; usually we assume that $w_k = 1$ for all k in practical applications.

Let the variables x_j and y_r denote the mean flows along direct links j and alternative routes $r \in \mathcal{R}$, respectively. Let us refer to these two flows as traffic of type one and two. For traffic of type one, we have the bound

$$x_j \le \nu_j(1 - E_B(\nu_j, C_j)), \tag{6.11}$$

where $E_B(\cdot)$ is the Erlang B-formula (5.39). The states are shown in Figure 6.8.

Routing of traffic of type two traffic is more expensive in terms of network resources. When type one traffic is prioritized by means of trunk reservation, the maximal acceptance rate of type two traffic can be expressed by a function of spare circuits $M(\nu_1, \nu_2, C; x)$, given that the mean acceptance rate of type one traffic is at least x.

Under the assumption of trunk reservation with parameter t, so that a call of type two is admitted onto a link only if there are at least t channels free, the link can be

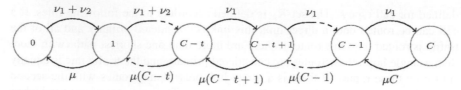

FIGURE 6.8

State diagram for a network resource with trunk reservation. Both direct and alternative flows are allowed up to the capacity $C - t$, where t is the trunk reservation parameter. When less capacity is available, only direct traffic is allowed.

described by the birth-death process shown in Figure 6.8. We then have the following expressions for the mean acceptance rates of type one and type two traffic, respectively,

$$x(\nu_1, \nu_2, C, t) = \tag{6.12}$$
$$\nu_1 G(\nu_1, \nu_2, C, t) \left[\sum_{s=0}^{C-t-1} \frac{(\nu_1 + \nu_2)^s}{s!} + ((\nu_1 + \nu_2)^{C-t} \sum_{s=C-t}^{C-1} \frac{\nu_1^{s-C+t}}{s!} \right]$$

and

$$y(\nu_1, \nu_2, C, t) = \nu_2 G(\nu_1, \nu_2, C, t) \left[\sum_{s=0}^{C-t-1} \frac{(\nu_1 + \nu_2)^s}{s!} \right] \tag{6.13}$$

where the normalization constant is given by

$$G(\nu_1, \nu_2, C, t) = \left[\sum_{s=0}^{C-t-1} \frac{(\nu_1 + \nu_2)^s}{s!} + (\nu_1 + \nu_2)^{C-t} \sum_{s=C-t}^{C} \frac{\nu_1^{s-C+t}}{s!} \right]^{-1}.$$

We wish to maximize the total amount of traffic in the network by selecting t appropriately. We may let $-C \le t \le C$, where positive values of t restrict the admittance of type two traffic, and negative values restrict the admittance of type one traffic. We now define the M-function by (the convex hull of) the set

$$\{x(\nu_1, \nu_2, C, t), y(\nu_1, \nu_2, C, t); t = -C, -C + 1, \ldots, C - 1, C\}, \tag{6.14}$$

where $x(\cdot)$ and $y(\cdot)$ are given by (6.12) and (6.13), and let $M = -\infty$ for $x > \nu_1$ $(1 - E_B(\nu_1, C))$.

In the context of the network model, assume that traffic arrives as independent Poisson processes with intensity ν_j to each link j. Then, for each $j \in E$, we define

$$M_j(x) = M(\nu_j, \sum_{k:j \in \mathcal{R}_k} \nu_k, C_j; x), \tag{6.15}$$

where $\sum_{k:j \in \mathcal{R}_k} \nu_k$ is the upper bound on the amount of traffic of type two (using alternative routes) going through link j.

Thus, each abscissa $M_x(x)$ of the M-function corresponds to an "Erlang bound" on carried traffic of type one, and the ordinate $M_y(x)$ to the "complement", the bound on type two traffic that can be carried on the link.

The function $M_j(x_j)$ is concave and piecewise linear in x_j, for $j \in E$ which allows the flow optimization problem to be formulated as the following linear program.

$$
\max \sum_k w_k f_k
$$

$$
f_k = x_k + \sum_{r \in \mathcal{R}_k} y_r \le v_k,
$$

$$
\sum_{r:j \in r} y_r \le M_k(x_k), \tag{6.16}
$$

$$
x_j \ge 0, \quad j \in E
$$

$$
y_r \ge 0, \quad r \in \mathcal{R}.
$$

We consider three network topologies: a fully connected network, a "cube" network, and a general network.

6.4.2 Symmetric networks

Consider a fully connected network with n nodes where each node pair is connected by a link. This network is called symmetric if it has equal capacity C and offered traffic v on each link. For any node pair, the preferred route is the direct link between the nodes. If this link is blocked, any two-link route connecting this node pair, via a third node, is a possible alternative route. Thus, each node pair has one direct and $(n-2)$ alternative routes.

Since the network is fully connected, we can index the demands and the links by the same set E, where $|E| = n(n-1)/2$. Let $v_k = v$, $C_j = C$ and $w_k = 1$ for all $j \in E$. Denote by $\mathcal{R}_k$ the set of $(n-2)$ two-link routes connecting the node pair identified by k. Then the problem can be formulated as the linear program

$$
\max \binom{n}{2} f
$$

$$
f = x + (n-2)y \le v
$$

$$
x \le M_x(x) \tag{6.17}
$$

$$
2(n-2)y \le M_y(x)
$$

$$
x, y \ge 0.
$$

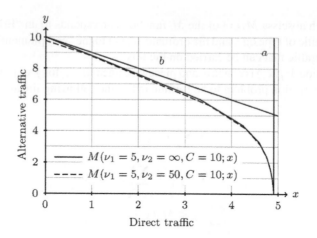

FIGURE 6.9

Different network bounds: (a) The Erlang bound, (b) the capacity bound, and the $M(\cdot)$ functions with exact (dashed line) and the approximation (solid line) with infinite alternative traffic intensity.

Under dynamic alternative routing, each node pair has $(n-2)$ possible alternative routes, and each link can be used in $2(n-2)$ alternative routes considering all node-to-node flows in the network. The traffic on the direct link (type one) is denoted x and traffic using alternative routes (type two) by y. By the symmetry of the network all flows are equal, we can drop the indices and only need to consider the flow variables x and y. The total flow is limited by the demand v, the direct traffic x by $M_x(x)$, and the alternative traffic on a link is allowed to utilize "spare capacity," given by the value of $M_y(x)$ (see Figure 6.9). The structure of this optimization problem makes it easy to solve by inspection, once the M-values are determined. By substituting the second side condition into the first gives the maximum flow $f = \min(M_x(x) + M_y(x)/2, v)$.

The M-function depends on the value of the trunk reservation parameter t, so this value should be determined with some care. We note that trunk reservation plays a role for traffic intensities close to the link capacity, $v \approx C$. For small v there is no blocking, so trunk reservation need not be utilized. For large v, on the other hand, the optimal routing is on the cheaper direct link, and the flow is limited by the Erlang B-formula. Too low t-values have little effect on protecting direct traffic, and a large portion is forced to alternative routes. Too large values, on the other hand, prevent alternative traffic. When $t = C$, no alternative traffic is allowed.

As an example, consider a fully connected symmetric network with $n = 12$ and $C = 120$ and suppose that we want to determine the optimal t-value for $v \approx 120$.

With the simple structure of the optimization problem (6.17), it is straightforward to solve it for different values of v and t. The optimal flows for $v = 110$ and some t-values are shown in Table 6.3. Comparing with the Erlang bound $v(1 - E_B(v, C))$ for the flow, it is apparent that too low t-values give a poor result. For $t \approx 0.1C$,

Table 6.3 Flow values corresponding to different values of the trunk reservation parameter for a symmetric network with $C = 120$ and $v = 110$

t	$M_x(x)$	$M_y(x)$	f
0	5.71	114.23	62.83
5	95.09	22.17	106.18
10	102.77	11.73	108.64
15	105.23	6.73	108.59
20	106.24	3.62	108.05
25	106.68	1.70	107.53
30	106.86	0.66	107.19

the flow achieves a maximum. The value of t where maximum is reached, however, is dependent on the offered traffic intensity v. As t increases, the flow approaches the Erlang bound. We notice that the gain in network efficiency is rather modest, less than 2%. The large benefit is in networks where local concentration of traffic causes overload situations which can be handled by using less utilized resources of the network. This common scenario is not captured by this model, since the traffic intensity is assumed to be equal for all node pairs.

The M-function can for symmetric networks be expressed as

$$M(x) = M(v, 2(n-2)v, C; x). \tag{6.18}$$

For simplicity, however, we may use

$$M(x) = M(v, \infty, C; x), \tag{6.19}$$

where Equation (6.12) and (6.13) reduce in the limit to

$$x(v_1, \infty, C; t) = v_1 \sum_{s=C-t}^{C-1} \frac{v_1^{s-C+t}}{s!} \bigg/ \left[\sum_{s=C-t}^{C} \frac{v_1^{s-C+t}}{s!} \right] \tag{6.20}$$

and

$$y(v_1, \infty, C; t) = v_1 \bigg/ \left[(C - t - 1)! \sum_{s=C-t}^{C} \frac{v_1^{s-C+t}}{s!} \right]. \tag{6.21}$$

Figure 6.9 shows a comparison of different bounds.

We illustrate the flow optimization problem for a symmetric network by the following example, borrowed from Gibbens and Kelly (1995).

Figure 6.10 illustrates the flows in a fully connected symmetric network with $n = 12$, $C = 120$, and $v \in [60, 140]$. The trunk reservation parameter is set to $t = 12$. The direct flow x and the alternative flow $2(n-2)y$ through a link are shown. The total flow $f = x + (n-2)y$ and the blocked flow vB, where B is the difference between offered v and carried f flow, are also shown.

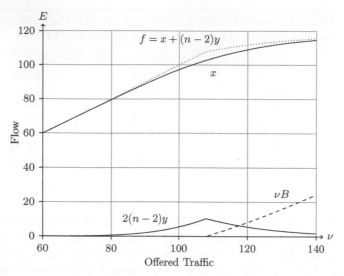

FIGURE 6.10

Traffic streams in a symmetric network. The solid lines show direct and alternative traffic, respectively. Also the total flow f and the blocked flow νB are shown.

With trunk reservation parameter $t = 12$, maximum gain in network efficiency is achieved for $\nu = 0.9C = 108$. At this point the fraction of alternative traffic peaks at 4.8%. This is also the point at which blocking sets in. For small ν, the blocking is zero. Then there are many optimal solutions to the linear program (6.17), ranging from zero to a positive amount of alternative traffic. In this region, we will take traffic on a direct link as the preferred route, so that no traffic on alternative routes are needed. When the traffic intensity approaches C, the amount of alternatively routed traffic increases. From the point where $\nu = 108$ and blocking starts to occur, the fraction of alternatively routed traffic decreases. In the limit, x approaches the Erlang bound and y zero.

6.4.3 The cube network

We now turn to an example of a sparsely connected network—the "cube" network in Figure 6.11. To simplify the discussion, we assume that the line capacities C are the same for all existing links (and zero for nonexisting links). There are 8 nodes, so we have $8 \cdot (8 - 1)/2 = 28$ node pairs, each pair assumed to have a traffic demand of ν.

In a sparsely connected network, some cuts may be heavier loaded than others. Taking such cuts into account, it is possible to formulate additional constraints on the problem.

To formalize the definition of a network cut, consider a subset J with sets of routes $\mathcal{R}_j$ such that for $j \in J$ and $r \in \mathcal{R}_j$ implies

$$|r \cap J| = 1, \tag{6.22}$$

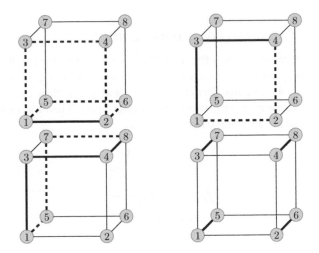

FIGURE 6.11

Routes in the cube network; Upper left shows a direct route between adjacent nodes (the solid line between nodes 1 and 2) and two alternative routes (dashed lines). Upper right shows the two shortest routes between nodes a distance two apart. Lower left shows two out of six possible shortest routes for nodes a distance three apart. Lower right shows the saturated network cut.

where $|r \cap J|$ is the number of links from J on route r. This defines the traffic that should be prioritized under high load. For example, a demand k which is not in J.

Note that J is a set of node pairs, not the cut itself. Thus, J forms a cut in the network, so that every call from the set J requires to use a single resource from J, however it is routed. Let

$$M_J(x) = M\left(\sum_{j \in J} v_j, \infty, \sum_{j \in J} C_j; x\right), \tag{6.23}$$

and consider the flow of traffic across the cut J. Under any dynamic routing scheme

$$\sum_{k \notin J} \sum_{r \in \mathcal{R}_k} |r \cap J| y_r \leq M_J(f_J), \tag{6.24}$$

where,

$$\sum_{j \in J} f_j = f_J. \tag{6.25}$$

From the "bottleneck" character of a cut, the additional constraints (6.24)–(6.25) should be added to the general problem (6.16).

Consider the cube network shown in Figure 6.11, and introduce a cut constraint for each of the three cut sets separating a face of the cube from the opposite face.

Then

$$M_J(f_J) = M(16v, \infty, 4C; f_J), \tag{6.26}$$

for each cut by symmetry. The offered traffic is $4 \cdot 4 = 16$ times v, the number of origin-destination pairs forced to use the cut. The cut, consisting of four links, has a total capacity of $4C$ circuits. Denote by f_1, f_2 and f_3 the flow carried of origin-destination pairs a distance one, two or three links apart. Then the problem (6.16) with the additional cut contraint can be written as

$$\max\ 12f_1 + 12f_2 + 4f_3, \tag{6.27}$$

subject to

$$f_1 = x_1 + 2y_1 \le v \tag{6.28}$$

$$f_2 = 2y_2 \le v \tag{6.29}$$

$$f_3 = 6y_3 \le v \tag{6.30}$$

$$6y_1 + 4y_2 + 6y_3 \le M(x_1) \tag{6.31}$$

$$16y_1 \le M_J(f_J) \tag{6.32}$$

$$4f_1 + 8f_2 + 4f_3 = f_J \tag{6.33}$$

over

$$x_1, y_1, y_2, y_3 \ge 0. \tag{6.34}$$

Let traffic case 1 denote traffic between nodes that are one edge apart, case 2 traffic between nodes that are a distance two apart, and case 3 traffic for nodes a distance 3 apart. For case 1, we have one direct route and two alternative routes. Denote the traffic on these routes by x_1 and y_1, respectively. Similarly, let y_2 be the traffic on the two routes in case 2, and y_3 the traffic on the six routes in case 3. Only one variable per traffic case is necessary due to the symmetry of the problem. This immediately gives the flow constraints for f_1, f_2, and f_3 above.

Next, we assign capacity to the links, which must be equal on all links due to symmetry. Let $v = 6$. The network has 12 links and therefore the total case 1 traffic is $12 \cdot 6 = 72$. Note that this is the preferred link and in cases of no blocking, the traffic $y_1 = 0$. There are two routes with two edges each for case 2 traffic, each carrying the traffic $v/2 = 3$. Summing over the 12 pairs, we obtain $2 \cdot 2 \cdot 3 \cdot 12 = 144$. Finally, for case 3 (four node pairs) we have six routes of length three each carrying the flow $v/6 = 1$, giving $6 \cdot 3 \cdot 1 \cdot 4 = 72$. The total nominal traffic carried by the network is, therefore, $72 + 144 + 72 = 288$, which divided by the number of edges yields $C = 24$ (see Table 6.4).

For the link constraints, note that the x_1 traffic is the cheapest and therefore preferred traffic type. The remaining capacity is then used by the other traffic streams (y_1, y_2, and y_3). For y_1 streams we have an alternative route with three edges, and two faces divided by each edge. Similarly, there are four case 2 streams (that is, node pairs) and six streams using a given edge for node pairs a distance three apart. Thus, we arrive at the constraint $6y_1 + 4y_2 + 6y_3 \le M(x_1)$.

Table 6.4   Cuts in the cube network

No. Nodes	No. Node Pairs	Traffic in Cut	Cut Capacity
1	7	42	72
2	12	72	96
3	15	90	120
4	16	96	96

We now look for a cut with equal flow and capacity f_J. We can look for cuts containing one, two, three, or four nodes. The result is that we find that the cut consisting of four links connecting two faces of the cube is a saturated cut with flow $f_J = 4f_1 + 8f_2 + 4f_3$. These the three (symmetric) cut constraints have the largest effect on the network performance.

Finally, in any of the cuts, spare capacity can be used for case 1 alternative routes y_1 for the flows with primary routes confined to any of the two sets separated by the cut. Thus, the constraint equals the product of the two sides, four node pairs, and two directions (to the other side across the cut, and back again), arriving at $16y_1 \leq M_J(f_J)$.

We can immediately conclude from Figure 6.12, that as the traffic intensity increases, the traffic between nodes a distance of three links apart—which is the most expensive in terms of resource utilization—quickly drops to zero. The traffic between nodes a distance two apart slightly exceeds the traffic a distance one apart, but then decreases, as it is more expensive than traffic on a direct link.

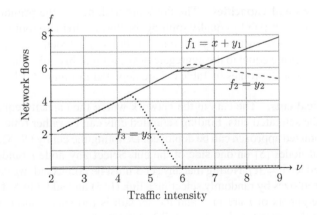

FIGURE 6.12

Flows in the cube network as a function of traffic intensity v : f_1 is the flow between nodes a distance 1 apart, f_2 the flow between nodes a distance 2 apart, and f_3 the flow between nodes a distance 3 apart.

6.4.4 General networks

The methodology presented in this section can be extended to general network topologies. The procedure is best illustrated by an example. We consider the topology from Section 4.4. For simplicity of the discussion, we let the traffic demand between each node pair be 12 (Erlangs). The general linear program (6.16) can be used to model this example. First, we need to find possible routes for each demand. Next, we assign nominal flows onto primary routes. Summing the nominal flows on each link gives the minimum link capacities. Finally, the saturated cuts need to be identified.

Identification of routes. The set of allowed routes consists of primary routes and alternative routes. The shortest routes between any node pair are selected as primary routes. It this route is unique (possibly the direct route), we allow the next shortest routes as alternative routes. If there are two or more shortest routes, these are used by equally splitting the traffic load between them.

Since the topology is given, it is usually sufficient to describe the length of a route by the number of hops, that is, the number of links that has to be traversed to connect an origin with a destination. Therefore, the link costs can simply be set to $c_{ij} = 1$ for all links (i, j). It is straightforward to find the shortest route for any node pair using for example Dijkstra's algorithm. To find alternative routes, however, is a little trickier. Suppose that a shortest route has been found. A simple heuristic method to accomplish a search for alternatives is to multiply the cost of each link in turn of the given route with a large number, and look for the shortest path in the network with modified costs. From the set of routes thus found, the shortest route is again selected and the search continued. An exhaustive search is time consuming for large networks with many links, but since routes are greedily selected, any alternative route of interest is likely to be found early in the search.

Assigning flows and capacities. The flows are assigned to the primary (shortest) routes between a node pair by equally splitting the flows onto these routes. Alternative routes are not assigned any flow. By summing the flows on all routes using a link (i, j), the link capacities are equal to the aggregate flow through this link. The capacities so calculated based on the routes in Table 6.5 are depicted in Figure 6.13.

Finding critical cuts. The cuts in the network where the capacity equals the flow demand are the saturated cuts. Finding saturated cuts can be rather time consuming. Here, a randomized approach can be deployed. Denoting the cut by $(S, \bar{S})$, we specify the number of nodes $|S|$ on one side of the cut. Select any node i randomly in the network G and let $S = \{i\}$. Noting that any set S must be connected, we can construct a set of required sizes by randomly selecting a link (i, j) and add j to S if not already in S. If all neighbors of i are in S and the set still is too small, another node in S is chosen and the procedure is repeated. When a set of specified sizes is found, the capacity of the cut is determined and compared with the required demand between S and $\bar{S}$. When these two quantities are equal, the cut is saturated. The saturated cuts in the example are shown in Figure 6.13.

Table 6.5 Routes for the commodities in Figure 6.13

s_k	t_k	Var.	Routes	Distance	Flow
1	2	x_1	$1-2$	1	12
		y_1	$1-3-2$	2	0
1	3	x_2	$1-3$	1	12
		y_2	$1-2-3$	2	0
2	3	x_3	$2-3$	1	12
		y_3	$2-1-3$	2	0
1	4	x_4	$1-3-4$	2	12
		y_{41}	$1-2-5-4$	3	0
		y_{42}	$1-2-3-4$	3	0
		y_{43}	$1-2-6-4$	3	0
2	4	y_{51}	$2-3-4$	2	4
		y_{52}	$2-5-4$	2	4
		y_{53}	$2-6-4$	2	4
3	4	x_6	$3-4$	1	12
		y_{61}	$3-2-5-4$	3	0
		y_{62}	$3-2-6-4$	3	0
1	5	y_7	$1-2-5$	2	12
		y_{71}	$1-3-2-5$	3	0
		y_{72}	$1-3-4-5$	3	0
		y_{73}	$1-2-6-5$	3	0
2	5	x_8	$2-5$	1	12
		y_8	$2-6-5$	2	0
3	5	y_{91}	$3-2-5$	2	6
		y_{92}	$3-4-5$	2	6
4	5	x_{10}	$4-5$	1	12
		y_{10}	$4-6-5$	2	0
1	6	y_{11}	$1-2-6$	2	12
		y_{111}	$1-3-2-6$	3	0
		y_{112}	$1-3-4-6$	3	0
		y_{113}	$1-2-5-6$	3	0
2	6	x_{12}	$2-6$	1	12
		y_{12}	$2-5-6$	2	0
3	6	y_{131}	$3-2-6$	2	6
		y_{132}	$3-4-6$	2	6
4	6	x_{14}	$4-6$	1	12
		y_{14}	$4-5-6$	2	0
5	6	x_{14}	$5-6$	1	12
		y_{141}	$5-2-6$	2	0
		y_{142}	$5-4-6$	2	0

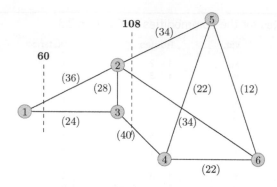

FIGURE 6.13

An example of general network topology with link capacities and two cuts.

Table 6.6 Maximum node-to-node flows for Figure 6.13			
Route	**Flow**	**Route**	**Flow**
1–2	11.06	3–5	12.00
1–3	12.00	4–5	11.09
2–3	12.00	1–6	8.17
1–4	10.94	2–6	12.00
2–4	12.00	3–6	7.48
3–4	11.21	4–6	10.01
1–5	12.00	5–6	12.00
2–5	12.00		

Determining trunk reservation parameter. The trunk reservation parameter is used as a parameter in the linear program. This parameter is typically a small integer. In principle, the linear program can be solved for different parameter values to find the optimal value. Now we can formulate a linear program for the example, which can be solved using a standard LP solver. The corresponding flows for this example are show in Table 6.6.

6.5 Simulation of loss networks

6.5.1 Discrete event simulation

In many situations, simulation is the only method to analyze a complex system. Since both arrivals to and departure from a queue are events that occur in an instantaneous point of time, the paradigm for simulation of such systems is discrete-event simulation. The fundamental steps in simulation are generation of the arrival and departure processes, bookkeeping of the queue states, and analysis of the results. Even if simulation is a very flexible method of analyzing complex systems, there are many pitfalls. The result can never be better than the quality of input and the implementation of the system. Some of the concerns may be the quality of the random number generator used and numerical stability of the implementation. Numerical overflow or underflow can give unusable results. Another factor is choosing the number of simulation runs.

Simulation results should be compared with approximate analytical methods whenever possible. We begin by simulating an Erlang link, which can be done relatively easily. The simulation results can then be compared with analytical results, which can be used to tune simulator parameters. The simulator is used to study a full mesh network under dynamic alternate routing with retries and is compared to a fixed-point equation in Section 6.6. Assuming Poisson arrivals of intensity λ, it follows that the time between consecutive arrivals is exponentially distributed with mean $1/\lambda$. Similarly, the time between consecutive departures is assumed to be exponentially distributed with mean $1/\mu$. The state of the queue can be represented by a gauge, which is stepped up when there is an arrival and the system has capacity to accommodate the call, and stepped down when there is a departure from the system. Naturally, a departure can only occur in a non-empty system.

In discrete-event simulation (DES), the operation of a system is represented as a chronological sequence of events. Each event occurs at an instant in time and marks a change of state in the system.

We refer to an implementation of a procedure for simulation as a *simulator*. A discrete event simulator has the following functions.

Master clock. The simulator generates events at different points in time and it is therefore necessary to have a clock that keeps track of the simulation time. The clock time is always stepped up with an amount of time equal to the next event. The clock value is therefore governed by the simulated event times rather than the other way around.

Events list. The simulated event types consist of a time when the next event is to be triggered, and a consecutive action. This information is formally kept in *event lists*. In case of a queue, the event types are job arrivals and job departures. One event list can then hold the next scheduled arrival, and some pointer to the logic to be executed when the arrival occurs (whether the job should enter the queue or be blocked). A second event list keeps the simulated service times of the current jobs in the queue and, again, some pointer to actions related to a service completion.

Bookkeeping of events keeps track of the event next in turn to be triggered, invokes the appropriate system logic and updates system counters, including the simulator clock.

Random-number generators. The event simulations are based on *pseudo-random numbers*. Since a computer is deterministic (it really is, even if it does not always appear so), it cannot produce real random numbers. Thereof the name "pseudo-random". A random number generator can be controlled by one or more *seeds*, and sometimes other parameters as well.

Most programming languages have routines to generate random numbers, but such built in functions may generate random numbers of low quality. If the random numbers driving the simulations are of inferior quality, the outcome is usually useless. It is advisable to test different random number generators and to perform the same simulation with different seeds and compare the outcomes for statistical properties.

The random numbers a generator produces is usually a *uniform* random variable, which needs to be transformed into the sought random variate. In queueing contexts, the by far most important variate (for simulating exponential inter-arrival times and service times) is the exponential variate.

Exponential variables are easily generated from a uniform random variable, which is the usual distribution of random numbers. Letting X denote a uniformly distributed random number, and μ^{-1} is the mean of the exponential distribution, an exponentially distributed variable is given by

$$Y = -\ln(X) \cdot \mu, \quad \text{for } X > 0.$$

Ending condition. An important issue is to determine an ending condition for the simulation. This may be at a certain system state, after a number of events have occurred or after a specified period of time has elapsed. The danger is to specify a too short simulation run. Typically, the simulation starts at a system state which is transient, and the simulation has to be run long enough so that a substantial part of the time the system is in steady state (which is assumed is of interest here). It may be necessary to let the simulator run for a *training period*, where no observations are recorded, in order to reach steady state before recording of system states can begin.

6.5.2 The Erlang link

In discrete-event simulation, we let the master clock keep track of the times at which either an arrival or a departure occurs. The queue state does not change between the events. The next event is therefore the closest event in time of arrivals and departures. To keep track of arrivals and departures requires bookkeeping as described. Each time an arriving call is accepted, we generate an exponentially distributed service time for this call and a new exponentially distributed time of the next arrival. A departure only changes the system state by the freed resource. At all times we keep the time of the next arrival and a vector of departure times for each call in progress. The vector of departure times has the size of the system capacity. The next event is the closest of the future arrival times and departure times, and when the next event is identified, its time value is deducted from the future times of all other events. In the simulator, we thus jump forward to the next event and change the state of the queue accordingly, while keeping track of the time spent in each state.

6.6　**Efficiency and stability of loss networks**

In this section, we will analyze network stability and efficiency properties with respect to dynamic routing, trunk reservation, and multiple route selections.

We consider a fully connected network and random alternate routing (RAR). By this strategy, each call has a first choice route, and a set of possible alternative routes. In a fully connected network the first choice route is the direct link between the origin and the destination. If possible, this route is used for connecting the call. Should there be a lack of resources the direct link, the routing strategy randomly selects an alternative route via a transit node. If the resources on the alternative route are insufficient to carry the call, it is lost.

By allowing alternative routes, any call would seem to be given a higher probability of being connected. However, the alternative routes are more expensive in terms of resources, and if the traffic load is high enough, a situation can occur where most calls will be forced to use alternative routes. Using more network resources for many calls leads to a higher blocking probability than if only the direct route is available to each call. Such a network can be shown to have several modes of operation under heavy traffic. It turns out that by using trunk reservation, the network stability can be greatly improved.

6.6.1　**Network model**

Consider a symmetric, fully connected network with n nodes, where every pair of nodes is connected by a link having a capacity of C circuits, which is the same for all links. Since all node pairs are connected the network has $K = n(n-1)/2$ links and demands. Calls between any two nodes u and v are assumed to arrive according to a Poisson process with rate v and all traffic streams are assumed to be independent. The holding times are exponentially distributed with mean one, and independent of all other processes. When there is free capacity on the direct link (u, v), this route is selected for the call. Should there be lack of resources, a third node w is selected at random as a transit node, and the call is routed on the links (u, w) and (w, v) if there is sufficient capacity on both these links. If not, the call is lost. If the alternate route is used by the call, one circuit on each link is occupied for the duration of the call.

Even in such an idealized network, analytical analysis by its Markov process becomes complex. Gibbens et al. (1990) suggest the following simplification to relax the dependence on the network topology. Let traffic between two nodes u and v arrive as a Poisson process with intensity v. If the first choice link (u, v) is blocked, two other links are chosen at random from the remaining $K - 1$ (omitting link (u, v)), say (s, w) and (y, z). If neither of these links are blocked, the call is set-up on this fictitious route, otherwise, the call is lost. Note, that in this approximation we may have $s \neq u$ and $z \neq v$.

Suppose that when used, each link is occupied independently for an exponential period of time with mean one. Thus, all circuits are held independently of each other instead of by the actual holding time of the call. These two assumptions decouple the dependence of the state from the topology, and each link is held independently of the others. The approximation can be expected to be good for large networks. The simplified model is much simpler to analyze than the original model.

It can be shown that the Markov process describing the simplified model defines a system of fixed-point equations, as the network size $K \to \infty$. Let x_j be the proportion of links with j occupied resources. Then the detailed balance equations are

$$(\nu + \lambda)x_j = (j+1)x_{j+1}, \quad j = 0, 1, \ldots, C-1, \tag{6.35}$$

where

$$\lambda = 2\nu x_C(1 - x_C). \tag{6.36}$$

To see this relationship, we note that the transition rate consists of both direct traffic ν and traffic from rerouted calls λ. The latter is a product of the probability of the direct link being blocked x_C and the probability of non-blocked links $(1 - x_C)$ in the alternative route, which is then fed with 2ν (since we have two links).

The amount of redirected traffic in Equation (6.36) may not be entirely clear. It is a consequence of the fully connected network. Figure 6.14 shows a four-node network. If a direct link is blocked, for example $(1, 3)$, then two alternative two-link routes are possible: $(1, 2)$ and $(2, 3)$; or $(1, 4)$ and $(4, 3)$. Any of these two is chosen with probability $\frac{1}{2}$. Now, consider link $(3, 4)$. This link may be used as the second leg in one out of two alternative routes whenever any of the links $(1, 3)$, $(1, 4)$, $(2, 3)$, or $(2, 4)$ are blocked. Thus, we have four cases which may choose link $(3, 4)$ for its alternative route with probability $\frac{1}{2}$. Link $(3, 4)$ may be used for redirected traffic only if a direct link (of the four cases above) is blocked and the link $(3, 4)$ itself admits traffic. This gives $4 \cdot \frac{1}{2}\nu x_C(1 - x_C)$. In a general fully connected network with n nodes, each blocked link has $n - 2$ alternative routes, and any link can be used in $2(n - 2)$ alternative routes. Since $\frac{2(n-2)}{n-2} = 2$, Equation (6.36) is valid for any fully connected network. Note, that if $x_C = 1$, there can be no alternate routes, and when $x_C = 0$, there is no need for it, so in either case, $\lambda = 0$.

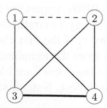

FIGURE 6.14

Illustration of alternative traffic.

6.6.2 Instability

Denoting the vector of state probabilities by **x**, it can be shown that a fixed point of **x** is of the form

$$x_j = \frac{\xi^j}{j!} \left(\sum_{i=0}^{C} \frac{\xi^i}{i!} \right)^{-1}, \quad j = 0, 1, \ldots, C,$$

where ξ is a solution to

$$\xi = v + 2v E(\xi, C)(1 - E(\xi, C)). \tag{6.37}$$

This is just the state probabilities for the $M/M/C/C$ queue given the offered traffic ξ, and

$$E(\xi, C) = \frac{\xi^C}{C!} \left(\sum_{i=0}^{C} \frac{\xi^i}{i!} \right)^{-1} \tag{6.38}$$

is the Erlang B-formula for the loss probability of a single link offered Poisson traffic at rate ξ. Equation (6.37) can be written

$$B = E(v + 2vB(1 - B), C), \tag{6.39}$$

where $B = E(\xi, C)$ is the blocking probability (by applying the Erlang B-formula to both sides of (6.37)). The parameter B corresponds to the link blocking probability, x_C.

Figure 6.15 shows multiple solutions for B, where C is large, and for a narrow range of the load v/C. The upper and lower solutions correspond to stable fixed points of the system, while the middle region corresponds to an unstable fixed point.

As the traffic load increases from a ratio $v/C = 0.92$, it follows the lower curve to the right. This corresponds to the situation where most of the traffic is routed through direct links. On the other hand, when load decreases from a load of approximately $v/C = 0.96$, it follows the upper curve to the left. Most traffic is the routed on

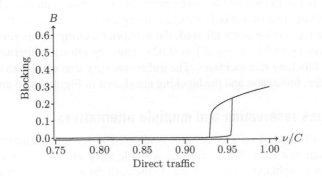

FIGURE 6.15

Hysteresis in a symmetric network with one retry.

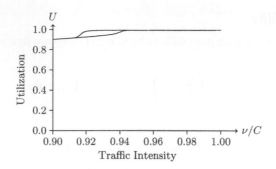

FIGURE 6.16

Network utilization.

alternative paths, and the network can then handle less traffic at the same blocking rate as compared to when traffic is routed directly.

The overall network loss probability is given by

$$L = x_C[1 - (1 - x_C)^2],\qquad(6.40)$$

and for the example simulated this value is 0.12. If alternative routing is not allowed, so that a call blocked on its direct link is lost, then the network loss probability is given by the Erlang B-formula (6.38) to be 0.05. Observe that allowing a blocked call to attempt a two-link alternative actually increases the loss probability of the network.

In order to solve the fixed-point equation, given a Poisson arrival rate v and link capacity C, we set B to some assumed value. The two curves correspond to two sets of initial values of B. We then iterate Equation (6.39) until convergence to a fixed point has been reached. The overall network blocking is then given by Equation (6.40).

Example 6.6.1. Consider the simplified model of a fully connected network with $C = 500$ and $v = 490$. The solution to the fixed-point equation (6.39) is $B = 0.274$, which corresponds to the link blocking rate when alternative routing is allowed. The network blocking rate (6.40) is $L = 0.129$, using the value of $x_C = B$. In contrast, when alternative routing is not allowed, the network blocking rate is simply the link blocking given by (6.38), $E_B(v, C) = 0.024$. Thus, by allowing alternative routing the network blocking rate increases. The traffic intensity that can be accommodated in the resource, utilization and the blocking are shown in Figures 6.16 and 6.17.

6.6.3 Trunk reservation and multiple alternatives

Suppose a call blocked on a direct link is allowed to try an alternative route. We call this situation a *retry*. Since allowing a single retry increases the network loss probability, we might expect the loss rate to increase even more if several retries are allowed before the call is blocked in the network. This is to say, that a call is allowed to try a maximum number $r > 1$ of alternative routes.

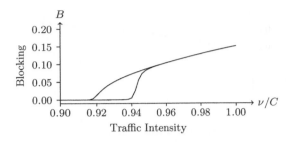

FIGURE 6.17

Network blocking.

When the proportion of traffic carried over alternative routes is high, calls that try to set up a connection on a direct link are likely to find the link congested, and so will try to find an alternative route. By trunk reservation, direct traffic can be prioritized over alternative traffic. We extend the simplified model for a fully connected network allowing each call r retries and where direct traffic is protected by trunk reservation with parameter t.

Consider a fully connected network with n nodes and $m = n(n-1)/2$ links. A call is first trying to connect over the direct link. If blocked, it tries two random links from the remaining $m-1$ links. If there are at least $C-t$ circuits free on both links, the call will be set up using these two links. If the free capacity is less than $C-t$ on any of the two links, the call will not be accepted onto this route, and the call will select two random links from the remaining $m-3$ links, and so on. If the call has failed to connect after a maximum of r retries, it is lost. On each link, trunk reservation with parameter t protects direct traffic. We assume that all holding times are independent, even on the two circuits making up an alternative route.

It can be shown that the system has a fixed point $\mathbf{x} = (x_0, x_1, \ldots, x_C)$ which satisfies

$$(j+1)x_{j+1} = (\nu + \lambda)x_j, \quad j = 0, 1, \ldots, C - t - 1 \tag{6.41}$$

$$(j+1)x_{j+1} = \nu x_j, \quad j = C - t, \ldots, C - 1, \tag{6.42}$$

where

$$\lambda = 2\nu B_1 (1 - B_2)^{-1} \left\{ 1 - \left[1 - (1 - B_2)^2 \right]^r \right\} \tag{6.43}$$

$$B_1 = x_C, \quad B_2 = \sum_{i=C-t}^{C} x_i. \tag{6.44}$$

The network loss probability corresponding to a solution to the fixed-point Equations (6.41)–(6.44) is

$$L = B_1 \left[1 - (1 - B_2)^2 \right]^r. \tag{6.45}$$

Equation (6.45) is a generalization of (6.40) and has a similar interpretation. A call is blocked on its direct route with probability B_1, and blocked on an alternative route

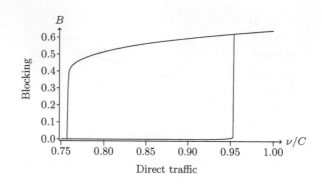

FIGURE 6.18

Hysteresis in a symmetric network with five retries.

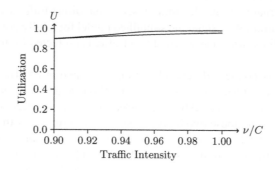

FIGURE 6.19

Network utilization.

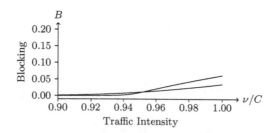

FIGURE 6.20

Network blocking.

with probability $(1 - (1 - B_2)^2)$. This can happen at most r times, where of the exponent r. The fixed-points, utilization and blocking are shown in Figures 6.18–6.20, respectively.

6.7 **Summary**

Some of the most important design methods for loss systems are presented—the Erlang fixed-point approximation for performance evaluation, Moe's principle for resource allocation, and network programming determining the optimal flow when alternative routing is deployed. The content of the chapter suggests a design method as follows. First, we determine a topology, using a shortest-path routing principle. Next, the Erlang fixed-point approximation is used to evaluate the network performance. The network programming method may be used to determine how alternative routing may improve the performance. Moe's principle can be used, together with the Erlang fixed-point approximation to adjust capacities until the blocking rates are within required limits.

Fixed-point methods are discussed in Kelly (1986, 1991), Moe's principle in Pirinen (2001) and Iversen (2013), and network programming in Gibbens and Kelly (1995).

Simple Packet Networks

This chapter and the next discuss design of packet networks (also known as packet-switched networks or store-and-forward networks). These differ from the loss networks analyzed in the previous chapter in that the traffic is broken up into chunks or packets rather than being sent in a continuous stream. These packets may take different routes from the origin to the destination during a conversation, referred to as a session. The advantage of packet networks compared to loss networks is that during periods of inactivity we do not need to occupy any network resources, which then can be used more efficiently.

As a consequence of not dedicating resources for a session is that different demands can compete for the same resources, which may lead to delay of the delivery of packets, and, in case of finite buffers, packet loss. We introduce the performance measures throughput, end-to-end delay and, for finite buffers, packet loss probability. Compared to loss networks, whose performance can be described by the loss probability only, we need several measures to describe the performance and efficiency of a packet network.

In this chapter, we will usually assume that we are given a set of n nodes that are required to communicate between themselves, a network topology (that is, a set of links) defined on this set of nodes, and an $n \times n$-matrix of traffic demands. The operational goals of packet network design are to fulfill certain performance requirements while minimizing the capital investments in link capacity. Since these objectives are conflicting, some sort of compromise will be necessary.

Usually, it is necessary to use a combination of various methods to solve such a design problem. One common approach is to divide the problem into subproblems, like dividing the network into core and access subnetworks. One may then try to apply different methods to each subproblem separately.

We will analyze resource allocation and flow optimization in packet networks subject to performance measures. A number of methods for design or modification of existing topologies are presented, as well as issues related to simulation of packet networks.

Under certain assumptions, a packet network can be described by a queueing network with state-space probabilities in so-called product-form, which can be solved analytically. In this way, a closed-form expression for optimal resource allocation is obtained. Flow optimization is achieved by using the flow deviation algorithm, which is closely related to the multicommodity flow algorithm in Section 3.3.

In case we have finite buffers, the state space probabilities of the queueing network are no longer in product-form and no closed-form expressions can be found, so we need to find another approach. One possible solution is to use a set of fixed-point equations to calculate packet loss probabilities.

Section 7.1 describes general properties of packet networks and some common routing principles. A number of performance measures are discussed, for which analytic expressions can be found for so-called open Jackson networks (Section 7.2). These performance measures, in particular end-to-end packet delay, are used in resource allocation (Section 7.3) and flow optimization (Section 7.4).

For queues with finite buffers, analysis is commonly performed using numerical methods (Section 7.6) or simulation (Section 7.8), as no exact analytic expressions for this case are known. A heuristic method for topology improvement is discussed in Section 7.7.

7.1 General properties of packet networks

To describe a packet network it is useful to consider an exchange of data of some kind between a pair of nodes. It may be a voice conversation or a transfer of some data files. The transport method in such a network is based on a digitally coded message that is broken up into packets, which are coded and sent out sequentially on the network, which provides control functions assuring that the message is properly delivered at the recipient end. The main feature of such a network is that the packets can be sent on different routes which is 'smoothing out' the load in the network. This becomes especially important when the traffic is bursty, that is, when large amounts of data are sent off at periods far apart. Another feature is that packets can be stored in memory of a sending device to be sent at a time when resources become available. The usefulness of this property is dependent on the type of service, however. In contrast to loss networks, where all conversations are treated similarly to voice traffic—all resources are occupied from session initiation to its end—a packet network may handle different types of conversations differently. For example, a voice conversation is sensitive to delay, but less sensitive to packet loss, whereas for a file transfer, the service is sensitive to packet loss and less sensitive to delay.

One of the consequences of providing the capability of handling various types of traffic is that control mechanisms have to be implemented in a packet network to protect the network from overload. Two such mechanisms are admission control (a term used to denote both flow and congestion control) and scheduling. Admission control is a way to limit the volume of traffic before it enters the system, whereas, scheduling is a way to prioritize the way packets are sent. These control methods are difficult to model mathematically. There are, however, approximate methods, such as fixed-point methods and the theory of effective bandwidth. This chapter introduces design and performance evaluation methods for networks without admission control.

We consider sending and receiving devices and links between them. These devices are called routers even if other technical solutions under other names exist. Each router is connected to other similar devices and may have a certain amount of buffer capacity dedicated per connected link for storage of data being sent over it.

There are two main types of packet transmission control. The first is an emulation of a loss network type of transmission, known as virtual channel (or virtual circuit), where the control function takes care that the message is delivered free from errors at the receiving end and protects the conversation from congestion by controlling the flow on the channel. In the second type, packets may take any route in the network, and the control functions need to handle reassembly of messages in addition to error detection, flow and congestion control. The packets are in this case often called datagrams. This is the mode of operation of the Internet, where the transmission control functions are handled by the Transmission Control Protocol (TCP).

In a virtual channel network, packets in a session are assigned to a virtual channel, which is flow controlled. Packets belonging to the same virtual channel are typically routed along the same path, thus assuring delivery of packets in sequence. The purpose of flow control (by means of some transmission protocols) is to protect the network against overload and to provide some guarantee of delivery of packets, subject to some performance criteria (such as high throughput, low delay, or low packet loss).

In a datagram packet network, each packet traverses the network independently from other packets, so that different packets may take different paths and therefore arrive out of sequence. The correct reassembly of the original message can be handled by a network transmission protocol or by the receiving node. During the transport of a packet from its source to its destination, resources are dynamically allocated; a packet is transmitted from one node to the next along the route, using a portion of the capacity of the link between the nodes and buffer space at the nodes at each end of the interconnecting link. Initially, the packet is stored in the buffer of the sending node and transmitted onto the link connecting with the next node on its route. If the packet is successfully received by the node at the end of the link, an acknowledgment (ACK) message is sent back to the sending node at the beginning of the link, and when the acknowledgment is received at the sending node, the buffer space is freed. If no such acknowledgment has been received after some time-out period, the packet is retransmitted.

This chapter discusses simple packet networks without flow or congestion control. Packet networks with flow or congestion control (for example by TCP) or virtual channel networks are referred to as flow-controlled packet networks, and the analysis of such networks is the topic of Chapter 8.

Packet networks without flow or congestion control are modeled by open Jackson networks. These are relatively easy to analyze. In this case, the throughput equals the input rate. By solving the related traffic equations, we can obtain the steady-state probabilities of the queues in the network and, hence, all performance indices.

Given a set of external traffic demands, the efficient utilization of a network's channel and buffer resources depends on the routing algorithm used in the network as well as its flow and congestion control functions. Measures of network performance typically include its throughput (in packets delivered per second) and some measure of the network transit delay. These performance measures may need to be characterized for all packets transported by the network or for individual classes of packets (for example packets between specific origin-destination pairs).

7.1.1 Routing

Routing is the process of selecting a suitable path along which a packet can be sent so that it will reach is destination. The 'suitability' may be dependent on the estimated total delay, which depends on the state of the network. Logically, the routing in datagram networks and virtual circuit networks functions differently. Two packets in a datagram network belonging to the same stream or chain may take different routes to reach their destinations, and a routing decision has to be made for each packet. In a virtual channel, the route is determined when the channel is set up, and packets only need to be associated with the virtual channel in order to be delivered to their destinations. It works like a pipe with the destination address as its end point. (A virtual channel can be rerouted, but the packets are still associated only with the channel, not with the links that are utilized to deliver packets to their destinations.)

In large networks, routing decisions are invariably decentralized, so that each node makes the necessary routing decisions based on the information it possesses about the network topology and the state of the network. In contrast, the logic of a centralized routing mechanism resides in a dedicated node which performs each routing decision (like a control tower at an airport). The disadvantage of a centralized routing logic is the risk of overload and failure, which would render the whole network defunct. The advantages of decentralized routing algorithms are resilience to faults and the possibility of making fast routing decisions, possibly with the risk of not always having all the information necessary about the network state available at the time of the decision. A potential problem with decentralized routing schemes without full network information is the risk for infinite loops. As an example, suppose node i finds, among possible routes for the packet, the link to node j the least congested, but does not have any further congestion estimates for the network. Node i then forwards the packet to node j. Node j, on the other hand, may have link i listed as a neighbor in a route for the packet. Should node j also find the link to node i the least congested, it would send the packet back to node i, causing a loop.

A robust routing scheme should be able to avoid links and nodes with failure or congestion, and rather select links with ample free capacity for the route. The mechanisms involved greatly affect network performance. An inefficient routing scheme could, for example, lead to an increase in network delay or packet loss by choosing a few, overloaded, links. Thus, network performance is closely linked to the routing scheme.

A routing scheme normally stores routing information in routing tables and this information may be static or it may be updated regularly as the state of the network changes (called adaptive schemes).

Thus, the routing and the flow-control schemes are intended to work so that, during high load, throughput (that is, network flow) is maximized with respect to an average allowable delay per packet, and during low loads, average delay per packet is minimized.

Routing schemes can be characterized subject to whether they adapt the routing depending on the prevalent traffic conditions or not, that is, into static or dynamic (adaptive) schemes. A static scheme does not change when traffic conditions change, it only changes routes as a result of link or node failure, whereas an adaptive scheme considers network congestion when making routing decisions.

Static Routing. In static routing, a table of available routes is kept in the node, and only in case of link or node failure, the route priority changes. This is the first routing method used in circuit switched networks. Since a path is set up by assigning resources on each path for an incoming call, a fault (link failure) is detected before any message is sent. In case of failure, the switch tries the alternative route in the routing table until the call can be connected or all possibilities have been exhausted.

An important feature of this routing strategy is that it can be modeled and its efficiency evaluated, which in general is not the case with adaptive routing strategies.

Random Routing. Random routing is based on a routing table with a number of routes to each destination. The traffic is transmitted onto a route on a random basis, either with equal or weighted probability, where the weight could be the route's distance to the destination, for example.

Shortest Path Routing. The concept of shortest path (see Section 2.4) provides a very versatile method of determining routes based on different criteria if we replace length (in the sense of distance) in the algorithm with some other criteria, such as congestion or delay. The result is a shortest (distance) path algorithm yielding a shortest delay path instead. The criterion used needs to be single-valued and approximately summable over the links (in other words, we make an assumption of independence of the links) for shortest path algorithms to work. More complex criteria, such as a combination of delay and packet loss, would then have to be represented by a single number. In practice, the Bellman-Ford algorithm (see Section 2.4) is often used to determine shortest paths in networks. A potential problem with shortest path routing schemes is the risk of oscillations. This can be compared with the stability problem of alternative routing in circuit-switched networks discussed in Section 6.6.

Optimal Routing. A limitation of shortest path routing is that it chooses a single link for routing traffic, when an optimal routing possibly would distribute traffic onto several links. Optimal routing of flows is discussed in Section 3.3 (subject to link congestion) and in Section 7.4, where the total network delay is used as performance measure. Optimal routing strategies also mitigate the risk for oscillations in the network.

When the offered load is high, a portion of the traffic will be rejected by the flow-control algorithm so that the throughput equals the offered load minus the rejected load,

$$\text{throughput} = \text{offered load} - \text{rejected load}.$$

The traffic accepted into the network will experience an average delay per packet that will depend on the routes chosen by the routing algorithm. However, throughput will also be greatly affected (if only indirectly) by the routing algorithm because typical flow-control schemes operate on the basis of striking a balance between throughput and delay (that is, they start rejecting offered load when delay starts getting excessive). Therefore, as the routing algorithm is more successful in keeping delays low, the flow-control algorithm can allow more traffic into the network.

Deflection Routing Schemes.   Another routing strategy may be to avoid packet losses due to buffer overflows. A node would then forward a packet to any link with free capacity instead of storing it in its buffer, even if that link, in other regards, is not optimal. This may be done randomly. If the node has d incident links and processes packets in discrete-time intervals, no more buffer space than the equivalent of $2d$ packets is necessary (receiving and storing d incoming packets and sending d packets per time interval). This may lead to long paths, even with cycles, and precautions should be taken so that packets are not transmitted in infinite loops.

7.2 Queueing networks

General queueing networks are difficult to analyze. There are, however, some networks that lend themselves to analytical solutions. Networks can be classified as open or closed, depending on whether jobs (or calls) are arriving from outside the network to some of the nodes in the network, or if the number of jobs within the network is fixed and no jobs enter or leave the network. A network can also be a mixture of the two types.

7.2.1 Open Jackson networks

We assume that each node is a $M/M/C$ queue (in steady-state), where $1 \leq C \leq \infty$. Thus, the arrival process to the queue is Poisson. Also note that the departure process from one node is the arrival process to another node. The fundamental observation is given by the following Theorem.

Theorem 7.2.1 (Burke's theorem).   *If $M/M/C$, $1 \leq C \leq \infty$ is a queue in steady-state with arrivals according to a Poisson process with intensity λ, then:*

- *The departure process is Poisson with intensity λ;*
- *At any time t the number of jobs in the queue is independent of the departure process prior to time t.*

Proof.   Viewing the queue as a Markov chain, the arrival times in the forward chain are identical to the departure times of the reversed Markov chain (see Section 5.3). Thus, the departure process is a Poisson process with intensity λ. In the forward direction, the arrival at time t is independent of the number of jobs after t. Thus, in the reversed process, the number of jobs at t is independent of departures taking place before t. $\square$

This result may be surprising since the departure process is independent of the service time $1/\mu$. As a consequence of this theorem, the theorem of Jackson follows.

Theorem 7.2.2 (Jackson's theorem).   *Consider a queueing network with n nodes satisfying the following conditions:*

(1)   Each node is an $M/M/C$ queue. Node i has C_i servers, and its average service time is $1/\mu_i$.
(2)   Jobs arrive from outside the network to node i according to a Poisson process with intensity γ_i. Jobs may also arrive to node i from other nodes following a fixed routing matrix $R = (r_{ij})$.
(3)   A job which has just finished service at node i is immediately transferred to node j with probability r_{ij} or leaves the network with probability $1 - \sum_{j=1}^{n} r_{ij}$.

The total average arrival intensity λ_i to node i is obtained by solving the traffic (or flow balance) equations

$$\lambda_i = \gamma_i + \sum_{j=1}^{n} \lambda_j r_{ji}. \tag{7.1}$$

Let $p(s_1, s_2, \ldots, s_n)$ denote the state space probabilities under the assumption of statistical equilibrium, that is, the probability that there are s_i customers at node i. Furthermore, we assume

$$\frac{\lambda_i}{\mu_i} < C_i.$$

Then the state space probabilities are given by the product form

$$p(s_1, s_2, \ldots, s_n) = \prod_{i=1}^{n} p_i(s_i),$$

where for node i, $p_i(s_i)$ is the state probabilities of the $M/M/C$ queueing system with arrival rate λ_i and service rate μ_i.

Assume that a network topology $G = (V, E)$ and an $n \times n$ traffic matrix A are given. Let the row indices represent traffic sources and columns indices sinks. For each row i of A, the row sum represents the traffic transported from node i throughout the network to other nodes. This quantity must also be equal to the external traffic into node i. Traffic may arrive into the network via any of the n nodes, and the row sums of A are denoted $\boldsymbol{\gamma} = (\gamma_1, \ldots, \gamma_n)$.

Thus, calls arrive to the network from outside according to Poisson processes with mean arrival rates γ_i to a node (a switch or router) i. However, we are interested in the links with their associated terminals, including buffers, which are part of the nodes. The links are assumed to be $M/M/1$ queues with infinite queue capacity for simplicity. How the traffic is formulated in the case of links—particularly the routing matrix—is described below. For the present discussion, nodes and links are conceptually very closely related - a node is an end point of a link which contains the queue - and the expressions therefore involve the same type of variables.

All servers at node i service a job according to an exponential distribution with mean μ_i. When a job completes service at node i, it goes next to node j with probability r_{ij} (independently of the state of the system). There is a probability r_{i0} that a job leaves the network at node i after completion of service.

Let λ_i be the total mean flow rate into node i, consisting of both external traffic and traffic from other nodes. In order to satisfy flow balance at each node, the traffic (flow balance) equations

$$\lambda_i = \gamma_i + \sum_{j=1}^{n} \lambda_j r_{ji} \tag{7.2}$$

must hold. In vector notation, this can be expressed as

$$\lambda = \gamma + \lambda R. \tag{7.3}$$

In order to remove the statistical dependence due to the packet length, Kleinrock (1976) introduced the assumption that packet lengths were exponentially distributed,

$$p(x) = \mu e^{-\mu x}, \quad x \geq 0,$$

where $1/\mu$ is the average packet length. In real networks, packets have a maximum length. The assumption of exponentially distributed packet length is an idealization, but necessary to obtain an analytic result.

Let $\rho_i = \lambda_i/\mu_i$ for $i = 1, 2, \ldots, n$. Then, for a network of n $M/M/1$ queues, the steady-state probabilities are given by

$$p(s_1, s_2, \ldots, s_n) = (1 - \rho_1)\rho_1^{s_1}(1 - \rho_2)\rho_2^{s_2} \cdots (1 - \rho_n)\rho_n^{s_n}. \tag{7.4}$$

The joint probability distribution in (7.4) is therefore a product of marginal distributions, one for each queue. Such a form is called a product form. The result can be interpreted as that each node in the network could be regarded as if it were an independent $M/M/1$ queue, with parameters λ_i and μ_i.

In general, however, the internal network flows are not Poisson. It can be shown that if there is any feedback (jobs returning to a node previously visited), the internal flows are not Poisson. The surprising result of Theorem 7.2.2 is that regardless of whether the internal flows are actually Poisson or not, Equation (7.4) still holds, and the network behaves as if the nodes were independent $M/M/1$ queues.

The traffic load ρ_i is obtained by solving the traffic Equation (7.3) for λ_i,

$$\lambda = \gamma(I - R)^{-1}, \tag{7.5}$$

where I is the $n \times n$ identity matrix. The matrix $(I - R)$ is invertible as long as there is at least one node from where traffic is routed to the outside of the network and no node is totally absorbing.

As jobs are routed differently in the network depending on its origin and destination, we sometimes wish to distinguish between the jobs. A job following a particular route is said to belong to a *routing chain*. The chains are indexed by k and the total number of chains is denoted K. When jobs are classified into chains, the state of a queue i is described by a vector $s_i = (s_{i1}, s_{i2}, \ldots, s_{iK})$ rather than the scalar $s_i = \sum_{k=1}^{K} s_{ik}$ denoting the state from the aggregate traffic over all chains. In the latter case, the aggregate load at queue i is

$$\rho_i = \sum_{k=1}^{K} \rho_{ik}, \tag{7.6}$$

where ρ_{ik} is the load contribution at queue i from chain k. We summarize the equilibrium probability distributions when considering each chain separately and the aggregate traffic at each queue:

$$p(\mathbf{s}) = \prod_{i=1}^{n} \frac{p_i(\mathbf{s}_i)}{G_i},$$

where

$$p_i(\mathbf{s}_i) = s_i! \prod_{k=1}^{K} \frac{\rho_{ik}^{s_{ik}}}{s_{ik}!},$$

and

$$G_i = 1/(1 - \rho_i),$$

when queue i is a first-come, first-served (FCFS) $M/M/1$ queue; and

$$p_i(\mathbf{s}_i) = \prod_{k=1}^{K} \frac{\rho_{ik}^{s_{ik}}}{s_{ik}!},$$

and

$$G_i = e^{\rho_i},$$

when server i is an infinite server (IS). The equilibrium probability of $\mathbf{s}$ when considering the aggregate traffic is

$$p(\mathbf{s}) = \sum_{i=1}^{n} \frac{p_i(s_i)}{G_i},$$

where

$$p_i(s_i) = \rho_i^{s_i}, \tag{7.7}$$

when server i is a first-come first-served $M/M/1$ queue; and

$$p_i(s_i) = \frac{\rho_i^{s_i}}{s_i!},$$

when server i is an infinite server.

If the network nodes have infinite or large buffers, the open Jackson network model can be used. Then the network throughput equals the arrival rate for each chain

$$\gamma_k^* = \gamma_k, \quad k = 1, 2, \ldots, K.$$

Otherwise, as a result of buffer overflow or flow control, the throughput is lower than the arrival rates, so that

$$\gamma_k^* < \gamma_k, \quad k = 1, 2, \ldots, K.$$

The difference $\gamma_k - \gamma_k^*$ is the rate at which arrivals of chain k are rejected. Since rejected packets can be retransmitted at a later time, the ratio γ_k / γ_k^* can also be interpreted as the number of times a packet needs to be transmitted on average in order to reach its destination successfully.

7.2.2 The routing matrix

When forming a routing matrix, it is instructive to separate direct traffic from transit traffic. We will also need to distinguish between traffic at nodes and on links. We can adopt a node view or a link view. In other words, a traffic matrix has different representations, just as the connectivity information for a graph in its adjacency matrix can be represented in different forms.

The node view may be a little simpler since only a single point (the node) need to be considered. In the link view, a pair of nodes needs to be analyzed. The row sums and the column sums of the traffic matrix A are the traffic into and out from a node, respectively.

We need to adopt the link view (since we are modeling links). We now define traffic as direct traffic on a link if it originates in either end point. Traffic just passing through the link or traffic having its origin at one end of the link, but does not terminate at the other end, but it passed onto another link is considered to be transit traffic. We denote originating traffic to a node i by γ_i and the traffic originating in either end point of a link by γ_{ij}. Transit traffic on a link is denoted τ_{ij}. The construction of the routing matrix is best illustrated by an example.

Example 7.2.1. Consider the network in Figure 7.1, and suppose that the traffic requirements are given by the matrix

$$A = \begin{pmatrix} 0 & 2 & 3 & 1 & 2 & 1 \\ 2 & 0 & 1 & 2 & 3 & 3 \\ 3 & 1 & 0 & 2 & 2 & 1 \\ 1 & 2 & 2 & 0 & 3 & 2 \\ 1 & 3 & 1 & 1 & 2 & 0 \end{pmatrix}.$$

Note that the sum of all the rows equals 58.

Table 7.1 shows the direct traffic as defined. For example, the link $(1, 2)$ has two end points and $\gamma_{12} = \gamma_{21} = 2$. The link is also included in the route 1–2–5. The flow from 1 terminates at 5. Thus, the direct traffic on this link is 6. Traffic originating at 5

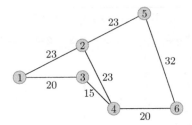

FIGURE 7.1

A capacity assignment problem. Letting the cost be proportional to the distance (the numbers without brackets) and the demand (resp. capacity) be represented by the figures in brackets, we can find an optimal solution.

Table 7.1 Direct traffic on links in Example 7.2.1

(i, j)	Shortest Path	Type	γ_i	1–2	1–3	2–4	2–5	3–4	4–6	5–6
(1,2)	1–2	D	2	4						
(1,3)	1–3	D	3		6					
(1,4)	1–3–4	T	1		1			1		
(1,5)	1–2–5	T	2	2			2			
(1,6)	1–3–4–6	T	1		1				1	
(2,3)	2–4–3	T	1			1		1		
(2,4)	2–4	D	2			4				
(2,5)	2–5	D	3				6			
(2,6)	2–4–6	T	3			3			3	
(3,4)	3–4	D	2					4		
(3,5)	3–4–2–5	T	2				2	2		
(3,6)	3–4–6	T	1					1	1	
(4,5)	4–2–5	T	3			3	3			
(4,6)	4–6	D	1						2	
(5,6)	5–6	D	2							4
Sum				**6**	**8**	**11**	**13**	**9**	**7**	**4**

is considered transit traffic for this link since it does not originate at either end of the link. Thus, we have $\lambda_{12} = 2\gamma_{12} + \tau_{12} = 6 + 2 = 8$. Note that

$$\sum_{(i,j)\in E} \gamma_{ij} = \sum_{i} \gamma_i,$$

that is, the sum of originating traffic on the links equals the originating traffic to the nodes.

The transit traffic figures are presented in Table 7.2. On the route 1–3–4, the link (3, 4) carries transit traffic from 1 aimed for node 4, and link (1, 3) for traffic from 4 to 1. On longer routes, such as 1–3–4–6, links (1, 3) and (3, 4) carry transit traffic to

Table 7.2 Transit traffic on links in Example 7.2.1

(i, j)	Shortest Path	Type	γ_{ij}	1–2	1–3	2–4	2–5	3–4	4–6	5–6
(1,2)	1–2	D	2							
(1,3)	1–3	D	3							
(1,4)	1–3–4	T	1		1			1		
(1,5)	1–2–5	T	2	2			2			
(1,6)	1–3–4–6	T	1		1			2	1	
(2,3)	2–4–3	T	1			1		1		
(2,4)	2–4	D	2							
(2,5)	2–5	D	3							
(2,6)	2–4–6	T	3			3			3	
(3,4)	3–4	D	2							
(3,5)	3–4–2–5	T	2			4	2	2		
(3,6)	3–4–6	T	1					1	1	
(4,5)	4–2–5	T	3			3	3			
(4,6)	4–6	D	1							
(5,6)	5–6	D	2							
Sum				2	2	11	7	7	5	0

Table 7.3 Constructing a routing matrix

(i, j)	1–2	1–3	2–4	2–5	3–4	4–6	5–6
(1,2)				2			
(1,3)					2		
(2,4)				5	3	3	
(2,5)	2		5				
(3,4)		2	3			2	
(4,6)		3			2		
(5,6)							

node 1; and (4, 6) and (3, 4) traffic from node 1 to 6. Thus, link (3, 4) carries transit traffic in both directions, and the traffic is therefore counted twice.

Next, the transit traffic figures need to be mapped onto the links. This is straight-forward using the figures in Table 7.2. The result is shown in Table 7.3. The final routing matrix is obtained by calculating the routing probabilities as τ_{ij}/λ_{ij} where τ_{ij} are taken directly from Table 7.3.

7.2.3 End-to-end delay

Let $\gamma = \sum_i \gamma_i$ be the total traffic load arriving to the whole network. Let the links (and the corresponding queues) be indexed by $l = (i, j), l = 1, 2, \ldots, m$. The average packet sending time is $\rho_l = 1/\mu C_l$. When $\mu = 1$, C_l equals the capacity in packets per second. Let:

$$T_l = \frac{1}{\mu C_l - \lambda_l} = \frac{1}{\mu C_l} \frac{1}{1 - \lambda_l/\mu C_l} = \frac{1}{\mu C_l} \frac{1}{1 - \rho_l}$$

by the $M/M/1$ queue assumption. This is the sojourn time at a link (delay plus transmission time). An alternative cost measure is the link utilization, defined as

$$\max_l \left\{ \frac{f_l}{C_l} \right\}. \tag{7.8}$$

This measure is used in the multicommodity flow algorithm in Section 3.3.

By Little's formula, the number of packets in link buffer l is

$$L_l = \lambda_l T_l.$$

The number of packets in the whole network is

$$\sum_l L_l = \sum_l \lambda_l T_l.$$

The total network delay is, therefore,

$$T = \frac{1}{\gamma} \sum_l L_l = \frac{1}{\gamma} \sum_l = \lambda_l T_l. \tag{7.9}$$

Assuming a network of $M/M/1$ queues, this expression can alternatively be deduced as follows. Let $\mathbf{E}(s_l)$ be the mean number of packets in queue l. Then, by Little's formula,

$$\gamma T = \sum_{l=1}^{m} \mathbf{E}(s_l).$$

The marginal probability distribution Equation (7.4) implies the mean queue length

$$\mathbf{E}(s_l) = \frac{\rho_l}{1 - \rho_l}, \quad l = 1, 2, \ldots, m,$$

so that,

$$\gamma T = \sum_{l=1}^{m} \frac{\rho_l}{1 - \rho_l}.$$

Since $\rho_l = \lambda_l/(\mu C_l)$, we have

$$T = \frac{1}{\gamma} \sum_{l=1}^{m} \frac{\lambda_l}{\mu C_l - \lambda_l},$$

which can be written as

$$T = \frac{1}{\gamma} \sum_{l=1}^{m} \frac{f_l}{C_l - f_l}, \tag{7.10}$$

where $f_l = \lambda_l/\mu$ is the flow on link l.

Example 7.2.2. Suppose that $\gamma = (6, 8, 11, 13, 9, 7, 4)$ and $C = (30, 30, 30, 30, 30, 30, 30)$ and $\mu = 1$.

The routing matrix is

$$R = \begin{pmatrix} 0 & 0 & 0 & 1/4 & 0 & 0 & 0 \\ 0 & 0 & 0 & 0 & 1/5 & 0 & 0 \\ 0 & 0 & 0 & 5/22 & 3/22 & 3/22 & 0 \\ 1/10 & 0 & 1/4 & 0 & 0 & 0 & 0 \\ 0 & 1/8 & 3/16 & 0 & 0 & 9/16 & 0 \\ 0 & 0 & 1/4 & 0 & 1/6 & 0 & 0 \\ 0 & 0 & 0 & 0 & 0 & 0 & 0 \end{pmatrix}.$$

Solving the traffic Equation (7.5) gives $\lambda = (8, 10, 22, 20, 16, 12, 4)$. $T = (0.05, 0.05, 0.12, 0.10, 0.07, 0.06, 0.04)$ by (7.10) and the network-wide delay (7.9) is $T = 0.131$.

7.3 Resource allocation

Two design-related tasks will be discussed next, resource allocation (that is, the problem of determining the link capacities) in this section, and flow optimization given a network with capacities assigned in the following section. By combining the two tasks, we have a more intricate problem, discussed in Section 7.5.

We assume that we have a network topology and a matrix of traffic demands given. For each link we determine the offered traffic under some routing scheme (for example, shortest path routing). The traffic offered to each link can be determined by solving the traffic equation (7.5) and proceed as described in Example 7.2.1. Denote the link traffic by $\lambda = (\lambda_{ij})$, where $|E| = m$ is the number of links.

Next, we need a cost model for the links. The simplest is the linear model. In this case, we can formulate closed-form expressions for the resource optimization problem. Some examples of cost functions are illustrated in Figure 7.2.

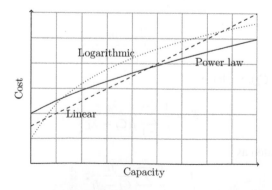

FIGURE 7.2

Different cost functions.

The resource allocation problem can be formulated in two ways (subject to network delay): either we minimize the delay given a specified budget constraint, or we minimize the network cost given a network delay constraint. Both formulations are equivalent in that the second problem is the dual of the first.

Assume that we are given a topology $G = (V, E)$ with n nodes and m links, modeled as $M/M/1$ queues, an $n \times n$ traffic matrix A, and a routing scheme $\mathcal{R}$. This routing scheme determines the flows f_{ij} on each link (i, j). Denoting the capacities of each link by C_{ij} and its cost by $c_{ij}(C_{ij})$, the resource allocation problem can be formulated as a nonlinear optimization problem,

$$\min \left\{ \sum_{(i,j) \in E} c_{ij} \right\}$$
$$T = \frac{1}{\gamma} \sum_{(i,j) \in E} \frac{f_{ij}}{C_{ij} - f_{ij}} \leq T_{\max} \qquad (7.11)$$
$$0 \leq \lambda_{ij} \leq C_{ij},$$

where c_{ij} denotes some cost function. The condition $T \leq T_{\max}$ ensures that the delay on any link does not exceed the pre-defined value $T_{\max}$.

This optimization problem can be solved without too much effort for three important classes of cost functions: linear costs, concave costs, and discrete costs. We look at each of these in turn. At the end of the section, a heuristic method is discussed.

The resource allocation problem is often used as a subroutine in more complex design procedures rather than in isolation. This is due to the fact that resource allocation and flow assignment are closely dependent on each other, and these two tasks should be solved simultaneously to yield a joint capacity and flow optimum.

7.3.1 Linear costs

Suppose that the cost can be expressed as

$$c_{ij} = \alpha_{ij} C_{ij} + \beta_{ij}, \qquad (7.12)$$

where β_{ij} is a fixed cost per link. Then the optimization problem can be written

$$\min \left\{ \sum_{(i,j) \in E} \alpha_{ij} C_{ij} + \beta_{ij} \right\}$$
$$T = \frac{1}{\gamma} \sum_{(i,j) \in E} \frac{f_{ij}}{C_{ij} - f_{ij}} \leq T_{\max}$$
$$0 \leq f_{ij} \leq C_{ij}.$$

Note that at optimum, it appears logical that $T = T_{\max}$. The optimization problem can be solved by the Lagrange method. The Lagrangian $\mathcal{L}$ is formed by multiplying the side condition by a multiplier, say ϑ, and adding it to the objective function,

$$\mathcal{L} = \sum_{ij}\left(\alpha_{ij}C_{ij} + \beta_{ij} + \vartheta\frac{1}{\gamma}\frac{f_{ij}}{C_{ij} - f_{ij}}\right),$$

where the abbreviated notation $\sum_{ij}$ is used for $\sum_{(i,j)\in E}$. The necessary conditions for an optimal solution are now that $\partial\mathcal{L}/\partial C_{ij} = 0$ and that the side condition is satisfied. Thus,

$$\frac{\partial\mathcal{L}}{\partial C_{ij}} = \sum_{ij}\left(\alpha_{ij} - \vartheta\frac{1}{\gamma}\frac{f_{ij}}{(C_{ij} - f_{ij})^2}\right) = 0.$$

Solving for C_{ij}, which we can do component-wise (ignoring the summation) yields

$$C_{ij} = f_{ij} + \sqrt{\frac{\vartheta f_{ij}}{\gamma\alpha_{ij}}}. \tag{7.13}$$

Substituting this expression into the side condition, we get

$$\frac{1}{\gamma}\sum_{ij}f_{ij}\sqrt{\frac{\gamma\alpha_{ij}}{\vartheta f_{ij}}} = \sum_{ij}\sqrt{\frac{\alpha_{ij}f_{ij}}{\vartheta\gamma}} = T_{\max}.$$

Now, solving for ϑ gives

$$\sqrt{\vartheta} = \frac{1}{T_{\max}}\sum_{ij}\sqrt{\frac{\alpha_{ij}f_{ij}}{\gamma}}$$

which, substituted back into (7.13), gives the optimal capacity and minimum cost

$$C_{ij} = f_{ij} + \frac{1}{\gamma T_{\max}}\sum_{kl}\sqrt{\alpha_{kl}f_{kl}}\sqrt{\frac{f_{ij}}{\alpha_{ij}}} \tag{7.14}$$

$$c_{\min} = \sum_{ij}\alpha_{ij}f_{ij} + \frac{1}{\gamma T_{\max}}\left(\sum_{kl}\sqrt{\alpha_{kl}f_{kl}}\right). \tag{7.15}$$

These equations can easily be used directly to calculate the optimal capacities in a network with given flows and under a linear link cost assumption. The linear cost case can give a first approximation to more complex scenarios. Including the average packet size μ in these expressions, we have

$$C_{ij} = \frac{\lambda_{ij}}{\mu} + \frac{1}{\gamma T_{\max}}\sum_{kl}\sqrt{\alpha_{kl}\lambda_{kl}/\mu}\sqrt{\frac{\lambda_{ij}}{\mu\alpha_{ij}}} \tag{7.16}$$

$$c_{\min} = \sum_{ij}\alpha_{ij}\frac{\lambda_{ij}}{\mu} + \frac{1}{\gamma T_{\max}}\left(\sum_{kl}\sqrt{\alpha_{kl}\lambda_{kl}/\mu}\right). \tag{7.17}$$

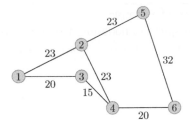

FIGURE 7.3
A capacity assignment problem where the cost is proportional to the distance.

The dual can be found by minimizing $T = \sum_{ij} \lambda_{ij} T_{ij}/\gamma$ subject to given link costs c_{ij} and a total budget $\hat{c}$. For the link capacities we have the constraint $\sum_{ij} c_{ij} C_{ij} \leq \hat{c}$. The minimum cost of the network is $\sum_{ij} c_{ij}\lambda_{ij}/\mu$. After investing this necessary cost, the excess funds $\bar{c} = \hat{c} - \sum_{ij} c_{ij}\lambda_{ij}/\mu$ are allocated to link capacities so as to minimize T. The optimal solution of the dual is found to be

$$C_{ij} = \frac{\lambda_{ij}}{\mu} + \bar{c} \frac{\sqrt{\lambda_{ij}/c_{ij}}}{\sum_{kl} \sqrt{\lambda_{kl} c_{kl}}}$$

$$T_{\min} = \frac{1}{\gamma \mu \bar{c}} \left(\sum_{ij} \sqrt{\lambda_{ij} c_{ij}} \right)^2 .$$

Example 7.3.1. Let the topology $G = (V, E)$ be as in Figure 7.3, also showing the link distances d_{ij}, and the traffic matrix A be

$$A = \begin{pmatrix} 0 & 2 & 3 & 1 & 2 & 1 \\ 2 & 0 & 1 & 2 & 3 & 3 \\ 3 & 1 & 0 & 2 & 2 & 1 \\ 1 & 2 & 2 & 0 & 3 & 2 \\ 1 & 3 & 1 & 1 & 2 & 0 \end{pmatrix} .$$

In the linear cost function, let $\alpha_{ij} = d_{ij}$ and $\beta_{ij} = 0$. First, it is necessary to define a routing scheme $\mathcal{R}$. Consider shortest path routing. Table 7.4 represents these flows in one direction. For example, the flows between node 1 and node 6 are routed on the path 1–3–4–6. The sum of the flows on each link is shown in the bottom of the table. Since the flow requirement is symmetric, these flow values have to be doubled.

Suppose $T_{\max} = 0.1$ and $\mu = 1$. We get $\gamma = 58$ from the traffic matrix A by summing all entries. This is the total external traffic arriving to the network. (In contrast, the sum $\sum_{(i,j) \in E} \lambda_{ij} = 92$, that is, the sum of the doubled flows in Table 7.4. Now, the optimal link capacities C_{ij} are given directly by (7.14) giving the total cost $c_{\min}$ given by (7.15).

Table 7.4 Table of flows in Example 7.3.1. The columns correspond to link flows and the rows correspond to node-to-node demands

Route	1–2	1–3	2–4	2–5	3–4	4–6	5–6
1–2	2						
1–3		3					
1–4		1			1		
1–5	2			2			
1–6		1			1	1	
2–3			1		1		
2–4			2				
2–5				3			
2–6			3			3	
3–4					2		
3–5			2	2	2		
3–6					1	1	
4–5			3	3			
4–6						1	
5–6							2
Sum	4	5	11	10	8	6	2

Table 7.5 Optimal capacities in Example 7.3.1

Link	λ_{ij}	d_{ij}	c_{ij}	$\lambda_{ij} T_{ij}/\gamma$
1–2	8	23	19.59	0.012
1–3	10	20	23.89	0.012
2–4	22	23	41.21	0.020
2–5	20	23	38.32	0.019
3–4	16	15	36.29	0.014
4–6	12	20	27.22	0.014
5–6	4	32	10.95	0.010
Sum	92	156	197.47	0.100

The result is shown in Table 7.5. Column three shows the optimal capacities, and column four the delay per each link. The sum of these is T_{max}, as required. The minimum cost link capacities are $c_{min} = 4197$.

7.3.2 Concave costs

The concave case can be solved iteratively by linearizing the costs and solving the linearized problem at each iteration. This method leads, in general, to local minima.

Consider the power law cost function

$$c = \sum_{ij} \alpha_{ij} C_{ij}^{\kappa}, \tag{7.18}$$

with $0 \leq \kappa \leq 1$. Note that for $\kappa = 0$, the cost is constant, and for $\kappa = 1$, we have the linear cost function. We therefore assume that $0 < \kappa < 1$. Let $\mu = 1$ for simplicity. It is straightforward to include μ in the calculations, if required.

Again, we need to solve (7.12), but with link costs $c_{ij} = \alpha_{ij} C_{ij}^{\kappa} + \beta_{ij}$. Formulating the Lagrangian, we arrive at

$$\mathcal{L} = \sum_{ij} \left(\alpha_{ij} C_{ij}^{\kappa} + \vartheta \frac{1}{\gamma} \frac{\lambda_{ij}}{C_{ij} - \lambda_{ij}} \right),$$

where ϑ is the Lagrange multiplier. Taking the partial derivative of $\mathcal{L}$ with respect to C_{ij}, we have

$$\frac{\partial \mathcal{L}}{\partial C_{ij}} = \kappa \alpha_{ij} C_{ij}^{\kappa-1} - \vartheta \frac{1}{\gamma} \frac{\lambda_{ij}}{(C_{ij} - \lambda_{ij})^2} = 0$$

for all links (i, j). Rearranging, taking the square root, and introducing a new variable g_{ij} for this quantity gives

$$g_{ij} = C_{ij}^{(\kappa-1)/2} (C_{ij} - \lambda_{ij}) = \sqrt{\frac{\vartheta}{\gamma} \frac{\lambda_{ij}}{\kappa \alpha_{ij}}}.$$

Thus, we have

$$C_{ij} = g_{ij} C_{ij}^{(1-\kappa)/2} + \lambda_{ij}, \tag{7.19}$$

which can be solved iteratively. In each iteration, ϑ is adjusted so that

$$T_{\max} = \frac{1}{\gamma} \sum_{ij} \frac{\lambda_{ij}}{C_{ij} - \lambda_{ij}} \tag{7.20}$$

is satisfied. Note that g_{ij} depends on ϑ, and C_{ij} depends on g_{ij}. The procedure requires a set of initial non-zero capacities to be given. The closer these initial estimates are to the final capacities, the faster the convergence. For numerical reasons, it may be more convenient to use a Lagrange multiplier $1/\theta'$ instead of ϑ. A fixed cost β_{ij} per link can be added to the final cost. Just as in the derivation of the linear cost case, it does not enter into the expression for the capacities. This can be summarized in the following algorithm (including μ):

Algorithm 7.3.1 (The Flow Deviation Algorithm).   Given a connected weighted graph $G = (V, E)$ with n nodes and m links, where link (i, j) has a fixed flow $f_{ij} = \lambda_{ij}/\mu$.

STEP 1:   (iterate)
Calculate

$$g_{ij} = \sqrt{\frac{\vartheta}{\gamma} \frac{\lambda_{ij}}{\kappa \mu \alpha_{ij}}}.$$

STEP 2:   (iterate)
Calculate

$$C_{ij}^{(N+1)} = \frac{\lambda_{ij}}{\mu} + g_{ij}(C_{ij}^{(N)})^{(1-\kappa)/2}$$

with ϑ adjusted so that

$$\frac{1}{\gamma} \sum_{ij} \frac{\lambda_{ij}/\mu}{C_{ij}^{(N+1)} - \lambda_{ij}/\mu} = T_{\max}.$$

Iterate until the capacities do not change. Calculate the link costs

$$c_{ij} = \alpha_{ij} \left(C_{ij}^{(N+1)} \right)^{\kappa}$$

and the total cost

$$c_{\min} = \sum_{ij} c_{ij}.$$

Output **C**, the optimal capacities and the optimal cost $c_{\min}$.

Example 7.3.2.   Consider the topology given by Figure 7.3. Let the traffic demand be the same as in Example 7.3.1. With shortest path routing, Table 7.4 describes the edge flows. Let $T_{\max} = 0.1$. Suppose we have a power law cost function (7.18) with $\kappa = 0.48$ and $\beta_{ij} = 3$.

Set initial link capacities $C_{ij} = 30$ for all links (i, j). By iterating Equation (7.19) and adjusting ϑ in each step so that Equation (7.20) is satisfied, the Equation (7.19) converges after a few iterations, giving the capacities (18.8, 23.8, 40.6, 41.0, 38.9, 27.7, 9.4) with a total link cost of $c_{\min} = 812$. The fixed costs β_{ij} are simply added to the final cost calculated using (7.18).

Using the same method, we can derive an expression for a logarithmic cost function,

$$c = \sum_{ij} \alpha_{ij} \log(\kappa C_{ij})$$

and using the same principle as for the power law cost function, we have

$$C_{ij} = \frac{\lambda_{ij}}{\mu} \left(1 + \frac{1}{2\gamma \vartheta \alpha_{ij}} + \left[\frac{1}{\gamma \vartheta \alpha_{ij}} + \left(\frac{1}{2\gamma \vartheta \alpha_{ij}} \right)^2 \right]^{1/2} \right),$$

where the Lagrange multiplier ϑ is adjusted in each iteration so that

$$T_{\max} = \sum_{ij} \left(\frac{1}{2\alpha_{ij}\vartheta} + \left[\frac{\gamma}{\alpha_{ij}\vartheta} + \left(\frac{1}{2\alpha_{ij}\vartheta} \right)^2 \right]^{1/2} \right)^{-1}$$

is satisfied.

7.3.3 Discrete costs

Discrete costs can be solved by fitting a continuous (concave) function to approximate the discrete values. After a solution has been found, the closest larger discrete values are selected for the solution. This method does not guarantee optimality, however.

7.3.4 Moe's principle

Variants of Moe's principle, used in Section 6.2 for heuristic resource allocation in loss networks, can also be used for packet networks. For $M/M/n$ systems, Erlang's C-formula for delay systems can be used to express the improvement function. The improvement when adding one server can be described by the decrease in proportion of total traffic that experience delay by

$$F(\lambda, n) = \lambda(E_C(\lambda, n) - E_C(\lambda, n+1)). \tag{7.21}$$

Each link is tested for improvement by adding a server to one link at a time. On the link which yields the largest improvement, a server is added in each iteration.

With a simple modification, the method can also be used for $M/M/1$ queues. In this case, we wish to allocate capacities to the links so that a target maximum network delay $T_{\max}$ is satisfied.

Let the improvement function be given by the total network delay

$$T = \frac{1}{\gamma} \sum_{(i,j)\in E} \frac{f_{ij}}{C_{ij} - f_{ij}},$$

so that

$$\Delta T = \frac{1}{\gamma} \left(\sum_{ij} \frac{f_{ij}}{C_{ij} - f_{ij}} - \sum_{ij} \frac{f_{ij}}{(C_{ij} + \Delta C) - f_{ij}} \right),$$

where $\Delta C > 0$ for some link (k, l) and zero otherwise.

The starting configuration is a network with capacities slightly larger than the flows, where the flows are kept fixed. This is necessary to get a finite positive value of the initial network delay. The procedure then checks the links in turn for the largest improvement in network delay by incrementing the capacity of one of the capacities by a fixed value. On the link which yields the largest delay improvement, say (k, l), the capacity is increased to $C_{kl} + \Delta C$. This is repeated until $T \leq T_{\max}$. One of the advantages of the Moe heuristic is that it is easy to use and to implement in a computer program, a disadvantage is that it converges rather slowly.

Table 7.6 Approximately optimal capacities in Example 7.3.3

Link	λ_{ij}	d_{ij}	C_{ij}	$\lambda_{ij} T_{ij}$
1–2	8	23	19	0.013
1–3	10	20	24	0.012
2–4	22	23	41	0.020
2–5	20	23	41	0.016
3–4	16	15	39	0.012
4–6	12	20	28	0.013
5–6	4	32	10	0.014
Sum	92	156	202	0.100

Example 7.3.3. Using the same input values as in Example 7.3.1, we may proceed as follows. Let the initial capacities be $C = (9, 11, 23, 21, 17, 13, 5)$, that is, $f_{ij} + 1$ where the flows follow the shortest paths, $\Delta C = 1$ and $T_{max} = 0.1$. The cost is given by a power law function with $\kappa = 0.48$ and $\beta_{ij} = 3$. After 102 iterations, $T \lesssim T_{max}$ and the final result is shown in Table 7.6. The final cost is 817.

7.4 Flow optimization

In a network with capacities assigned to the links we may wish to optimize the flow through the network in order to minimize delay, for example. This is referred to as the flow optimization (or optimal routing) problem. With optimal routing we here mean optimal with respect to delay. The problem is then a multicommodity flow problem with the delay used as distance metric in the network. The algorithm discussed here is known as the flow deviation method.

In the previous section, the flows are assumed to be assigned to some routes and capacities are optimally determined based on these flows. If, on the other hand, capacities are given, then the flows may be optimally assigned to some routes, based on the capacities. This case is the topic of this section.

Optimal routing is difficult to achieve practically in a large network. The routing mechanism in one part of the network needs information about the states in other parts. Sending such information across a network adds load to the network and the information may be obsolete when it reaches another part of the network.

With most performance objectives, optimal routing can be formulated as a shortest path problem with an appropriate weight defined on the links making up a route. The distance metric is usually the (estimated) mean delay of a link (The mean delay includes both the expected waiting time in the queue and the transmission time. In practice, a fixed bias term is also included to reduce looping behavior.) At each node, a packet is routed to an outgoing link on the path with the shortest (estimated) mean delay to its destination. At any such node, the routing decision is made independently of the origin of the packet. This routing strategy minimizes the (estimated) mean delay of each individual packet. It can be shown, however, that such individual flow optimization does not necessarily lead to an overall optimal network flow distribution. The reason is that the mean transit delay T for all packets transported by the network is not optimized by this routing strategy. There is an efficient method to optimize T globally, which is described next. It successively redistributes flows on a network in order to minimize the network delay.

7.4.1 The flow deviation method

The flow deviation method is an algorithm for solving nonlinear optimization problems related to network flows. It is a version of the Frank-Wolfe method. It reduces the value of a special cost function by successively rerouting flows in the network.

We denote by $f_{ij} = \lambda_{ij}/\mu$ the flow on a link (i, j). A link flow is called feasible if

$$0 \leq f_{ij} \leq C_{ij}.$$

Similarly, we denote by f_r an end-to-end flow along some route r. The collection of routes available to an origin-destination pair (chain) k is denoted $\mathcal{R}_k$. Also, denote the set of all routes in the network by $\mathcal{R} = \cup_k \mathcal{R}_k$.

The optimal routing problem can be formulated

$$\min \left\{ \sum_{(i,j) \in E} c \left(\sum_{r:(i,j) \in r} f_r \right) \right\} \tag{7.22}$$

$$\sum_{r \in \mathcal{R}_k} f_k = \gamma_k$$

$$f_r \geq 0 \text{ for all } r \in \mathcal{R}_k, k \in K,$$

where $c(\cdot)$ is a cost function. We assume that the cost $c(f)$ is a differentiable function defined in an interval $[0, C_{ij})$, where C_{ij} is the capacity on link (i, j). The derivative of this function defines the search direction for finding an optimal routing (7.22).

In order to characterize an optimal route we let the cost of a link be the packet delay (omitting constants)

$$c(f) = \frac{f}{C - f}. \tag{7.23}$$

The derivative is

$$\frac{dc(f)}{df} = \frac{C}{(C - f)^2}. \tag{7.24}$$

This derivative, evaluated at a current flow f_r, defines a 'length' of a link, that we want to minimize. This corresponds to the 'fastest' direction to find an optimal solution. Denote the route length by l_r. The minimum route length is known as the *minimum first derivative length*. It is clearly dependent on the flow f_r along r. The larger the flow, the longer the route becomes. Denote this derivative with respect to a flow f_r on a route r by

$$\frac{\partial c(f)}{\partial f_r} = \sum_{r:(i,j) \in \mathcal{R}_k} \frac{C}{(C - f)^2}.$$

For a direct route (i, j), formally written $r = \{(i, j)\}$, the length with respect to a route flow f_r is

$$l_r = \frac{\partial c(f)}{\partial f_r} = \frac{C_{ij}}{(C_{ij} - f_{ij})^2}.$$

If, for example, a route r of an origin-destination pair $k(r \in \mathcal{R}_k)$ is $r = \{(v, w), (w, y), (y, z)\}$, the length of the route is

$$l_r = \frac{\partial c(f)}{\partial f_r} = \frac{C_{vw}}{(C_{vw} - f_{vw})^2} + \frac{C_{wy}}{(C_{wy} - f_{wy})^2} + \frac{C_{yz}}{(C_{yz} - f_{yz})^2}.$$

For an optimal flow f_r^* we have that $f_r^* > 0$ implies that

$$\frac{\partial c(f^*)}{\partial f_q} \geq \frac{\partial c(f^*)}{\partial f_r} \text{ for all } q \in \mathcal{R}_k.$$

Thus, an optimal path flow is positive only on paths with a minimum first derivative length.

The idea is to start from any feasible flow and search for a suboptimal flow, say f_q, and an alternative route of minimum length f_r. That is, a flow for which $l_q > l_r$. First, we reroute all the flow f from q to r.

Next, let $\alpha \in [0, 1]$ be the amount of flow to reroute from q to r. The optimal amount to reroute is such that

$$c(f_q + \alpha(f_r - f_q)) \tag{7.25}$$

is minimized. That is

$$c(f_q + \alpha^*(f_r - f_q)) = \min_{\alpha \in [0,1]} c(f_q + \alpha(f_r - f_q)). \tag{7.26}$$

Once α^* has been determined, the new path flows are defined as

$$\bar{f} = f_q + \alpha^*(f_r - f_q) \text{ for all } q, \quad r \in \mathcal{R}_k.$$

The alternative flow need not be feasible. The cost function effectively penalizes flows on routes where links are heavily loaded.

The optimal amount to reroute, α^*, is found by optimizing (7.26). There are many possible ways to accomplish this. A simple and reasonably accurate expression for α^* is obtained by making a second order Taylor expansion of $c(\cdot)$ around $\alpha = 0$ and minimizing with respect to α:

$$\alpha^* = \min \left\{ 1, -\frac{(f_r - f_q)(\partial T/\partial f)}{(f_r - f_q)^2(\partial^2 T/\partial f^2)} \right\},$$

where $\partial T/\partial f$ and $\partial^2 T/\partial f^2$ are evaluated at f_q.

The procedure is iterated until no suboptimal flow can be found or until the total cost (delay) does not decrease further. Should the cost increase, the iteration is stopped.

Algorithm 7.4.1 (The Flow Deviation Algorithm).   Given a connected weighted graph $G = (V, E)$ with n nodes and m links, where link (i, j) has capacity C_{ij}.

STEP 0:   (initialize)

Identify a set of routes $\mathcal{R}_k$ for each demand k. Find an initial feasible flow $\tilde{f}$. Calculate the length l_r for each route. Calculate the cost (delay) $T = \sum_{ij} \tilde{f}_{ij}/(C_{ij} - \tilde{f}_{ij})$. Set the tolerance ϵ.

STEP N: (iterate)

while $T^{(N-2)} - T^{(N-1)} > \epsilon$ (there are suboptimal flows) **do**

reroute all flows onto cheaper alternative routes, the new flow is denoted $\hat{f}$;

determine α so that $\bar{f} = \tilde{f}_{ij} + \alpha(\hat{f} - \tilde{f}_{ij})$ gives minimum total cost (delay) $T = \sum_{ij} \tilde{f}_{ij}/(C_{ij} - \tilde{f}_{ij})$. Recalculate the length l_r for each route.

Let $\tilde{f}_{ij} \leftarrow \bar{f}$;

end;

Output $\bar{f}$, the optimal flow.

Example 7.4.1.   Consider the network in Figure 7.3 and suppose we are given the same traffic matrix as in Example 7.3.1. Let all capacities $C_{ij} = 30$. As initial feasible flow, we use the flows on the shortest path routes in Table 7.7.

Next, we need to list alternative routes and compute the route lengths that result from the delays induced by the current flows. Let the alternative routes be the second shortest paths, using the geometric distances. For simplicity, only one alternative route is considered. Using Equation (7.24), the route lengths l_r can be computed. Table 7.7 summarizes routes, their geometric distances, and lengths caused by delays for the initial routes.

Note that the physical distance between nodes is the sum of geometric distances between the nodes, called distance, and the derivative of the delay along a route is called length.

The initial edge flows are given in Table 7.8. Suboptimal flows can be identified by looking up a source-destination pair, then mark the one with shorter distance, which is the original path. If the alternative route has a shorter length made up by the derived delay metric, rerouting of flow is desirable.

Entries in boldface un Table 7.7 represent routes that could accommodate more flow from the shorter ones. Thus, route 2–4–3 (with distance 38) is longer than route 2–1–3 (with distance 43). Therefore, we wish to reroute some of the flow from 2–4–3 to 2–1–3. Similarly, some of the flow between the origin-destination pairs (2,4), (2,6), (3,5) and (4,5) could be rerouted. (Route 1–3–4–6 has not been considered in the first step due to the relative by small differences between the distance of the original and alternative routes).

Note that the total flow is double the column sums, since the routing table shows flows in one direction only. Next, we need to find the α minimizing the cost for the convex combination (7.25). Table 7.9 shows the flows f_q before rerouting, the flows f_r after rerouting, the new flows $\bar{f} = f_q + \alpha^*(f_r - f_q)$, and the cost (7.23). The

Table 7.7 Primary and alternative routes in Example 7.4.1 with respective lengths

Route	Distance	$l_r^{(1)}$	$l_r^{(2)}$	$l_r^{(3)}$	$l_r^{(4)}$	Length
1–2	23	0.062				0.062
1–3–4–2	58	0.075	0.153	0.469		0.697
1–3	20	0.075				0.075
1–2–4–3	61	0.062	0.469	0.153		0.684
1–3–4	35	0.075	0.153			0.228
1–2–4	46	0.062	0.469			0.531
1–2–5	46	0.062	0.300			0.362
1–3–4–2–5	81	0.075	0.153	0.469	0.300	0.997
1–3–4–6	55	0.075	0.153	0.093		0.321
1–2–4–6	66	0.062	0.469	0.093		0.623
2–4–3	38	0.469	0.153			0.622
2–1–3	43	0.062	0.075			**0.137**
2–4	23	0.469				0.469
2–1–3–4	58	0.062	0.075	0.153		**0.290**
2–5	23	0.300				0.300
2–4–6–5	75	0.469	0.093	0.044		0.606
2–4–6	43	0.469	0.093			0.561
2–5–6	55	0.300	0.044			**0.344**
3–4	15	0.153				0.153
3–1–2–4	66	0.075	0.062	0.469		0.606
3–4–2–5	61	0.153	0.469	0.300		0.922
3–1–2–5	66	0.075	0.062	0.300		**0.437**
3–4–6	35	0.153	0.093			0.246
3–1–2–4–6	86	0.075	0.062	0.469	0.093	0.698
4–2–5	46	0.469	0.300			0.769
4–6–5	52	0.093	0.044			**0.137**
4–6	20	0.093				0.093
4–2–5–6	78	0.469	0.300	0.044		0.813
5–6	32	0.044				0.044
5–2–4–6	66	0.300	0.469	0.093		0.861

proportion of flows to reroute in the first iteration is $\alpha^* = 0.37$ by which the cost decreases from 8.18 to 6.43.

By iterating this procedure until no suboptimal flows longer can be found, an optimal flow assignment is achieved. The final result is shown in Table 7.10. The table shows the link delays for the initial and the final flows. Note how the flow deviation algorithm "levels out" the delays, so that at optimum they have about the same values. By increasing the total flow from 92 to 96.3 through rerouting, the maximum network delay T is decreased from 0.131 to 0.105. Figure 7.4 shows how the flows converge to their final values in the iterations.

Different distance metrics can be used, in the algorithm. It may, for example, be more appropriate to use simply the number of links (or hops) as route distances. Note

Table 7.8 Table of flows in Example 7.4.1. The columns correspond to link flows and the rows correspond to node-to-node demands

Route	1–2	1–3	2–4	2–5	3–4	4–6	5–6
1–2	2						
1–3		3					
1–4		1			1		
1–5	2			2			
1–6		1			1	1	
2–3	1	1	0		0		
2–4	2	2	0		2		
2–5				3			
2–6			0	3		0	3
3–4					2		
3–5	2	2	0	2	0		
3–6					1	1	
4–5			0	0		3	3
4–6						1	
5–6							2
Sum	**9**	**10**	**0**	**10**	**7**	**6**	**8**

Table 7.9 Table of flows in Example 7.4.1. The columns correspond to link flows and the rows correspond to node-to-node demands

Link	f_q	f_r	$\alpha^*(f_r - f_q)$	Cost
1–2	8	18	11.7	0.64
1–3	10	20	13.7	0.84
2–4	22	0	13.9	0.86
2–5	20	20	20	2.00
3–4	16	14	15.3	1.04
4–6	12	12	12	0.67
5–6	4	16	8.4	0.39
Sum	**92**	**100**	**95**	**6.43**

that the algorithm reroutes flows in bulk or chunks, that is, several flows together, so some flows may partially have to be rerouted back to its original route at a later stage. A strategy to differentiate between flows may be to start with flows on routes which show the largest deviations from average length, and then successively refine the result.

Example 7.4.2. Using the same problem as in Example 7.4.1, we can compare the flow deviation algorithm with the multicommodity flow algorithm from Section 3.3. The result is shown in Table 7.11. The flow deviation method gives slightly lower maximum link load as compared to the multicommodity algorithm. In principle, they work similarly. The main differences are the cost functions (exponential or rational)

Table 7.10 Initial and final flows in Example 7.4.1

Initial Flow	T_i	Final Flow	T_i
8	0.045	15.52	0.069
10	0.050	13.00	0.059
22	0.125	12.00	0.055
20	0.100	16.08	0.072
16	0.071	11.44	0.054
12	0.056	15.92	0.071
4	0.038	12.36	0.057

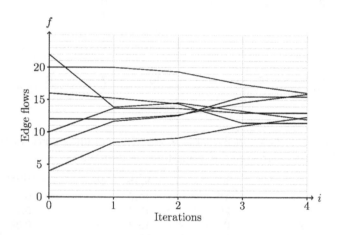

FIGURE 7.4

The convergence of the flow deviation algorithm.

Table 7.11 Initial and final flows in Example 7.4.1

Link	Flow Deviation		MC Flow		Shortest Path	
	Flow	Load	Flow	Load	Flow	Load
1–2	8.65	0.29	15.52	0.51	8	0.27
1–3	10.65	0.36	13.00	0.43	10	0.33
2–4	18.64	0.62	12.00	0.40	22	0.73
2–5	18.65	0.62	16.08	0.54	20	0.67
3–4	16.65	0.56	11.44	0.38	16	0.53
4–6	12.70	0.42	15.92	0.53	12	0.40
5–6	7.36	0.25	12.36	0.41	4	0.13

and the fact that the flow deviation method reroutes several flows simultaneously, whereas the multicommodity flow algorithm (in this presentation) reroutes one flow at a time.

7.5 Simultaneous resource and flow optimization

Resource allocation and flow assignment are very closely tied together, so it is desirable to do both optimizations together. There are two possible approaches—either we can iterate the methods in the previous sections, alternating between flow optimization and resource allocation, or use the flow deviation algorithm with modified costs and lengths, when the costs are assumed to be linear.

It is possible to combine the resource allocation method of Section 6.2 under the assumption of linear costs (7.12) and the flow deviation method of Section 7.4. Thus, for each link (i, j), the cost to be minimized is defined as

$$c_{ij} = \alpha_{ij} f_{ij} + \beta_{ij} + \frac{\left(\sum_{kl} \sqrt{\alpha_{kl} f_{kl}}\right)^2}{\gamma T_{\max}}, \qquad (7.27)$$

and the link lengths as

$$l_{ij} = \alpha_{ij}\left(1 + \frac{\sum_{kl} \sqrt{\alpha_{kl} f_{kl}}}{\gamma T_{\max}} \frac{1}{\sqrt{\alpha_{ij} f_{ij}}}\right). \qquad (7.28)$$

Note that the capacities C_{ij} do not enter into these expressions, but the cost parameters are used instead. The flow deviation method works in the same way with these changes.

Example 7.5.1.   To illustrate the method, consider again the network in Figure 7.3 with the same traffic matrix as in Example 7.3.1. No assumptions are made on the capacities other than that they should be sufficient to carry the initial flows. As initial feasible flow, we now use the flows on the shortest path routes, but with the $(1, 4)$ flow changed as in Table 7.12. The initial flow analysis is presented in Table 7.13, where the redefined length function (7.28) is used. The shorter route for $(1, 4)$ (both in terms of distance and length) is marked in boldface. By rerouting all the flow to this route ($\alpha^* = 1$) the cost is minimized, and no further suboptimal flow can be found. The optimal capacities are then found using (7.16)–(7.17) with the flows thus found.

When the costs are nonlinear or discrete, the iterative approach can be used. Starting from a set of fixed capacities, the flow is optimized. Then, keeping the flows fixed, an optimal capacity allocation is performed. This is repeated until the solution converges.

It should be noted that both in the linear and in the nonlinear case, the problem has many local optima and is difficult to optimize. The solution obtained depends upon the initial flow $\mathbf{f}^{(0)}$. The optimization should therefore be repeated with different initial flows. If these flows are generated randomly, it is possible to formulate a bound on the probability of having found a global optimum. If N optimizations are performed with

Table 7.12 Table of flows in Example 7.5.1. The columns correspond to link flows and the rows correspond to node-to-node demands

Route	1–2	1–3	2–4	2–5	3–4	4–6	5–6
1–2	2						
1–3		3					
1–4	1	0	1		0		
1–5	2			2			
1–6		1			1	1	
2–3			1		1		
2–4			2				
2–5				3			
2–6			3			3	
3–4					2		
3–5			2	2	2		
3–6					1	1	
4–5			3	3			
4–6						1	
5–6							2
Sum	4	5	11	10	8	6	2

random initial flows and k of these have the lowest cost in the sample, the probability p that there exists a cheaper solution is approximately $p < k/N$.

Another aspect of the simultaneous flow and capacity optimization is that at cost minima, there is a tendency of the flow and therefore also the capacity, to be zero. Once a link is assigned a zero flow and capacity, these will remain zero for all subsequent iterations. This property can be useful, modifying the topology by removing uneconomical links, but it may also affect resilience aspects of the topology, so care should be taken in interpreting the results.

In the case of discrete costs, it is possible to interpolate these by a continuous (concave) function. After a solution has been found, the closest larger discrete values are chosen as the capacities. This procedure does not guarantee optimality, however, even if the solution to the continuous problem is optimal.

7.6 Finite buffers

In reality, the buffers in servers are, of course, finite, meaning that under operation, under heavy traffic, or even under normal traffic in an optimal network, the buffers may from time to time overflow, leading to packet loss. Networks consisting of queues with limited buffers cannot be solved analytically, because the loss probability makes the queues statistically dependent, and so no product-form solution exists.

In other words, a loss in one queue affects the traffic further down in the network. A packet sent, for example by a queue i, may experience congestion in a queue j further down its route, and then be discarded. It would, however, contribute to the work load

Table 7.13 Primary and alternative routes in Example 7.5.1 with respective lengths

Route	Distance	$l_r^{(1)}$	$l_r^{(2)}$	$l_r^{(3)}$	$l_r^{(4)}$	Length
1–2	23	52.82				52.82
1–3–4–2	58	51.09	35.35	42.25		128.70
1–3	20	51.09				51.09
1–2–4–3	61	52.82	42.25	35.35		130.43
1–3–4	35	51.09	35.35			**86.45**
1–2–4	46	52.82	42.25			95.07
1–2–5	46	52.82	44.09			96.91
1–3–4–2–5	81	51.09	35.35	42.25	44.09	172.79
1–3–4–6	55	51.09	35.35	45.39		131.83
1–2–4–6	81	52.82	42.25	45.39		140.46
2–4–3	38	42.25	35.35			77.61
2–1–3	43	52.82	51.09			103.92
2–4	23	42.25				42.25
2–1–3–4	58	52.82	51.09	35.35		139.27
2–5	23	44.09				44.09
2–4–6–5	75	42.25	45.39	87.62		175.26
2–4–6	43	42.25	45.39			87.64
2–5–6	55	44.09	87.62			131.71
3–4	15	35.35				35.35
3–1–2–4	66	51.09	52.82	42.25		146.17
3–4–2–5	61	35.35	42.25	44.09		121.69
3–1–2–5	66	51.09	52.82	44.09		148.00
3–4–6	35	35.35	45.39			80.74
3–1–2–4–6	86	51.09	52.82	42.25	45.39	191.55
4–2–5	46	42.25	44.09			86.34
4–6–5	52	45.39	87.62			133.01
4–6	20	45.39				45.39
4–2–5–6	20	42.25	44.09	87.62		173.96
5–6	32	87.62				87.62
5–2–4–6	66	44.09	42.25	45.39		131.73

of all servers before j, but not to servers after j. The effect of lost packets "along the way" works as a simple form of congestion control. No discrepancy, however, is made on from which chain a packet is discarded.

In the case of finite buffers, we have basically two approaches available, fixed-point methods and simulation. The former approach will be discussed here, the latter in Section 7.8.

Again, we will assume that a network topology $G = (V, E)$ is provided, so that we can calculate the consolidated traffic on each link. For simple queues, like $M/M/C/b$, the blocking can be determined analytically. A fixed-point equation is an iterative method to determine the actual load on links subject to blocking. The effect of blocking gives rise to thinning of the traffic, meaning that when a packet is blocked and subsequently disappears from the system, the load further down the route is diminished.

The state probabilities of an $M/M/1$ queue with buffer b are

$$p(s) = \frac{(1-\rho)\rho^s}{1-\rho^{b+1}}. \tag{7.29}$$

The time blocking probability is given by the probability $p(b)$ of the queue of being in state $s = b$, where b is the size of the buffer.

The fixed-point method consists of two sets of equations, an estimate of traffic offered to each queue $\hat{\lambda}$ and the blocking in the queues, given by (7.29). The offered traffic is the external traffic γ arriving to the queue plus transit traffic corrected for blocking in preceding queues. Initially, the blocking can be set to zero. The equations are iterated until the fixed point is reached, whereby the variables do not change further.

The blocking B_{ij} in queue (i, j) is given by Equation (7.29) with $\hat{\rho}_{ij} = \hat{\lambda}_{ij}/(\mu C_{ij})$. The throughput is $\hat{\lambda}_{ij}(1 - B_{ij})$ and the delay is given by

$$T_i = \frac{\sum_{s=0}^{b-1}(1+s)p_i(s)}{C\sum_{s=0}^{b-1}p_i(s)},$$

where $p_i(s)$ is given by (7.29).

An important effect is that the delay decreases with increasing blocking. In essence, due to blocking, fewer packets are competing for the same resources, thus the decrease in delay.

Example 7.6.1.   Consider the network in Example 7.3.1 with shortest paths flows only. The traffic offered to the queues, taking blocking into account, is given by the equation:

$$\hat{\lambda}_{12} = \gamma_{12} + \tau_{12}^{(51)}(1 - B_{25})$$

$$\hat{\lambda}_{13} = \gamma_{13} + \tau_{13}^{(41)}(1 - B_{34}) + \tau_{13}^{(61)}(1 - B_{46})(1 - B_{34})$$

$$\hat{\lambda}_{24} = \gamma_{24} + \tau_{24}^{(32)}(1 - B_{34}) + \tau_{24}^{(54)}(1 - B_{25}) + \tau_{24}^{(62)}(1 - B_{46}) + \tau_{24}^{(53)}((1 - B_{25})$$
$$+ (1 - B_{34}))$$

$$\hat{\lambda}_{25} = \gamma_{25} + \tau_{25}^{(15)}(1 - B_{12}) + \tau_{25}^{(45)}(1 - B_{24}) + \tau_{35}(1 - B_{34})(1 - B_{42})$$

$$\hat{\lambda}_{34} = \gamma_{34} + \tau_{34}^{(14)}(1 - B_{13}) + \tau_{34}^{(23)}(1 - B_{24}) + \tau_{34}^{(63)}(1 - B_{46}) + \tau_{34}^{(16)}((1 - B_{13})$$
$$+ (1 - B_{46})) + \tau_{34}^{(53)}(1 - B_{25})(1 - B_{42})$$

$$\hat{\lambda}_{46} = \gamma_{46} + \tau_{46}^{(26)}(1 - B_{24}) + \tau_{46}^{(36)}(1 - B_{34}) + \tau_{46}^{(16)}(1 - B_{13})(1 - B_{34})$$

$$\hat{\lambda}_{56} = \gamma_{56},$$

where $B_{ij} = p_{ij}(b)$. After iterating these equations until reaching the fixed point, the performance metrics can be calculated. Figures 7.5, 7.6, and 7.7 show the blocking,

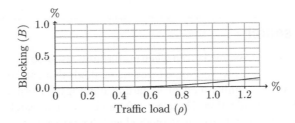

FIGURE 7.5

The blocking in a network of $M/M/1$ queues with finite buffers.

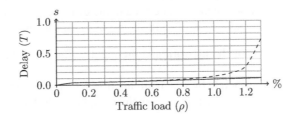

FIGURE 7.6

The waiting time in a network of $M/M/1$ queues with infinite and finite buffers.

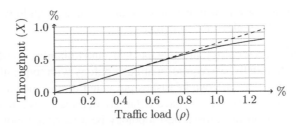

FIGURE 7.7

The throughput in a network of $M/M/1$ queues with infinite and finite buffers.

delay, and throughput for the network with buffer sizes $b = 5$ and $b = 30$, respectively. The larger buffer is used to approximate an infinite buffer. Each server has capacity $C = 30$. Each performance metric is plotted against traffic load ρ. At a load of $\rho \approx 0.6$, blocking starts to occur in the network with the smaller buffer. No blocking occurs in queues with large buffers. Due to packet loss, the delay in the former network is relatively constant for high traffic load, whereas it increases dramatically in the large buffer case. The throughput is limited by the blocking and levels out for the network where $b = 5$.

7.7 Local search

Usually, the topology is assumed given when performing resource allocation and flow optimization. There are, however, some heuristic methods that can be used to search for topologies in the neighborhood of a starting topology, incorporating resource allocation and flow optimization. These are referred to as local search methods. Two such methods are the branch exchange method and the cut-saturation method. Both start with an initial topology to which changes are gradually made, with the hope of finding a solution closer to optimum.

These methods start with a network topology and change one or more links at a time. We assume that any initial topology has capacities allocated to its links and that there is some routing principle that can be used in the evaluation of a new topology.

In general terms, we make the following assumptions for a local search procedure.

(1) The node locations of the network and the input traffic for each node pair are known.
(2) There is a routing principle by which all link flows f_{ij} can be determined, given the link capacities C_{ij}. The routing principle can be such that it minimizes the average packet delay.
(3) There is a maximum delay constraint T_{max} that has to be met.
(4) There is a reliability constraint that has to be met. A simple constraint is to require that the network is 2-connected, that is, each node has links to at least two other nodes.
(5) A cost function is given, with which the total cost of any topology can be calculated and used for ranking the topologies.

The goal is to find a topology, meeting the delay and reliability constraints, which has minimum cost. Initially, we require an initial topology with (typically) at least the reliability constraint met. The two heuristics presented do not increase or decrease the number of links, so the search space is largely defined by the initial topology. At each iteration, we let the current best topology be a feasible topology, that is, a topology satisfying the delay and reliability constraints, with the lowest cost. From this topology, a test topology is generated, which is evaluated and compared to the current best topology. These steps are summarized in Algorithm 7.7.1.

Algorithm 7.7.1.   Given a feasible network solution $G = (V, E)$, a routing principle and a cost function.

STEP 1: (Generate a Test Topology). A test topology is generated by using some heuristic to change some links and/or link capacities in the current best topology.
STEP 2: (Assign Flows). Assign flows f_{ij} in the test topology by means of some routing principle.
STEP 3: (Check Delay). Calculate the average packet delay T for the test topology. If the delay constraint is satisfied, go to Step 4. Otherwise go to Step 1.

STEP 4: (Check Reliability). Test whether the trial topology meets the reliability constraints. If the constraints are met, go to Step 5; else go to Step 1.

STEP 5: (Check Cost Improvement). If the cost of the trial topology is less than the cost of the current best topology, replace the current best topology with the trial topology. Return to Step 1.

Output the best topology found.

A local search algorithm is usually run until no substantial decrease in cost seems likely, or when no more test topologies can be generated. There is no guarantee that the final solution is optimal. It is often advisable to try different starting topologies, some of which may lead to better final solutions. In particular, when the number of links remain unchanged throughout the search, topologies with different number of links should be tested.

7.7.1 The branch exchange method

The branch exchange method - first introduced in Section 4.3 - is a conceptually simple local search method for topologies. It can easily be extended to search for an optimal topology taking also link capacity and flows into account. Starting from an initial topology, it makes local transformations by selecting two links and switching their end points, as shown in Figure 4.3. An important property of this transformation is that 2-connectivity is maintained. However, the links must be chosen so that they are completely disjoint, that is, not having any end point in common.

Algorithm 7.7.2 (The Branch Exchange Algorithm).   Given an initial topology $G = (V, E)$ with n nodes and m links and minimum degree 2, where link (i, j) has capacity C_{ij}, and a routing principle $\mathcal{R}$.

STEP 1: Perform a local transformation: Select two disjoint links and swap their end points maintaining the 2-connectivity.

STEP 2: Determine capacities and flows in the new topology using the methods in Sections 7.3–7.5; evaluate the cost. If there is a cost improvement, then the new topology is accepted. Otherwise, it is rejected.

STEP 3: If all local transformations have been explored, stop. Otherwise, go to Step 1.

Output the best topology found.

7.7.2 The cut-saturation method

The cut-saturation method is an improvement of the branch exchange method. Instead of selecting links randomly, it removes an underutilized link and adds a link where it can off-load existing links. The method thus identifies a partition of the nodes into two sets S_1 and S_2 (or equivalently, a cut) such that the links between S_1 and S_2 are heavily utilized, and adds a link to this cut. The method can be summarized as follows.

Algorithm 7.7.3 (The Cut-Saturation Algorithm). Given an initial topology $G = (V, E)$ with n nodes and m links, where link (i, j) has capacity C_{ij}, and a routing principle $\mathcal{R}$.

STEP 1: Create a list of all links, sorted in decreasing utilization, calculated by f_{ij}/C_{ij}.

STEP 2: Identify the link (k, l) such that (a) if all links above (k, l) on the list are removed, the network remains connected, and (b) if (k, l) is removed together with all preceding links in the list, the network is disconnected into two sets, S_1 and S_2.

STEP 3: Remove the most underutilized link in the network, and randomly add a link to the (S_1, S_2)-cut, so that one end point is in S_1 and the other in S_2. This link must not coincide with any existing link.

Calculate capacities and flows in the new topology using the methods in Sections 7.3–7.5; evaluate the cost. If there is a cost improvement, then the new topology is accepted. Otherwise, it is rejected. Go to Step 1.

Output the best topology found.

There are some subtleties to consider when implementing the cut-saturation algorithm. The capacities of the new link can be chosen in different ways. One possibility is to relocate the capacity from the removed link. Another possibility is to recalculate all capacities when a new topology has been created. The method tends to promote direct links, which may lead to a suboptimal topology. Links have to be selected so that 2-connectivity is maintained. Should a node have a single link only, it will never be considered in any transformation if we require the network to remain connected. Should this link be underutilized, it is a waste of resources.

These effects make the cut-saturation algorithm a suggestive method, and the result may require manual modification. A direct link, for example, might be merged with an existing path, increasing the efficiency.

Example 7.7.1. The cut-saturation algorithm applied to a random initial topology on the seven nodes shown in Figure 7.8 resulted in the feasible topology shown on the left-hand side. The demands in the form (s_k, t_k, d_k) are $(1, 6, 5)$, $(3, 5, 2)$, and $(2, 7, 2)$. The maximum utilization (congestion) is denoted φ in the figure. By merging links and maintaining the total network capacity, the network can be made cheaper, while maintaining feasibility. The result is shown in Figure 7.8 to the right.

7.8 Simulation of general packet networks

Simulation of single queues is rather straightforward. Simple networks, in particular single chains, can also be simulated with a reasonable effort. Simulation of large, realistic networks is much more complex, the main reason is the difficulty of representing the routing scheme when there are many routes available. In a real network, each packet carries its destination address, which each node then uses for making a decision on which route to send the packet next. A simulator modeling individual packets and the behavior of routing protocols is beyond the scope of this text. We use the discrete-event simulation principles introduced in Section 6.5.

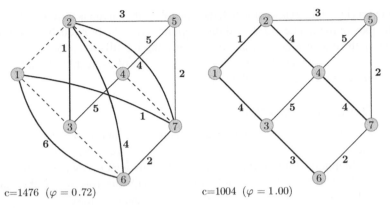

$c=1476 \ (\varphi = 0.72)$ $c=1004 \ (\varphi = 1.00)$

FIGURE 7.8

A result from iterating the cut-saturation heuristic. Direct (long) links are promoted, so a manual modification may be required.

The time to the next event is the time closest in the future when anything will happen that changes the state of the network - an arrival or a departure. A simulator clock variable is used to keep track of the progress of the simulation. It simplifies the programming task delegating bookkeeping of each queue's internal states and statistics to the queues themselves. This is conveniently implemented using an object-oriented programming language.

An object instance then keeps track of its own states and other metrics. In essence, an $M/M/1/b$ queue can be implemented as an object with only the size b of the buffer and service rate μ.

It is useful to equip the object with some internal gauges, counters and other variables. The queue can for example have a gauge that reflects the state of the queue and changes whenever an event occurs.

7.8.1 The single link

We consider the simulation of a $M/M/1/b$ link.

Since we use discrete-event simulation, there are two sets of pending events: the next scheduled arrival of a packet to the queue, and the next scheduled departure from the server. The jobs in the queue can be ordered sequentially in a vector of size b.

Each time service of a job is completed, a new service time for the next job in the queue is generated. This job is at this time moved from the buffer to the server.

The blocking rate at the queue is easily found by using a counter that is stepped up each time the buffer is full at the time of a job arrival. Of interest is also measuring the delay (which may or may not include service time). The delay is defined per job and is dependent on its position in the buffer. It is therefore convenient to let the delay be represented by a vector of size b. Each time an event occurs, all the entries in the vector corresponding to jobs in the buffer are updated with the inter-event time interval. It is important to keep track of the order of the jobs, since delays are accumulating. Usually we have a first-come first-served queueing policy. Then, each time the service of a job is completed, the first job in the queue is moved to the server and the jobs in the queue are shifted one step forward. The previously occupied last position in the queue is now vacant until an arriving job fills it again.

One of the advantages of simulation is the flexibility of using arbitrary inter-arrival and/or inter-departure time distributions.

7.8.2 Queueing networks

In case of modeling networks with random routing, the routing mechanism needs to be implemented as well. We can use a routing matrix like in Section 7.2, including the traffic which terminates at the own node. Then, by generating a uniform random number, scale it to the row total and compare it with the row where this random number falls, we can make a decision on where to route the packet.

Note, that we have no way of identifying where the packet is addressed to, or where in the queue it is. This microscopic behavior has to be captured on a macroscopic level by the routing matrix and the associated random routing. If the node decides to send the packet onto a different route due to link state issues, the receiving node would have to have its routing table altered.

One difficulty in simulating queueing networks is the following. If we do not want to put a tag on each simulated packet—which makes the simulation cumbersome and complicated—we can only route traffic on a probabilistic basis. This means that we can only form a routing matrix proportional to incoming traffic, and thus, transit traffic cannot be represented in this manner.

It is rather straightforward to implement a network simulator based on the routing matrix R using discrete-event simulation and random routing. Let the traffic arriving to the links be as described in Section 7.2. If the traffic matrix is symmetric, the amount of terminating traffic equals the amount of originating traffic. Therefore, using the routing matrix R including the terminating traffic γ_{ij} on the matrix diagonal, random sampling on this matrix will represent the operation of the network. One drawback, however, is that some jobs will not terminate at its destination, but be routed further in the network. This leads to an overestimation of the traffic in the network and an underestimation of performance metrics.

7.9 Summary

Simple packet networks are packet networks without any flow or congestion control functions. It is an idealization of a packet network which lends itself nicely to analytical treatment. Thus, a simple packet network can often be designed optimally, and this turns out to be of practical importance also for more complex networks.

In essence, what we call simple networks are networks that can be modeled as open Jackson networks. There are many references of importance, some of which are Bertsekas and Gallager (1987), Kleinrock (1976), and Gerla and Kleinrock (1977). Network optimization problems for a range of cost functions is analyzed in Kleinrock (1970) and Gerla and Kleinrock (1977). An introduction to nonlinear optimization is given in, for example, Luenberger (1989).

Flow-Controlled Packet Networks

Flow control in a network is a mechanism that controls the amount of work entering the network and synchronizes the input rate with the admittance rate at the sink to prevent overload. The effect is, that during high load, the amount of work in the network is kept approximately constant.

This is necessary both to ensure quality of service, which is related to the quality that the jobs experience from the delivery by the network, and to protect the network itself from overload. For example, if a file in transfer experiences large packet loss as a result of high load, the file may be corrupt and unusable at the recipient end. Then, it may be better to wait until the network load decreases. Another example, may be a voice call that experiences large delays and therefore becomes incomprehensible to the people involved. Admitting these services onto a heavily loaded network would just be a waste of network resources. At the same time the network itself would suffer from being loaded by services that it cannot handle while maintaining the requirements that services demand. As a result, other services would suffer poorer delivery as well.

The mechanism is referred to as flow control when a service is to be guaranteed a certain quality, and congestion control when an overall overload protection of the network is intended. The two mechanisms can be implemented in similar ways.

One way of accomplishing this is by statically limiting the total number of jobs in the network, called window-based flow control, where the number of jobs is referred to as the window size. This can be done on a network basis or on a traffic stream basis. A virtual channel is a stream with a fixed window size.

Another way is to dynamically adjust the window size for a stream based on the performance of its end-to-end connection, such as delay and packet loss. Such a mechanism is implemented in the Transmission Control Protocol (TCP).

The analysis of flow-controlled networks is often modeled by closed queueing networks, because the number of jobs is kept approximately constant in the network. Flow control can be accomplished by using permits to access the network, so that the number of jobs never can exceed the number of permits. The analysis of closed queueing networks is more difficult than for open ones due to the computation of the normalizing constant. Some traditional methods for analyzing such networks are convolution, mean value analysis, decomposition (Norton's theorem for queues), fixed-point methods for flow-controlled networks and simulation.

The main difficulty of an efficient flow control is that it should be adaptive. That is, the network flow changes depending on the statistical variations of arriving traffic, and the flow control should react accordingly. In TCP, in particular, the algorithm is difficult to model due to its statistical dependence and strongly variable window size. It tries to squeeze as much as possible through the network without disturbing the calls or sessions in progress too much, and when a threshold is reached, it decreases the throughput to a minimum level, starting to increase it again should performance metrics show acceptable values.

Performance measures are typically nonlinear, so some type of statistical measures are needed to assess the performance of the network, subject to flow control. In this case, the mean of the throughput is rather unrevealing as a measure, so the median is used instead. Other common measures that are used are more or less pessimistic percentiles.

8.1 Flow control and congestion control

In this chapter, we distinguish between two types of packet networks: virtual channel networks, for which flow control is achieved by permits; and packet networks with TCP, which adaptively allows traffic into the network based on the quality the service the traffic experiences.

The fundamental difference between these two types is the way routing is performed. A virtual channel is similar to a circuit-switched connection in that a route is identified at call set-up, and this route is the preferred route under normal operating conditions. Each packed contains a virtual channel identifier, which is used for routing.

In a TCP controlled network, on the other hand, each packet contains address fields that make it possible to send it onto any route available at any instant. This makes such networks more efficient in dynamically utilizing network resources, but the overhead of routing information is larger, and the service quality is typically less reliable than for a virtual channel.

8.1.1 Virtual channels

A virtual channel (VC), also known as a virtual connection or a virtual circuit, is a transport technology similar to circuit-switched transmission, but based on packet switching. The similarity to circuit switching is the necessary establishment of a virtual channel between two nodes before communication can start. This transport method is referred to as connection-oriented. The characteristics of the method are that packets are delivered in sequence, and the signaling volume can be kept low by defining a transmission path in the set-up phase of a communication session that is used throughout the session.

In contrast to circuit switching, however, a virtual channel uses packet-switching technology, which makes more efficient utilization of transmission resources. In circuit switching, resources are reserved during periods of inactivity as well as when data needs to be sent. A virtual channel, however, is only using resources when needed by the application. The quality of service provided depends on packet lengths, bit rates, and load distribution between applications sharing the resources.

A virtual channel can be implemented in different ways, or on different network layers described by the Open Systems Interconnection (OSI) model. On a lower network level, data is always following the same physical path through the network. Examples of technologies realizing virtual channels on lower levels (network layer or datalink layer) are X.25, Asynchronous Transfer Mode (ATM), and General Packet Radio Service (GPRS).

In contrast to the technology based on virtual channels, datagram (IP) packet networks can use any available paths for transmission, a method which is also referred to as a connectionless service. In a datagram network, more signaling overhead is required, as each packet is transported independently of other packets, and more sophisticated methods for message re-assembly are necessary at the recipient node. A virtual channel can be implemented using higher-level (transport layer) protocols like the Transmission Control Protocol (TCP), which then handles division of messages into packets, error correction, and message reassembly at the destination.

The advantages of virtual channel implementation on lower levels are that resources (bandwidth) can be reserved for a service, which can be used to emulate circuit-switched services, and that less signaling overhead follows from the set-up of a dedicated transmission path so that only a small virtual channel identifier (VCI) is required to route a packet correctly instead of the full destination address. Since the routing of packets is simplified by a pre-determined path, the nodes handling the traffic are faster than in the case of datagram-based technologies where each packet is routed separately. In datagram (IP) networks, Multi-protocol Label Switching (MPLS) provides a similar functionality to the virtual channel concept inherent in Asynchronous Transfer Mode (ATM).

In a virtual channel, the traffic flow is protected from overload by using a limited number of access permits. Such a permit is required by any job entering the network. A job first requests a permit from the network, and if such a permit is available, it is attached to the job until delivered and a delivery acknowledgment has been sent from the receiving node back to the originating node. When the transfer is completed, the permit is freed and a new job can be accepted into the network.

8.1.2 Transmission Control Protocol (TCP)

The Transmission Control Protocol (TCP) is a protocol in the transport layer of the OSI model. It breaks up a data stream into segments (or chunks), and adds a TCP header. Such a segment is called a protocol data unit (PDU). TCP provides a reliable transport of data, including error correction and retransmission in case of packet loss. The protocol is used for many applications such as the World Wide Web (WWW), e-mail, file transfer, and some streaming media.

TCP is intended for reliable delivery rather than transport of delay sensitive services, such as voice services. Due to the features implemented to obtain reliable

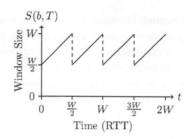

FIGURE 8.1

The TCP congestion window size can approximately be described by a sawtooth function.

delivery, large delays can sometimes be incurred. For delay sensitive services, proto-cols like Real-time Transport Protocol (RTP) are commonly used instead.

From a design point of view, one of the most important features of TCP is the sliding window mechanism, which prevents overload of the network by limiting the number of jobs admitted. This can be done both for an end-to-end connection, such as a virtual channel, as well as for an entire network. The former is referred to as flow control and the latter congestion control. The term sliding window is reserved for flow control, and the similar mechanism for congestion control is known as congestion window.

The purpose of flow control is to synchronize the transmission speed between the sending node and the receiving node so that it complies with the optimal capabilities at both ends. By doing so, it can also enforce some quality-of-service requirements, such as limits on packet loss probability and, to a certain degree, delay. The size of the sliding window specifies how much data the source is allowed to send before it has to wait for an acknowledgment from the receiving node that the transmitted data has been received properly. In case of congestion, when large packet loss probability and/or delay is experienced, the window size is decreased. If, on the other hand, these performance metrics indicate a good connection, the window size is increased. Figure 8.1 shows the window size dependence on delay (or round-trip time (RTT)).

In contrast to using a fixed number of permits—a static number which then has to be specified—TCP adjusts its window size so that acceptable service performance is achieved and at as high throughput as possible. Under stationary traffic conditions, it typically reaches a size for which the maximum possible number of jobs can be sent while maintaining sufficient transmission quality. After that, a new job is admitted whenever an acknowledgment is received. Thus, the sliding window works similarly to the permit paradigm of virtual channel technologies, in that, whenever an acknowl-edgment is received back at the sending node, the mechanism will allow a new job into the network, that is, the window is 'sliding' forward in time.

Consider a route between an origin node and a destination node with constant round-trip time and packet loss probability. Let the congestion window have maxi-mum size W. The congestion avoidance algorithm of TCP increases the window size by one packet (or, maximum segment size (MSS)) when no congestion is detected. In case of congestion, the window is re-sized to half the current size.

If the recipient end acknowledges all packets, transmitted periodically, the window will be increased by one packet per round-trip, until packet loss occurs (which by assumption is at window size W). In this idealized situation, each cycle is $W/2$ round trips long and the congestion window can be described by a sawtooth function, as shown in Figure 8.1.

8.2 Closed queueing networks

In a closed queueing network (also known as a closed Jackson network or a Gordon-Newell network) a number of jobs are circulating within the network without any exchange with the outer world.

Closed networks are common models of flow-controlled or congestion-controlled networks—particularly when permit based—since the number of jobs during heavy traffic is kept approximately constant. The number of jobs in a closed network is commonly referred to as the population of the network.

Since the number of jobs in a closed network is limited, the state probabilities have to be calculated using a normalization constant. This constant is cumbersome to calculate, since all feasible states contribute, and the number of states grows exponentially with the size of the network and the number of jobs.

A closed queueing network is a Gordon-Newell network (or closed Jackson network) if the following criteria are met:

(1) The network is closed, meaning that no jobs can enter or leave the network.
(2) The service times are exponentially distributed and the queue discipline is first-come, first-served (FCFS).
(3) The utilization of each queue is less than unity.
(4) Each queue has a sufficiently large buffer so that no job is lost.
(5) A job finishing service at queue i is routed to queue j with probability r_{ij}. The routing probabilities are described by a routing matrix R where all rows have sum 1. This matrix is found by solving the traffic equation corresponding to the network.

A closed network can be thought of as having a number of access permits allowing jobs to use it. The number of jobs that circulate in the network thus equals the number of permits. Suppose we have a traffic stream arriving from outside the network and jobs leaving the network upon service completion. Then any new job is accepted only when another job leaves. If the jobs arrive to a virtual channel k at intensity γ_k, then the relation between this arrival rate and the throughput γ_k^* of the virtual channel is $\gamma_k \leq \gamma_k^*$. The quantity γ_k is called the offered load to the virtual channel. If a job is blocked due to lack of free permits, it is usually scheduled for retransmission at a later time. Therefore, the ratio γ_k/γ_k^* estimates the number of times a job needs to be retransmitted in order to gain access to the network.

8.2.1 The product-form solution

Let a closed network have m servers, and index the virtual channels (also known as chains) in the network by $k = 1, 2, \ldots, K$. Then, the actual arrival rates at server i contributed by chain k are

$$\lambda_i^{(k)} = \sum_{j=1}^{m} \lambda_j^{(k)} r_{ji}^{(k)}, \quad i = 1, 2, \ldots, m. \tag{8.1}$$

The solution $\boldsymbol{\lambda}$ to (8.1) can only be determined up to a multiplicative constant due to the fact that $R^{(k)}$ is a stochastic matrix (which is a consequence of the network being closed). Two methods to determine $\boldsymbol{\lambda}$ are discussed in the next subsection.

We can now formulate a general solution for the state probabilities of this network. Let N_k be the population size of chain k and the population vector in the network having K chains and m queues

$$\mathbf{N} = (N_1, N_2, \ldots, N_K).$$

It can be shown (see for example Gross and Harris, 1998) that the equilibrium network state probability has a product form. Denoting by $\mathbf{s}_i$ the K-vector of the number of jobs in queue i from chains $k = 1, 2, \ldots, K$, so that $\mathbf{s}_i = (s_{i1}, s_{i2}, \ldots, s_{iK})$. The state of the network is defined by $S = \{\mathbf{s}_1, \mathbf{s}_2, \ldots, \mathbf{s}_m\}$. Then

$$\mathbf{P}(S) = \frac{1}{G(\mathbf{N})} \prod_{i=1}^{m} p_i(\mathbf{s}_i), \tag{8.2}$$

where $p_i(\mathbf{s}_i)$ is given by

$$p_i(\mathbf{s}_i) = s_i! \prod_{k=1}^{K} \frac{\rho_{ik}^{s_{ik}}}{s_{ik}!}$$

for a FCFS server, with $s_i = s_{i1} + s_{i2} + \cdots + s_{iK}$, and

$$p_i(\mathbf{s}_i) = \prod_{k=1}^{K} \frac{\rho_{ik}^{s_{ik}}}{s_{ik}!} \tag{8.3}$$

for an IS server. In these expressions,

$$\rho_{ik} = \lambda_{ik}/\mu C_i,$$

where C_i is the server rate and μ the job length (holding time).

The normalization constant $G(\mathbf{N})$ is defined by

$$G(\mathbf{N}) = \sum_{S \in \Omega} \prod_{i=1}^{m} p_i(\mathbf{s}_i), \tag{8.4}$$

where

$$\Omega = \left\{ S : \sum_{i=1}^{m} \mathbf{s}_i = \mathbf{N} \right\}$$

is the total state space. In addition to the difficulty of calculating the normalizing constant $G(\cdot)$ is that since λ is determined up to a scaling factor, which is absorbed in $G(\cdot)$, this normalization constant can become very large or very small, leading to numerical difficulties with floating point overflow or underflow. The subsequent sections discuss some efficient methods to determine the performance metrics for closed queueing networks.

For a particular network state **s**,

$$\mathbf{P(s)} = \frac{1}{G(\mathbf{N})} \prod_{i=1}^{m} q_i(s_i), \tag{8.5}$$

where $q_i(s_i)$ is given by

$$q_i(s_i) = \rho_i^{s_i}$$

for FCFS servers, and

$$q_i(s_i) = \frac{\rho_i^{s_i}}{s!}$$

for IS servers.

8.2.2 Solving the traffic equation

The routing matrix and the solution of the traffic equation (8.1) for a closed network can be determined as follows. Let $R^{(k)} = r_{ij}^{(k)}$ be a routing matrix for chain k. In the following discussion, we omit the indexing by k for brevity. Since the number of jobs is constant, that is, no job can enter or leave the system, the routing matrix R must be a stochastic matrix having each row summing to one. We also decompose the closed network into subnetworks so that each routing matrix in a subnetwork corresponds to an irreducible Markov chain (where each state is reachable from all other states in the chain). The assumption of irreducibility is important for the numerical methods to work.

We will present two methods for solving the traffic equation (8.1), called inverse iteration and removal of an equation, presented in Stewart (1994). The closed network follows a continuous-time Markov process with $Q = I - R$, determined by the routing matrix R, where Q is the infinitesimal generator of the process. By collecting all terms of (8.1) on the left hand, we can express the equation as

$$\lambda(I - R) = \lambda Q = \mathbf{0}.$$

The equation for the steady-state probabilities of the network is, therefore, the solution to the equation $\lambda Q = \mathbf{0}$. However, it can be shown that Q has rank $n - 1$, where n is the number of nodes, so that the system is underdetermined. The solution can therefore be determined only up to a multiplicative constant.

8.2.2.1 *Inverse iteration*

Equation (8.1) can be seen as an eigenvalue problem. In fact, the solution λ is the right-hand eigenvector corresponding to the zero eigenvalue of Q^T.

It is well known that when Q is irreducible it has an LU-factorization $Q^T = LU$. The inverse iteration builds on LU-factorization of Q, which first must be found. This can easily be done using a software package with routines for linear algebra. Suppose then that a factorization has been found so that $Q^T = LU$. Let $\mathbf{e}_i = (0, \ldots, 0, 1, 0, \ldots)$ be a vector with a one in position i and zeros elsewhere. Next, we modify the upper-diagonal matrix $U = (u_{ij})$ of the LU-factorization by setting $u_{ii} = \epsilon$, where ϵ is a small number. Calling the modified upper-diagonal matrix $\bar{U}$ and solving the system of equations

$$\bar{U}\bar{\lambda} = \mathbf{e}_n \tag{8.6}$$

for $\bar{\lambda}$, we get the traffic weights by normalizing, so that

$$\lambda = \frac{\bar{\lambda}}{\sqrt{\bar{\lambda}^T \cdot \bar{\lambda}}}. \tag{8.7}$$

These are then the relative steady-state offered traffic rates to each queue (that is, offered rates determined up to the multiplicative constant).

Example 8.2.1. Suppose the network in Example 7.2.1 is closed instead of open. The routing matrix constructed in Example 7.2.1 now needs to be modified in two ways. First all rows have to sum to one (since the number of jobs is constant), so that the r_{ii} entry is set to equal one minus the sum of the rest of row i. Secondly, we require the resulting Markov chain to be irreducible. By restricting R to include only queues 1 to 6, it is easily seen that each queue can be reached from each other queue, and hence, this subnetwork is irreducible. Then, we have

$$R = \begin{pmatrix} 3/4 & 0 & 0 & 1/4 & 0 & 0 \\ 0 & 4/5 & 0 & 0 & 1/5 & 0 \\ 0 & 0 & 1/2 & 5/22 & 3/22 & 3/22 \\ 1/10 & 0 & 1/4 & 13/20 & 0 & 0 \\ 0 & 1/8 & 3/16 & 0 & 9/16 & 1/8 \\ 0 & 0 & 1/4 & 0 & 1/6 & 7/12 \end{pmatrix}. \tag{8.8}$$

The LU-factorization is

$$LU = \begin{pmatrix} 1 & 0 & 0 & 0 & 0 & 0 \\ 0 & 1 & 0 & 0 & 0 & 0 \\ 0 & 0 & 1 & 0 & 0 & 0 \\ -1 & 0 & -0.455 & 1 & 0 & 0 \\ 0 & 0 & -0.273 & -0.5 & 1 & 0 \\ 0 & -1 & -0.273 & -0.5 & -1 & 1 \end{pmatrix}$$

$$
\cdot \begin{pmatrix}
0.25 & 0 & 0 & -0.1 & 0 & 0 \\
0 & -0.2 & -0.136 & 0 & 0.438 & -0.167 \\
0 & 0 & 0.5 & -0.25 & -0.188 & -0.25 \\
0 & 0 & 0 & 0.136 & -0.085 & -0.114 \\
0 & 0 & 0 & 0 & -0.219 & 0.292 \\
0 & 0 & 0 & 0 & 0 & 0
\end{pmatrix} \cdot
$$

Note that the last row in the upper-diagonal matrix has all zero entries. This is a result of the rank of Q being $n - 1$. Next, let the entry $u_{66} = \epsilon$ with $\epsilon = 0.001$, say. Now solving (8.6) for $\bar{\lambda}$ gives

$$
\bar{\lambda} = \begin{pmatrix} 667 \\ 833 \\ 1833 \\ 1667 \\ 1333 \\ 1000 \end{pmatrix},
$$

and after normalization

$$
\lambda = \begin{pmatrix} 0.210 \\ 0.263 \\ 0.578 \\ 0.526 \\ 0.420 \\ 0.315 \end{pmatrix}. \tag{8.9}
$$

It is strictly speaking not necessary to perform the normalization at this stage when using the traffic as input to a closed network problem, since they only represent the relative traffic in that context. However, it may be useful to avoid numerical instability, in particular if the convolution method is to be used.

8.2.2.2 *Removal of an equation*

An alternative approach is to partition the matrix Q^{T} as

$$
Q^{\mathrm{T}} = \begin{pmatrix} Q_1 & \mathbf{q}_2 \\ \mathbf{q}_3 & q_4 \end{pmatrix},
$$

where Q_1 is of order $n - 1$. For an irreducible Markov chain, $\mathbf{q}_2 \neq \mathbf{0}$ and Q_1 is non-singular. Now we can remove the last equation and solve

$$
Q_1 \bar{\lambda}_1 = -\mathbf{q}_2
$$

for $\bar{\lambda}_1$. By appending a 1 to the last equation, we get

$$
\bar{\lambda} = \begin{pmatrix} \bar{\lambda}_1 \\ 1 \end{pmatrix}.
$$

After normalization, as in (8.7), we get the steady-state offered traffic to each queue.

8.3 Convolution

Two important methods to analyze closed queueing networks are the convolution method (Buzen, 1973) and mean value analysis (Reiser and Lavenberg, 1980). As mentioned earlier, the analysis of closed queueing networks is complicated by the necessity of calculating the normalization constant $G(\cdot)$. The convolution method gives the normalization constant $G(\mathbf{N})$ and the state probabilities directly, and per-formance metrics can then be calculated based on these. Mean value analysis, on the other hand, gives rather the queue performance metrics directly, without the need to compute $G(\mathbf{N})$ explicitly. However, the state probabilities can also be computed along the way. The convolution method is presented in this section and mean value analysis in the next.

Consider a closed queueing network with m queues and a fixed number N of jobs circulating in the network. The state of queue is determined by the number s of jobs currently in the queue, where $0 \le s \le N$. If there are K chains in the network, and we distinguish between the jobs belonging to different chains, the state is represented by a vector $\mathbf{s}_i = (s_{i1}, s_{i2}, \dots, s_{iK})$ and the set of jobs is also a vector $\mathbf{N} = (N_1, N_2, \dots, N_K)$ with $N = \sum_{k=1}^{K} N_k$. For simplicity of the present discussion, we assume that we have a single chain only, or equivalently, that the jobs belonging to different chains are indistinguishable. The generalization of the convolution method to K chains or traffic classes is discussed in Section 11.2.

In the convolution method, the state probabilities $p_i(s)$ in (8.4) for each queue i are successively calculated for an increasing number of jobs s by means of convolution. Given two functions with scalar arguments, say $q_1(s)$ and $q_2(s)$, defined on some domain. Then the *discrete convolution* $g(s)$ of the two functions defined as

$$g(s) = \sum_{v=0}^{s} q_1(v) q_2(s - v), \quad 0 \le s \le N, \tag{8.10}$$

is a function on the same domain.

In shorthand notation, this is usually written as

$$g(s) = (q_1 * q_2)(s) = (q_2 * q_1)(s).$$

The normalization constant $G(N)$ can be obtained by successive convolution of the functions where the argument is the population size N,

$$g_{(1,2,\dots,m)}(N) = (q_1 * q_2 * \cdots * q_m)(N).$$

For a network with m queues and N circulating jobs, we can describe the network state by the vector $\mathbf{s} = (s_1, s_2, \dots, s_m)$, where s_i is the number of customers present at the ith queue and $\sum_{i=1}^{m} s_i = N$.

Assume that the service time for a customer at the ith queue is given by an exponentially distributed random variable with mean $1/\mu_i$, and that the probability a customer will proceed to the jth queue is equal to r_{ij} for $i, j = 1, 2, \ldots, m$. It then follows that the equilibrium distribution of customers in the network is

$$\mathbf{P}(s_1, s_2, \ldots, s_m) = \frac{1}{G(N)} \prod_{i=1}^{m} (\rho_i)^{s_i}, \tag{8.11}$$

where $(\rho_1, \rho_2, \ldots, \rho_m)$ is the real positive solution to the traffic equations

$$\mu_j \rho_j = \sum_{i=1}^{m} \mu_i \rho_i r_{ij}, \quad 1 \le j \le m,$$

where $\rho_i = \gamma_i / mu_i$ and $G(N)$ is a normalizing constant defined so that all the $\mathbf{P}(s_1, s_2, \ldots, s_m)$ sum to one. That is

$$G(N) = \sum_{s \in S(N,m)} \prod_{i=1}^{m} p_i(s_i), \tag{8.12}$$

or, for an $M/M/1$ queue with $p_i(s_i) = \rho_i^{s_i}$

$$G(N) = \sum_{s \in S(N,m)} \prod_{i=1}^{m} (\rho_i)^{s_i}$$

where

$$S(N, m) = \left\{ (s_1, s_2, \ldots, s_m) | \sum_{i=1}^{m} s_i = N, s_i \ge 0 \right\}. \tag{8.13}$$

Note that we can substitute for the state probabilities for an $M/M/n$ queue in (8.12). The summation in (8.13) is taken over all $\binom{N+m-1}{N}$ possible system states $(s_1, s_2, \ldots, s_m)$.

First, we note that for any integer $s \le N$

$$\mathbf{P}(s_i \ge s) = (\rho_i)^s \frac{G(N-s)}{G(N)}. \tag{8.14}$$

It then follows that

$$\mathbf{P}(s_i = s) = \frac{(\rho_i)^s}{G(N)} (G(N-s) - \rho_i G(N-s-1)), \tag{8.15}$$

where it is assumed that $G(s) = 0$ for $s < 0$. Note that (8.14) is of interest in its own right since, for $s = 1$, it yields the probability that the ith queue is busy (i.e., non-empty). It also follows directly from (8.14) that $\mathbf{E}(s_i)$, the expected number of customers present at the ith queue, is given by

$$\mathbf{E}(s_i) = \sum_{s=1}^{N} (\rho_i)^s \frac{G(N-s)}{G(N)}. \tag{8.16}$$

Hence, once the values of $G(1)$, $G(2)$, ..., $G(N)$ have been calculated it is possible to use Equations (8.11), (8.14)–(8.16) to calculate the state probabilities and from these derive network performance measures.

Considering the normalization constant

$$G(N) = \sum_{s_1 + \cdots + s_m = N} \prod_{i=1}^{m} q_i(s_i), \qquad (8.17)$$

where $q_i(s_i)$ is the probability of an independent queue i of being in state s_i. We can define an auxiliary function such that for any $1 \leq s \leq N$ and $1 \leq w \leq m$,

$$g_w(s) = \sum_{s_1 + \cdots + s_w = s} \prod_{i=1}^{w} q_i(s_i). \qquad (8.18)$$

The function $g_w(s)$ would equal $G(N)$ if $w = m$ and $s = N$; in other words, it is a normalizing constant for a system with s customers and w nodes. This function can now be used in a recursive procedure for computing $G(N)$.

Consider $g_w(s)$ and let $s_w = i$ be fixed in the summation, so that we have

$$
\begin{aligned}
g_w(s) &= \sum_{i=0}^{s} \left(\sum_{s_1 + \cdots + s_{m-1} + i = s} \prod_{i=1}^{w} q_i(s_i) \right) \\
&= \sum_{i=0}^{s} q_w(i) \left(\sum_{s_1 + \cdots + s_{m-1} + i = s - i} \prod_{i=1}^{w-1} q_i(s_i) \right) \qquad (8.19) \\
&= \sum_{i=0}^{s} q_w(i) g_{w-1}(s - i), \quad s = 0, 1, \ldots, N.
\end{aligned}
$$

We have the initial values $g_1(s) = q_1(s)$ and $g_w(0) = 1$, so that we can use (8.19) recursively to calculate $G(N) = g_w(N)$.

In principle, we can find the marginal distribution at queue i, namely $p_i(s) = \mathbf{P}(N_i = s)$. Let

$$S_i = s_1 + s_2 + \cdots + s_{i-1} + s_{i+1} + \cdots + s_m. \qquad (8.20)$$

Then we have

$$
\begin{aligned}
p_i(s) &= \sum_{S_i = N - s} p_{s_1, \ldots, s_m} = \sum_{S_i = N - s} \frac{1}{G(N)} \prod_{i=1}^{m} q_i(N_i) \\
&= \frac{q_i(s)}{G(N)} \sum_{S_i = N - s} \prod_{\substack{j = 1 \\ j \neq i}}^{m} q_i(s), \quad s = 0, 1, \ldots, N
\end{aligned} \qquad (8.21)
$$

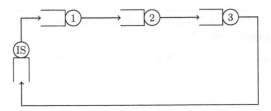

FIGURE 8.2
Virtual channel.

This expression is difficult to compute. For queue m, however, the expression simplifies to

$$p_m(s) = \frac{q_m(s)}{G(N)} \sum_{S_m=N-s} \prod_{i=1}^{m-1} q_i(s) = \frac{q_m(s)g_{m-1}(N-s)}{G(N)}, \quad s = 0, 1, \ldots, N,$$

(8.22)

which easily can be evaluated. To find the marginal distributions for another queue $p_i(s), i \neq m$, we can make a permutation to change the order of the queues in the network so that queue i of interest becomes queue m. After the permutation, some of the functions $g_w(s)$ will need to be recomputed.

Example 8.3.1. Consider a virtual channel, as depicted in Figure 8.2. Let the window size be small, $N = 3$. Then each server can be in the states $s = 0, 1, 2,$ or 3. Note that queue four is a delay (infinite) server, modeling the delay of the acknowledgments sent back from the destination to the origin. Thus, we have

$$q_i(s) = \mu_i^s, \quad \text{for } i = 1, 2, 3 \text{ and } s = 0, 1, 2, 3.$$

and

$$q_4(s) = \frac{\mu_4^s}{s!}, \quad \text{for } s = 0, 1, 2, 3.$$

The values are shown in the first four columns of Table 8.1. We use the auxiliary notation of $q_i(s)$, since these values are not normalized to the state probabilities $p_i(s)$

Table 8.1   Unnormalized state probabilities and two convolutions in Example 8.3.1

s	q_1	q_2	q_3	q_4	q_{12}	q_{23}
0	1	1	1	1	1	1
1	$2.63 \cdot 10^{-2}$	$2.94 \cdot 10^{-2}$	$2.78 \cdot 10^{-2}$	$6.45 \cdot 10^{-3}$	$5.57 \cdot 10^{-2}$	$5.72 \cdot 10^{-2}$
2	$6.93 \cdot 10^{-4}$	$8.65 \cdot 10^{-4}$	$7.72 \cdot 10^{-4}$	$2.08 \cdot 10^{-5}$	$2.33 \cdot 10^{-3}$	$2.45 \cdot 10^{-3}$
3	$1.82 \cdot 10^{-5}$	$2.54 \cdot 10^{-5}$	$2.14 \cdot 10^{-5}$	$4.48 \cdot 10^{-8}$	$8.68 \cdot 10^{-5}$	$9.36 \cdot 10^{-5}$

Table 8.2 Convolutions of three and four probabilities, and state probabilities for queue 1 in Example 8.3.1

s	q_{123}	q_{234}	q_{1234}	$p_1(s)$
0	1	1	1	0.447
1	$8.35 \cdot 10^{-2}$	$6.36 \cdot 10^{-2}$	$9.00 \cdot 10^{-2}$	0.302
2	$4.65 \cdot 10^{-3}$	$2.84 \cdot 10^{-3}$	$5.21 \cdot 10^{-3}$	0.178
3	$2.16 \cdot 10^{-4}$	$1.11 \cdot 10^{-4}$	$2.48 \cdot 10^{-4}$	0.074

yet, which of course must sum to unity. Next, the convolutions $q_{ij}(s) = (q_i * q_j)(s) = \sum_{l=0}^{s} q_i(l)q_j(s - l)$ are calculated. The first of these are the values of

$$q_{12}(0) = q_1(0)q_2(0),$$
$$q_{12}(1) = q_1(0)q_2(1) + q_1(1)q_2(0)$$
$$q_{12}(2) = q_1(0)q_2(2) + q_1(1)q_2(1) + q_1(2)q_2(0)$$
$$q_{12}(3) = q_1(0)q_2(3) + q_1(1)q_2(2) + q_1(2)q_2(1) + q_1(3)q_2(0).$$

The values of q_{12} and q_{23} are shown in columns five and six in Table 8.1. Continuing in this way, we can calculate

$$q_{123}(s) = (q_{12} * q_3)(s) = (q_1 * q_{23})(s)$$
$$q_{234}(s) = (q_{23} * q_4)(s)$$
$$q_{1234}(s) = (q_{123} * q_4)(s) = (q_1 * q_{234})(s).$$

The normalizing constant $G(N)$ is now given by $q_{1234}(N)$ where $N = 3$. Now, suppose that we wish to find the steady-state marginal probabilities for queue 1. These are found as

$$p_1(0) = q_1(0) \cdot q_{234}(3)/q_{1234}(3)$$
$$p_1(1) = q_1(1) \cdot q_{234}(2)/q_{1234}(3)$$
$$p_1(2) = q_1(2) \cdot q_{234}(1)/q_{1234}(3)$$
$$p_1(3) = q_1(3) \cdot q_{234}(0)/q_{1234}(3),$$

where we multiply the un-normalized state probabilities $q_1(s)$ by the compliment $q_{234}(N - s)$ in the network and normalize to get the actual probabilities. These values are shown in Table 8.2. Note that in order to calculate $p_1(s)$ we need $q_{234}(s)$, that is, the state probability of the complement of queue one in the network. Similarly, if we want to calculate $p_2(s)$, we need to calculate q_{134} first, and so on.

8.4 Mean value analysis

The solution to closed queueing network as described in Section 8.2 requires the computation of the normalizing constant $G(S)$. This can be done using the convolution method. Another method which does not require the explicit calculation of the normalizing constant is mean value analysis (MVA) by Reiser and Lavenberg (1980). We assume initially that all nodes are $M/M/1$ queues.

The method is built on the following two principles:

(1) The queue length observed by a job arriving to a queue i is the same as the mean queue length in a network with one job less, and waiting time with the arriving job is therefore

$$T_i(s) = \frac{1 + L_i(s-1)}{\mu_i}, \tag{8.23}$$

where s is the number of jobs in the network, μ_i is the service rate, and $T_i(s)$ is the delay at queue i in a network with s jobs.

(2) Little's formula is applicable throughout the network.

The first principle states that the mean waiting time when a new job arrives is the mean waiting time given by the queue length ahead as seen by the arriving job plus the service time of the job itself.

The second principle gives the relation

$$L_i(s) = \lambda_i(s)T_i(s), \tag{8.24}$$

where $\lambda_i(s)$ is the throughput at queue i with s jobs in the network.

We also note that the sum of the mean queue lengths in a network with m queues must equal the total number of jobs, that is

$$s = \sum_{j=1}^{m} L_j(s), \quad j = 1, 2, \ldots, m. \tag{8.25}$$

Equation (8.25) states the conservation of jobs in the network. We do not yet know $\lambda_i(s)$, but we can get v_i, the relative arrival rates given by the traffic equation (8.1). Since $\lambda_i(s)$ is proportional to v_i (by definition) for all s, we have

$$L_i(s) = s \frac{v_i T_i(s)}{\sum_{j=1}^{m} v_j T_j}, \quad j = 1, 2, \ldots, m, \tag{8.26}$$

which can be seen as an application of Little's formula to the whole network, appropriately scaled for the v_i to make the sum of the average queue length equal to s. Since the v_i are relative throughputs, we do not need to normalize the vector as in Section 8.3. The normalization step performed there may, however, be beneficial in order to avoiding numerical instability.

Noting that $L_i(0) = 0$ and $T_i(1) = 1/\mu_i$ for all i, we can formulate a recursive procedure to compute the queue lengths L, the delays T, and the throughputs λ for all queues i and increasing job populations s.

We can now summarize the mean value analysis procedure for computing performance metrics for a network with m single-server queues and a fixed number of N circulating jobs as follows.

Algorithm 8.4.1 (Mean Value Analysis (Single-Server Queues)). Given a network of m $M/M/1$ queues with service rates μ_i, a routing matrix R, and a fixed number of jobs N.

STEP 0: (initialize)

Solve the traffic equation (8.25) for v_i. Let $L_i(0) = 0$ for all queues i.

STEP $1 - N$: (recursion)

> **for** $i = 1$ to m **do**
>> Compute $T_i(s) = (1 + L_i(s - 1))/\mu_i$;
>>
>> Compute $L_i(s) = s \dfrac{v_i T_i(s)}{\sum_j v_j T_j(s)}$, $j = 1, 2, \ldots, m$.
>>
>> Optionally, compute $\lambda_i(s) = L_i(s)/T_i(s)$;
>
> **end**;

Output $L_i(N)$, $T_i(N)$, and (optionally) $\lambda_i(N)$.

Example 8.4.1. Consider three queues in series, which could be a model of a virtual channel, for example. The network is depicted in Figure 8.2. The nodes 1–3 are assumed to be $M/M/1$ queues with $\mu_1 = 38$, $\mu_2 = 34$, and $\mu_3 = 36$. In addition, the acknowledgment delay is modeled by an infinite server with $\mu_{IS} = 155$. This is a simple network, and we do not have to solve the traffic equation. It is sufficient to let $v_1 = v_2 = v_3 = v_{IS} = 1$, since all jobs can only take one route. The result for $N = 3$ is shown in Table 8.3. Note that $L_1(s) + L_2(s) + L_3(s) + L_{IS} = s$ for each s, and that all the throughputs $\lambda_i(s) = L_i(s)/T_i(s)$ are equal. The round-trip time is $T_{RTT} = \sum_{i=1}^3 T_i = 0.14$ and the weighted average delay $\bar{T} = \sum_{i=1}^3 \lambda_i T_i / \sum_{i=1}^3 \lambda_i = 0.036$.

Example 8.4.2. Consider again the closed network in Example 7.2.1 with the routing matrix given by (8.8). The solution to the traffic equation is given by (8.9). Suppose we wish to compare the throughput of this closed network with its open equivalent in Example 7.2.1. Applying the mean value analysis, now using the v_i values from the traffic equation, we find the throughputs as listed in Table 8.4 for successive s. For the closed network, the total throughput is $\lambda = 96$ with a weighted average delay of $\bar{T} = 0.073$. For the corresponding open network with similar throughput, the weighted average delay is 0.13—much higher. The reason for this is that the variability in input rate can cause very long delays, which cannot occur in a closed network which has a fixed number of jobs.

Delay of acknowledgment is usually modeled by an infinite server. Such a delay can be included in the mean value analysis. Such a queue has infinitely many servers

Table 8.3 Mean value analysis of three queues in series in Example 8.4.1

s	L_1	T_1	L_2	T_2	L_3	T_3	L_{IS}	T_{IS}
0	0	—	0	—	0	—	0	—
1	0.315	0.026	0.352	0.029	0.333	0.028	0.072	$6.45 \cdot 10^{-3}$
2	0.621	0.035	0.714	0.040	0.665	0.037	0.111	$6.45 \cdot 10^{-3}$
3	0.919	0.043	1.086	0.050	0.996	0.046	0.136	$6.45 \cdot 10^{-3}$

Table 8.4   Mean value analysis of the network in Example 8.4.2

s	λ_{12}	λ_{13}	λ_{24}	λ_{25}	λ_{34}	λ_{46}
0	0	0	0	0	0	0
1	2.73	3.41	7.50	6.82	5.46	4.09
2	4.60	5.74	12.64	11.49	9.19	6.89
3	5.93	7.42	16.31	14.83	11.86	8.90
4	6.92	8.65	19.03	17.30	13.84	10.38
5	7.67	9.58	21.08	19.17	15.33	11.50
6	8.24	10.30	22.67	20.61	16.49	12.36
7	8.70	10.87	23.91	21.74	17.39	13.04
v_{ij}	0.210	0.263	0.578	0.526	0.420	0.315

and so the only difference from the equations used for $M/M/1$ queues is that $T_{IS}(s) = \frac{1}{\mu_{IS}}$ for all s. The equations for queue length (or perhaps better, system size) L and throughput λ remain the same as for an $M/M/1$ queue.

The marginal steady-state probabilities for each queue can also be calculated by recursion. Denoting states by j, and the number of jobs by s, we have

$$p_i(j, s) = \frac{\lambda_i(s)}{\mu_i} p_i(j-1, s-1), \quad (j, s \geq 1), \qquad (8.27)$$

where $p_i(j, s)$ is the marginal probability at queue i of state j in a network with s jobs. The recursion starts with $p_i(0, 0) = 1$. The marginal probabilities are thus computed from higher states to lower, and $p_i(0, s) = 1 - \sum_{j=1}^{s} p_i(j, s)$. The sequence of computation for queue i is therefore $p_i(0, 0)$, $p_i(1, 1)$, $p_i(0, 1)$, $p_i(2, 2)$, $p_i(1, 2)$, $p_i(0, 2), \ldots$

Mean value analysis can also be used to analyze multiple-server queues. In this case, principle (8.23) for an arriving job is no longer valid, since several servers are working simultaneously reducing the length of the queue. The expression for $T_i(s)$ for queue i now becomes

$$T_i(s) = \frac{1}{\mu_i} + \frac{1}{C_i \mu_i} \sum_{j=C_i}^{s-1} (j - C_i + 1) p_i(j, s-1).$$

In case there are $j > C_i$ customers at queue i when a job arrives, it has to wait until $j - C_i + 1$ have been served at rate $C_i \mu_i$ in order to be serviced. This can be simplified to

$$T_i(s) = \frac{1}{C_i \mu_i} \left(C_i + \sum_{j=C_i}^{s-1} j p_i(j, s-1) - (C_i - 1) \sum_{j=C_i}^{s-1} p_i(j, s-1) \right)$$

$$= \frac{1}{C_i \mu_i} \left[C_i + L_i(s-1) - \sum_{j=0}^{C_i-1} j p_i(j, s-1) - (C_i - 1) \right.$$

$$\times \left(1 - \sum_{j=0}^{C_i-1} p_i(j, s-1) \right) \Bigg]$$

$$= \frac{1}{C_i \mu_i} \left(1 + L_i(s-1) + \sum_{j=0}^{C_i-2} (C_i - 1 - j) p_i(j, s-1) \right).$$

Thus, in the multiple-server case, the marginal state probabilities $p_i(j, s-1)$ have to be calculated in any case (or at least, for $j = 0, 1, \ldots, C_i - 2$). These are calculated recursively as

$$p_i(j, s) = \frac{\lambda_i(s)}{\alpha_i(j) \mu_i} p_i(j-1, s-1),$$

where

$$\alpha_i(j) = \begin{cases} j & j \le C_i, \\ C_i & j \ge C_i. \end{cases} \tag{8.28}$$

The mean value analysis procedure for multiple-server queues can now be summarized as:

Algorithm 8.4.2 (Mean Value Analysis [Multiple-Server Queues]). Given a network of m $M/M/1$ queues with service rates μ_i, a routing matrix R, and a fixed number of jobs N.

STEP 0: (initialize) Solve the traffic equation (8.25) for v_i. Let $L_i = N/m(i = 1, 2, \ldots, m)$, $p_i(0, 0) = 1$; $p_i(j, 0) = 0$, $(j \ne 0)$ for all queues i.

STEP $1 - N$: (recursion)
  for $i = 1$ to m **do**
    Compute $T_i(s) = \frac{1}{C_i \mu_i} \left(1 + L_i(s-1) + \sum_{j=0}^{C_i-2} (C_i - 1 - j) p_i(j, s-1) \right)$;
    Compute $L_i(s) = s \frac{v_i T_i(s)}{\sum_j v_j T_j(s)}$, $j = 1, 2, \ldots, m$.
    Compute $\lambda_i(s) = L_i(s)/T_i(s)$;
    Compute $p_i(j, s) = \frac{\lambda_i(s)}{\alpha_i(j) \mu_i} p_i(j-1, s-1)$, $(j = 1, 2, \ldots, s)$. **end**;

Output $L_i(N)$, $T_i(N)$, and $\lambda_i(N)$.

Proofs of validity of the equations for multiple servers can be found in, for example, Gross and Harris (1998).

8.4.1 Approximate mean value analysis

For large networks with a large number of jobs, the convolution method and MVA are computationally hard. In such situations, an approximate mean value analysis procedure can be used. Many different approximations have been suggested. We will present the algorithm proposed by Bard and Schweitzer, which, together with other approximations, can be found in Cremonesi et al. (2002).

Table 8.5   Approximate MVA for Example 8.4.3

Iteration	L_1	T_1	L_2	T_2	L_3	T_3	L_{IS}	T_{IS}
0	0.75	–	0.75	–	0.75	–	0.75	–
1	0.899	0.039	1.005	0.044	0.949	0.042	0.147	$6.45 \cdot 10^{-3}$
2	0.883	0.042	1.030	0.049	0.951	0.045	0.135	$6.45 \cdot 10^{-3}$
3	0.875	0.042	1.039	0.050	0.951	0.045	0.135	$6.45 \cdot 10^{-3}$
4	0.873	0.042	1.042	0.050	0.950	0.045	0.135	$6.45 \cdot 10^{-3}$
5	0.871	0.042	1.044	0.050	0.950	0.045	0.135	$6.45 \cdot 10^{-3}$

We assume that we have a closed network where all queues are $M/M/1$ (that is, single server) or infinite server systems.

In the Schweitzer-Bard approximation, we fix the number of jobs N and construct a system of fixed-point equations that are iterated until convergence is reached. Let

$$L_i(N-1) \approx L_i(N) - \frac{L_i(N)}{N},$$

so that Equation (8.23) becomes

$$T_i(N) = \frac{1 + L_i(N)\left(1 - \frac{1}{N}\right)}{\mu_i}. \tag{8.29}$$

Since N is fixed, we drop the argument and let $L_i(N) = L_i$, $T_i(N) = T_i$, and $\lambda_i(N) = \lambda_i$. Initially, we solve the traffic equation (8.1) for v_i. For initial values of L_i, we can set $L_i = N/m$, where m is the number of queues. Now Equations (8.29), (8.26), and (8.24) are iterated until the fixed point is reached. Notice that the approximation has the property that $\sum_i L_i = N$ throughout the iterations. The approximation is believed to be asymptotically exact as N grows to infinity.

Example 8.4.3.   If the problem in Example 8.4.1 is solved by approximate mean value analysis, we get the result in Table 8.5. After the fifth iteration, the values do not change with respect to the number of significant digits. The approximation gives a reasonable result even for this small value of N.

8.5 Closed network approximations

Most design methods for queueing networks are formulated for open networks, which are more tractable. It is therefore of interest to understand how well a closed network can be approximated by an open network. In general, a closed network has 'better performance' when some of the defining parameters or performance metrics are of the same order. This means that if we design an open network and 'make it closed,' the resulting closed network should not perform worse than the open network, using the same design. It turns out that open models can be used to approximate both mixed and closed networks.

We will compare two strategies of approximating a closed network with an open network. One is suggested in Lam and Lien (1982) and the other in Whitt (1984). The difference is by what metric the open network is approximated to the closed network. In the former model, the arrival rates λ_i are adjusted to fit those in the closed network, whereas in the latter, the number of jobs in the network (the sum of the queue lengths) is adjusted to correspond.

8.5.1 The equal throughput method

In the first approximation method, here called the equal throughput method and proposed by Lam and Lien (1982), the traffic equations of an open network (7.2) are solved for λ_i and then a scaling factor w is introduced to adjust the arrival rates so that they correspond to that of the closed network. It is therefore necessary to create a closed queueing network (this can be, for example, a virtual channel or a subnetwork) to determine the throughputs (and mean delays) using, for example, mean value analysis.

By constructing such an open queueing network approximation, methods described in Chapter 7 can be used for optimizing flow assignment and resource allocation. After such optimization methods have been applied to the open network approximation, the results should be verified in the closed network model, so that the two models are used somewhat in parallel.

A description of the method goes as follows. Suppose we want to design a flow-controlled network and decide its window size. We begin by creating an open network approximation. The closed network is then analyzed by, for example, mean value analysis, to determine the window size so that delay requirements are fulfilled. The analysis of the closed network will most likely have different throughput rates, so the throughputs in the open model (there, they are equal to the arrival rates) are adjusted to correspond to the closed network by scaling them down until they match. The open model can now be analyzed or optimized further, and whenever there is a change in the open model that carries over to the closed model, the latter is again analyzed to adjust the performance values.

The open network is a poorer approximation when the load is high than at low-to-moderate load. This should be expected since the flow-control limits the effects of heavy traffic on the network. In Lam and Lien (1982), it was found that the accuracy of the open model improved with increasing window size N (number of jobs).

The delay estimates in the open network model are usually higher than in the closed network. This is (most likely) an effect of the number of arriving jobs not being limited in the open network: the delay increases rapidly with the load, so that the delay distribution in an open network suffers from long tails. This effect is not so pronounced in the closed network, since the number of jobs is limited. This also implies that the open model gives somewhat pessimistic estimates for the closed network.

8.5.2 The fixed-population mean method

In another approach, proposed by Whitt (1984), called the fixed-population mean method, one node in the network is identified as an exit-entry node, through which jobs can enter and leave the network. The approximating network is thereby open, and the exit-entry node can be supplied by a Poisson process with a certain rate, so that the expected number of jobs in the closed network and the open approximation are the same.

For product-form closed networks, any node can serve as exit-entry node, and the equilibrium distribution of the corresponding open model is independent of this choice. The throughput of the approximating model is found by solving the traffic equation for an open network (7.2), with R suitably modified.

In the approximating network thus obtained, the ratio between the throughputs to the throughputs in the closed network for each queue is constant. By trial and error, it is straightforward to find the arrival rate γ_0 which yields approximately the same number of expected jobs in the open network as in the closed network. When this matching has been performed, the approximating network can be used for analysis and simulation.

Note that the arrival rates λ_i in the closed network are determined up to a multiplicative constant, whereas the rates calculated for the open network are determined exactly, relative to the external arrival rate γ_0.

An additional measure of the closeness of the two models can be obtained by considering the population variability. Whitt (1984) suggests taking estimates of the population variance into account when selecting the most suitable model. The measure proposed is the squared coefficient of variation, which is the variance divided by the square of the mean. Since we assume $M/M/1$ queues throughout the network, the number of jobs at each queue has a geometric distribution, and so

$$\mathbf{E}(N_i^{(o)}) = \rho_i/(1 - \rho_i),$$

$$\mathrm{Var}(N_i^{(o)}) = \rho_i/(1 - \rho_i)^2,$$

$$c^2(N_i^{(o)}) = 1/\rho_i,$$

where $\rho_i = \lambda_i/C_i$ is the traffic intensity at queue i with capacity C_i. The superscripts (o) and (c) are used to distinguish between values on the open and the closed network. The population variability of the entire network is

$$\mathbf{E}(N^{(o)}) = \mathbf{E}(N_1^{(o)}) + \cdots + \mathbf{E}(N_m^{(o)}), \tag{8.30}$$

$$\mathrm{Var}(N^{(o)}) = \mathrm{Var}(N_1^{(o)}) + \cdots + \mathrm{Var}(N_m^{(o)}),$$
$$c^2(N^{(o)}) = \mathrm{Var}(N^{(o)})/[\mathbf{E}(N^{(o)})]^2.$$

In the closed network, the population variability is zero, so finding an open model with minimum variability might suggest a suitable model unless the expected number of jobs N in the open network differs too much from that in the closed network.

In networks with few nodes, the difference in variability is usually rather large, but it tends to decrease with increasing number of nodes. To see this, let all $\rho_i = \rho$ be equal for all i. Then, $c^2(N^{(o)}) = 1/m\rho$, which suggests that the variability decreases as m increases. Thus, using the population variability as a goodness-of-fit measure is more applicable to large networks than small ones. The expression also indicates that ρ not should be too small, that is, the queues should not be too lightly loaded. The population variability can also be used to judge whether an open or a closed model is more appropriate for a problem.

An interesting application of the method is that it can be used to approximate an open network with finite buffers. By letting the network population for the closed model N represent the buffer size in the open model, each queue in the open model can have at most N jobs.

To determine the queue lengths L_i in a network with limited buffers, we create a model corresponding to a closed network with N jobs, and assume that the product form of the open system holds. Then, we modify the queue length distribution by truncating the distribution at N jobs and renormalize to get the distribution of a finite buffer queue.

Denote by $\bar{N}_i$ the equilibrium number of jobs at queue i in the open model. For an $M/M/1$ queue, the mean is $\mathbf{E}(N_i) = \rho_i/(1 - \rho_i)$ and the utilization is $U_i = \mathbf{P}(N_i > 0) = \rho_i$, where $\rho_i = \lambda_i/\mu_i$ is the traffic intensity at queue i. Then, using the well-known formulas for the truncated queue $M/M/1/N$, where N is the number of jobs in the network and letting $w(\rho_i, N)$ be the normalization constant,

$$\mathbf{E}(\bar{N}_i) = w(\rho_i, N)\mathbf{E}(N_i),$$

$$w(\rho_i, N) = \left(1 - (N + 1)\rho_i^N + N\rho_i^{N+1}\right) / \left(1 - \rho_i^{N+1}\right),$$

$$\bar{U}_i = \mathbf{P}\left(\bar{N}_i > 0\right) = \left(\rho_i - \rho_i^{N+1}\right) / \left(1 - \rho_i^{N+1}\right),$$

provided that $\rho_i \neq 1$. Since $w(\rho_i, N) \leq 1$, we have $\mathbf{E}(\bar{N}_i) < \mathbf{E}(N_i)$ and $\bar{U}_i < U_i$. We can also adjust γ_0 so that $\mathbf{E}(\bar{N}_i) = \mathbf{E}(N_i)$. Then $\bar{U}_i > U_i$.

Example 8.5.1. Let the closed network in Example 8.4.1 consist of $M/M/1$ with the same traffic matrix. By using MVA, we have the exact values for the performance metrics of the network.

In the first model, we adjust the arrival rates given by solving the traffic equations (7.2) and find a suitable scaling parameter so that the throughputs in the two models agree. The scaling parameter was found to be $w = 0.958$, and the results are shown in Table 8.6, in the second section. The exact figures are included in the top of the table for comparison.

To find the second approximating open network, we let the first node connect to the outside. The open model then has the arrival rate (and throughput)

$$\gamma = \begin{pmatrix} \gamma_0 \\ 0 \\ \vdots \\ 0 \end{pmatrix},$$

Table 8.6 Comparison of a closed network and the approximating open networks

Parameter	Link 1	Link 2	Link 3	Link 4	Link 5	Link 6
Queue length	0.328	0.439	1.518	1.275	0.874	0.566
Delay	0.043	0.046	0.072	0.067	0.057	0.049
Throughput	7.67	9.58	21.08	19.17	15.33	11.50
Queue length	0.343	0.469	2.364	1.769	1.045	0.622
Delay	0.089	0.049	0.112	0.092	0.068	0.054
Throughput	7.67	9.58	21.08	19.16	15.33	11.50
Queue length	0.293	0.396	1.658	1.309	0.831	0.516
Delay	0.043	0.047	0.089	0.077	0.061	0.051
Throughput	6.79	8.51	18.71	17.00	13.61	10.21

where γ_0 is to be determined so that $\sum_i L_i = 5$, as in Example 8.4.1. As routing matrix, the matrix (8.8) from Example 8.2.1 is used, where the first row is modified. Since we let traffic out of the system, the element in $(1, 1)$ is set to zero. Thus, we have

$$
R = \begin{pmatrix}
0 & 0 & 0 & 1/4 & 0 & 0 & 0 \\
0 & 4/5 & 0 & 0 & 1/5 & 0 & 0 \\
0 & 0 & 1/2 & 5/22 & 3/22 & 3/22 & 0 \\
1/10 & 0 & 1/4 & 13/20 & 0 & 0 & 0 \\
0 & 1/8 & 3/16 & 0 & 9/16 & 1/8 & 0 \\
0 & 0 & 1/4 & 0 & 1/6 & 7/12 & 0
\end{pmatrix}.
\tag{8.31}
$$

Solving the traffic equation with (8.31) and $\gamma_0 = 0.849 \cdot 6 = 5.094$ gives the system parameters in the third section of Table 8.6. Note that we are not restricted to the irreducible Markov chain as in the closed network case. The last link is actually not contributing to the final result—the queue length is zero since it is unaffected by the external traffic inserted into link 1. Since the closed network solution does not include link 7, this is omitted altogether in Table 8.6.

The mean squared errors $\sum_i (T_i^{(o)} - T_i^{(c)})^2/m$ of the delays for the first and the second models give $7.49 \cdot 10^{-4}$ for the first model and $6.83 \cdot 10^{-5}$ for the second.

For the latter model, the population variability was found to be 0.43. The network is rather small and the traffic rather light, but the model can possibly be accepted as a rough model of the closed network.

8.5.3 Virtual channels

It is quite straightforward to analyze a virtual channel, avoiding analysis of an entire network, which can be a daunting task. By making the assumption that traffic is unidirectional (that is, traffic is flowing in only one direction) and compensating for that by simply doubling the offered traffic, the performance analysis of the virtual channel is simplified further. Figure 8.2 shows a virtual channel consisting of three queues and an infinite server modeling the feedback of permits.

When analyzing the performance of a virtual channel in isolation it is useful to take the longest chain and consider the traffic that is carried on this chain. A network can therefore be decomposed into such chains. The reason for this is that we would like to know the end-to-end performance metrics, which are poorer for long chains, since other streams are using the same resources. We then determine the window size N for such chains in the network.

The queues in the virtual channel are usually carrying traffic between other source-destination pairs than that of the virtual channel, which should be handled in some way. The subnetwork carrying such traffic not using the virtual channel, is typically modeled as an open queueing network. In effect, then, the model is a mixed queueing network.

We refer to traffic streams between different origin-destination pairs as chains, and call such a chain open if it does not belong to the virtual channel, and closed otherwise. Since traffic from open chains is carried by the queues in the virtual channel, the load in each queue must first be determined and the capacity of the queue be reduced by this amount. Then, we can create a closed queueing model and analyze it with the reduced queue capacities.

8.6 Decomposition

There is an interesting analogy between closed queueing networks and Norton's theorem for electrical networks. Suppose a certain queue or group of queues are of interest for analysis. The principle is to reduce the rest of the network into a single flow-equivalent server (FES) in order to facilitate the analysis.

The method, suggested by Chandy et al. (1975), can be shown to be exact for closed networks with $M/M/1$ queues. It can also be used to approximate networks with non-exponential service distributions and multiple job classes, but we will only consider networks of $M/M/1$ queues with a single class of jobs.

8.6.1 Norton's theorem

An outline of the method is as follows. First, a queue or group of queues are identified for analysis. These queues are called the designated network. These queues are 'removed' from the network by short circuiting them. The remaining nodes, called the aggregate, are replaced by a single flow-equivalent server. The properties of this server can be found by mean value analysis. In the aggregate, the service rates equal the throughputs in the short circuited network, called the local balance interface. This leads, in general, to state-dependent service rates at the FES, which makes the analysis of the decomposed network a little trickier. To analyze such a network, a modified version of mean value analysis can be used. Note that the service rates at the FES therefore depend both on the total number of jobs in the network and the number of jobs in the aggregate.

Example 8.6.1.   Consider the network depicted in Figure 8.3, and suppose that we wish to analyze queue 4 in the network with $N = 5$ jobs. First, we short circuit queue 4, as shown in Figure 8.4, to obtain the subnetwork that is to be aggregated. This is equivalent to setting the service rate to zero. Solving for the flow-equivalent server using MVA, we first solve the traffic equation for the short circuited network to get $v = (1, 0.5, 1)$. The service rate of the FES equals the throughput from the aggregate, $\lambda_3/2$ for $N = 5$. Thus, $\mu(s) = (0, 0.141, 0.205, 0.239, 0.259, 0.272)$, corresponding to the states of 0–5 jobs in the aggregate.

Aggregation gives exact results for queueing networks which have product-form solutions. It may be more computationally expensive than mean value analysis on small network problems. The method is useful when the effect of a single or small number of queues is to be studied in a larger network. Another advantage of decomposition is the possibility of approximately analyzing queueing networks with no product form solution. It is then common to put the queues which do not have a product form probability distribution into the designated network and aggregate queues having a product form solution. The decomposed network can then be analyzed by approximate methods (see Chandy et al. 1975) or by simulation. When simulating large networks, decomposition can reduce simulation time significantly (see Figure 8.5).

The decomposition method consists of two steps: construction of the flow-equivalent server, and solving the resulting reduced network. This network cannot

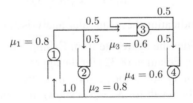

FIGURE 8.3

Network example.

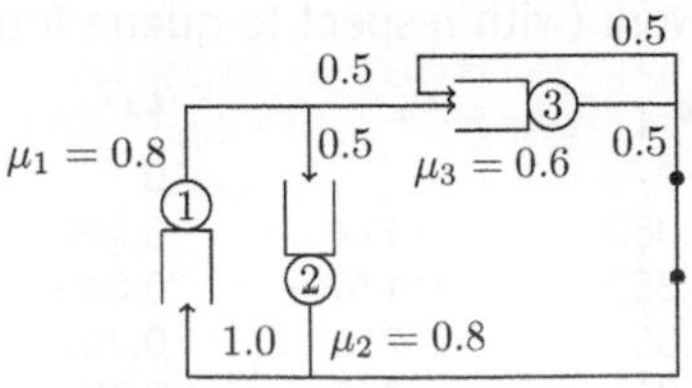

FIGURE 8.4

Network example with a short circuit.

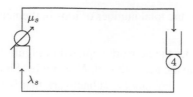

FIGURE 8.5

The equivalent network with state-dependent service rates.

be analyzed using the mean value analysis discussed in Section 8.4. The reason is that the flow-equivalent server has state-dependent service rates, where each state is an integer, but the actual mean queue lengths are real numbers. Two ways of tackling this are by linear interpolation and a modified MVA method called conditional mean value analysis.

8.6.2 Linear interpolation

A simple approximate method is to interpolate the monotonically increasing service rate values. Given the service rates $\mu = (0, \mu_1, \ldots, \mu_N)$, a linear approximation can be formed as $\mu(x) \approx \mu_{\lfloor x \rfloor} + (x - \lfloor x \rfloor)(\mu_{\lceil x \rceil} - \mu_{\lfloor x \rfloor})$, where $\lfloor x \rfloor$ is the largest integer smaller than x and $\lceil x \rceil$ is the smallest integer larger than x. This function then replaces the constant μ_i in Equation (8.23) for the aggregate, so that

$$T_a(s) = \frac{1 + L_a(s - 1)}{\mu_a(1 + L_a(s - 1))}, \tag{8.32}$$

where the subscript a is indicating the mean value analysis parameters for the aggregate.

Example 8.6.2.   Consider the aggregate from Example 8.6.1 with the service rates computed there. By using Equation (8.32) for the aggregate, mean value analysis can be applied to approximately solve for performance metrics of the two-component decomposed network. The result for $N = 5$ is shown in Table 8.7.

Table 8.7 Example of mean value analysis adapted by linear interpolation for state-dependent service rates. Parameters with subscript c denote values for the complementary network (with respect to queue four)

n	L_c	T_c	λ_c	L_4	T_4	λ_4
0	0	–	–	0	–	–
1	0.810	7.083	0.114	0.190	1.667	0.114
2	1.651	9.382	0.176	0.349	1.984	0.176
3	2.515	11.66	0.216	0.485	2.248	0.216
4	3.402	14.09	0.242	0.598	2.475	0.242
5	4.311	16.66	0.259	0.689	2.663	0.259

Note, that this is an approximation due to the linear interpolation of the service rate. A modification of mean value analysis can be used to solve such problems exactly, which is discussed next.

8.6.3 Conditional mean value analysis

The mean value analysis can be adapted to cope with the load dependent service rates that result from decomposition. We consider a closed network in product form with N jobs (of a single class) and m queues. The first $m - 1$ queues have service rates of constant mean, and we use the variable $D_i = 1/\mu_i$ for $1 \le i \le m - 1$ to denote the service intensity. The remaining queue, m, is a flow-equivalent server with $D_m = 1$ and load dependent service rates $\boldsymbol{\mu}$, that is represented as a vector of service rates. The service intensity is the product between the mean service time per job and the arrival rate of jobs.

We introduce an auxiliary parameter, the rate shift t, that is used to select a feasible service rate from $\boldsymbol{\mu}$. Thus, if $t = j$, the flow-equivalent server operates with the rate $\mu(j + s_m - 1)$ when there are s_m jobs in the queue. For $t = 1$ the rate is simply $\mu(s_m)$. Just as with the original mean value analysis, we start with $n = 0$, successively increase s until $s = N$ and using relations between queue length, delay, and throughput adapted to the load dependent service rate of the FES.

Let $\mathbf{s} = (s_0, s_1, \ldots, s_m)$ be an allocation of jobs across the network, where $\sum_{i=0}^{m} s_i = N$ and $s_i \ge 0$. The equilibrium state probabilities are then given by

$$\mathbf{P}(\mathbf{s}; N, t) = \frac{\prod_{i=0}^{m} D_i^{s_i}}{G(N, t) \prod_{s=1}^{s_m} \mu(t + s - 1)}, \tag{8.33}$$

where $G(N, t)$ is the normalizing constant making the state probabilities sum to unity.

The marginal probabilities of having $j \ge 1$ jobs in queue i are given by

$$\mathbf{P}(s_i = j; N, t) = \begin{cases} D_i \lambda(N, t) \mathbf{P}(s_i = j - 1; N - 1, t), & 1 \le i \le m - 1, \\ \frac{D_m}{\mu(j+t-1)} \lambda(N, t) \mathbf{P}(s_i = j - 1; N - 1, t), & i = m, \end{cases} \tag{8.34}$$

in which $\lambda(N, t) = G(N - 1, t)/G(N, t)$ denotes the throughput as measured at queue m assumed as reference for the number of job completions. The probability that queue i is idle is

$$\mathbf{P}(s_i = 0; N, t) = G_{-i}(N, t)/G(N, t) = 1 - \sum_{j=1}^{N} \mathbf{P}(s_i = j; N, t) = 1 - U_i(N, t), \tag{8.35}$$

where $U_i(N, t) = \sum_{j=1}^{N} \mathbf{P}(s_i = j; N, t), 0 \le i \le m$, is the utilization of queue i and $G_{-i}(N, t)$ is the normalizing constant of a model with queue i removed.

The delay T_i at queue i is

$$T_i(N, t) = \begin{cases} D_0 & i = 0, \\ D_i(1 + L_i(N - 1, t)) & 1 \le i \le m - 1, \\ D_m(N, t)(1 + L_m(N - 1, t + 1)) & i = m, \end{cases} \tag{8.36}$$

where $D_m(N, t) = U_m(N, t)/\lambda(N, t)$ is the mean service demand of the flow-equivalent server. This is equal to $D_m(1, t) = D_m/\mu(t)$ for $N = 1$ and for larger population is computed recursively as

$$D_m(N, t) = \frac{\lambda(N - 1, t)}{\lambda(N - 1, t + 1)} D_m(N - 1, t), \quad N \geq 2. \tag{8.37}$$

We can now summarize this procedure in an algorithm.

Algorithm 8.6.1 (Conditional Mean Value Analysis).
Input: Service demands D_i, $0 \leq i \leq m$, population N,
flow-equivalent server rates $\mu(k)$, $1 \leq k \leq N$.
Let $L_i(0, t) = 0$ for $t = 1, 2, \ldots, N + 1$.
  for $n = 1$ **to** N
    for $t = 1$ **to** $N - n + 1$
    Compute $D_m(n, t)$ using (8.37) if $n \geq 2$,
    otherwise set $D_M(1, t) = D_m/\mu(t)$ if $n = 1$.
    Compute $T_i(n, t)$ using (8.36), for $i = 0, 1, \ldots, m$.
    Compute by Little's formula the throughput $\lambda(n, t) = N / \sum_{i=0}^{m} T_i(n, t)$.
    Compute $L_i(n, t) = \begin{cases} D_i \lambda(n, t)(1 + L_i(n - 1, t)) & 1 \leq i \leq m - 1, \\ D_m(n, t)\lambda(n, t)(1 + L_m(n - 1, t + 1)) & i = m, \end{cases}$
  end
  end
Output: Throughput $\lambda(N) = \lambda(N, t = 1)$ and mean queue lengths $L_i(N) = L_i(N, t = 1)$, $i = 1, 2, \ldots m$.

A delay server can also be included in the procedure without any difficulty, but we will not consider that here.

Example 8.6.3. We can now solve the problem in Example 8.6.1 exactly. Following the procedure in Algorithm 8.6.1 with $\boldsymbol{\mu} = (0.141, 0.205, 0.239, 0.259, 0.272)$ gives the values shown in Table 8.8. Initial values are $D_a = 1$ and $D_4 = 1/\mu_4 = 1.667$. The final result for queue 4 is $L_4 = 0.691$, $T_4 = 2.669$, and $\lambda_4 = 0.259$.

8.7 TCP controlled networks

As opposed to permit-based virtual channels, a TCP controlled connection dynamically adjusts its window size (that is, the external input rate) to the prevailing network conditions. The number of jobs in a TCP controlled network may therefore vary in time.

Due to complex interdependencies, adaptive controls (routing or, as in this case, flow control) are notoriously difficult to analyze analytically. One way of solving this, however, is by using Monte Carlo simulation and formulating a set of fixed-point equations as described in Gibbens et al., 2000.

Table 8.8 Example of conditional mean value analysis for state-dependent service rates

s	t	$D(s,t)$	L_a	T_a	λ_a	L_4	T_4	λ_4
0	1	–	0	–	–	0	–	–
	2	–	0	–	–	0	–	–
	3	–	0	–	–	0	–	–
	4	–	0	–	–	0	–	–
	5	–	0	–	–	0	–	–
	6	–	0	–	–	0	–	–
1	1	7.083	0.810	7.083	0.114	0.190	1.667	0.114
	2	4.878	0.745	4.878	0.153	0.255	1.667	0.153
	3	4.181	0.715	4.181	0.171	0.285	1.667	0.171
	4	3.857	0.698	3.857	0.181	0.302	1.667	0.181
	5	3.678	0.688	3.678	0.187	0.312	1.667	0.187
2	1	5.298	1.647	9.246	0.178	0.353	1.984	0.178
	2	4.359	1.563	7.475	0.209	0.437	2.091	0.209
	3	3.950	1.516	6.708	0.226	0.484	2.142	0.226
	4	3.732	1.488	6.300	0.236	0.512	2.170	0.236
3	1	4.513	2.510	11.565	0.217	0.490	2.256	0.217
	2	4.032	2.427	10.144	0.239	0.573	2.395	0.239
	3	3.780	2.375	9.405	0.253	0.625	2.473	0.253
4	1	4.094	3.399	14.031	0.242	0.601	2.483	0.242
	2	3.820	3.324	12.892	0.258	0.676	2.622	0.258
5	1	3.846	4.309	16.631	0.259	0.691	2.669	0.259

The main idea is to use fixed-point equations to determine network performance parameters, such as packet loss and end-to-end delay, calculate the TCP window size, and finally determine the adjusted input rate by Monte Carlo simulation.

To characterize the system to be modeled, we formulate some quantities of interest. Let the network support two traffic classes, having high and low priority, respectively. Such traffic classes could be implemented on the DiffServ classes assured forwarding (AF) and best effort (BE), for example. These services are indexed by 1 and 2 in the respective variables.

Consider a network with m resources. Let r be a direct route in the network and the offered traffic rates of the high and low priority classes be λ_{1r} and λ_{2r} on this route. Let the loss probabilities be B_{1i} and B_{2i}, where i is any resource $i = 1, 2, \ldots, m$. Assuming independence, the reduced load at resource i is

$$v_{1i} = \sum_{i:i\in r} \lambda_{1r} \prod_{i'\in r: p(i',r)<p(i,r)} (1 - B_{1i'}), \tag{8.38}$$

$$v_{2i} = \sum_{i:i\in r} \lambda_{2r} \prod_{i'\in r: p(i',r)<p(i,r)} (1 - B_{2i'}), \tag{8.39}$$

where $p(i, r)$ gives the position of resource i on route r. The losses, in turn, are usually dependent on the (reduced) loads via some functions,

$$B_{1j} = B_{1j}(v_{1j}, v_{2j}), \tag{8.40}$$

$$B_{2j} = B_{2j}(\nu_{1j}, \nu_{2j}). \tag{8.41}$$

8.7.1 Performance metrics

We assume that the queue is an $M/M/1/b$, with capacity C and a finite buffer b. The prioritization in the queue is of trunk reservation type. Let t be a number of jobs (or packets) such that, if there are t or more jobs already in the queue, a low priority job is rejected and lost. This mechanism reserves a portion of the buffer for high priority jobs. Note, that there is a slight difference between the trunk reservation of Chapter 6; the trunk reservation parameter in the present case represents the number of jobs present in the buffer, as compared to the number of free channels in the loss network case.

Let s be the job occupancy of the buffer, and let the arrival rates of high and low priority jobs be λ_1 and λ_2, respectively. If the queue has the capacity of C jobs per time unit, the state probabilities of the queue follow a Markov chain with transition rates

$$
\begin{aligned}
q(s, s+1) &= (\nu_1 + \nu_2 : 0 \leq s < t; \nu_1 : t \leq s < b), \\
q(s, s-1) &= (C : 1 \leq s \leq b)
\end{aligned}
\tag{8.42}
$$

that is, $q(s, s+1)$ is the transition rate from state s to $s+1$, dependent on t, and $q(s, s-1)$ is the transition rate from state s to $s-1$, dependent only on C. The equilibrium distribution for state s is

$$\pi(s) = \pi(0) \prod_{j=1}^{s} \frac{q(j-1, j)}{q(j, j-1)}, \tag{8.43}$$

where

$$\pi(0) = \left(1 + \sum_{s=1}^{b} \prod_{j=1}^{s} \frac{q(j-1, j)}{q(j, j-1)}\right)^{-1}. \tag{8.44}$$

This gives two different loss probabilities for the high and low priority jobs;

$$B_1 = B_1(\nu_1, \nu_2) = \pi_b, \tag{8.45}$$

$$B_2 = B_2(\nu_1, \nu_2) = \sum_{s=t}^{b} \pi_s. \tag{8.46}$$

The mean delay of each traffic type admitted into the queue (when there are n jobs in the system) is given by

$$\mathbf{E}(T_1(\nu_1, \nu_2)) = \frac{\sum_{j=0}^{b-1}(1+j)\pi(j)}{C \sum_{j=0}^{b-1} \pi(j)} \tag{8.47}$$

$$\mathbf{E}(T_2(\nu_1, \nu_2)) = \frac{\sum_{j=0}^{t-1}(1+j)\pi(j)}{C \sum_{j=0}^{t-1} \pi(j)}, \tag{8.48}$$

where the expressions are normalized to include only non-blocking states.

A consequence of the prioritization mechanism is that low priority jobs are less likely to be admitted into the queue, and since they are admitted at a buffer occupancy $t < b - 1$ they experience shorter queue lengths than high priority jobs on average. Therefore, low priority jobs experience higher packet loss but lower delays than high priority jobs. Thus,

$$B_2(\nu_1, \nu_2) \geq B_1(\nu_1, \nu_2)$$

and

$$\mathbf{E}(T_1(\nu_1, \nu_2)) \geq \mathbf{E}(T_2(\nu_1, \nu_2)).$$

8.7.2 The TCP model

The TCP dynamically controls the rate at which traffic is admitted to the network. Therefore, a model of TCP is required, which in turn depends on the packet loss probabilities B_{jr} and the round-trip time T_{jr}, where $j = 1, 2$, indicates the traffic type and r is an outgoing route.

There are several models of TCP that can be used, depending on the level of detail one wishes to incorporate. The more elaborate model, the more parameters need to be estimated. In the present context, the parsimonious model described in Mathis et al. (1997), can be expected to be sufficiently accurate,

$$S(B_{jr}, T_{jr}) = \frac{1}{T_{jr}\sqrt{2B_{jr}/3}}, \tag{8.49}$$

In practice, it is convenient to modify this equation to avoid division by zero when no packet loss occurs. It is also convenient to limit the window size to the link capacity, since the window size can be very large when b_{jr} and T_{jr} attain low values. We then may use the model

$$S(B_{jr}, T_{jr}) = \min\left\{\frac{1}{T_{jr}\sqrt{2(B_{jr} + \epsilon)/3}}, C\right\}, \tag{8.50}$$

where $\epsilon > 0$ is a small constant.

In order to estimate packet loss rates and delays on which $S(\cdot)$ and therefore the offered traffic rates λ depend, we use Monte Carlo simulation. In the absence of TCP, we can assume that traffic arrives according to a Poisson process, and use the arrival intensity (which is just the mean rate). In a TCP controlled connection, however, the instantaneous arrivals affect the loss and delay, on which TCP recomputes the window size $S(\cdot)$, which in turn affects the offered load.

Omitting any subscripts for the sake of clarity, let γ be the arrival intensity of a Poisson process, and s the instantaneous number of arriving jobs, sampled from this Poisson distribution. Then, the instantaneous offered traffic is $\lambda = sS(B, T)$. Now, the loss rate B can be calculated from (8.46) for high and low priority traffic, and the delay T from (8.47) and (8.48), respectively.

The model (8.50) is nonlinear and depends on some 'worst-case' performance during a transmission, rather than the mean loss and delay values. To estimate blocking and delay, we can sample $\lambda = s\,S(B, T)$ and calculate B and T a large number of times, and take as $B_{\max}$ and $T_{\max}$ the 95 percentile of the sampled B- and T-values, for example. Thus, let $B_{\max} = B_{95}$ and $T_{\max} = T_{95}$ where B_{95} and T_{95} are the 95 percentiles, and recalculate the window size $S(B_{95}, T_{95})$.

The actual window sizes are increased gradually, so it is not very realistic to use the so-obtained window sizes as calculated directly from (8.50). This usually leads to violent oscillations which prevent convergence of the iterations. Instead, we linearize the new window size by damping. If $0 < \alpha < 1$ is a damping parameter, the new window size is simply

$$S_k^{(d)} = \alpha S_{k-1}^{(d)} + (1 - \alpha) S_k, \tag{8.51}$$

where k is the sequence of iterations and the superscript (d) denotes the damped window size. There are other possible ways of damping the window size, but numerical experiments show that linear damping works quite well. The damping parameter α should not be too small in order to suppress oscillations, and not too large either, which would lead to unnecessary slow convergence. A damping parameter of $0.6 \le \alpha \le 0.9$ seems appropriate from numerical experiments. The described procedure can now be iterated until convergence is reached.

Example 8.7.1. Consider a single TCP controlled $M/M/1/b$ queue with capacity $C = 155$ and buffer size $b = 10$. There are two traffic streams, a high priority traffic class with Poisson arrival rate $\gamma_1 = 10$ and a low priority with arrival rate $\gamma_2 = 20$. The job lengths are exponentially distributed with means $\mu_1 = \mu_2 = 1$. Further, the parameter $t = 8$, so that low priority traffic is admitted only when the buffer contains less than t jobs.

By using Monte Carlo simulation with 1000 samples, the 95 percentiles of B_j and T_j, $j = 1, 2$, are used as performance metrics and arguments to the window size function $S(\cdot)$. The damping parameter α is set to 0.8 and the simulation is performed 100 times. The initial values of blocking and delay are $B_1 = B_2 = 0$ and $T_1 = T_2 = 1/C$, respectively.

The results of this simulation are shown in Figure 8.6, showing the packet loss, and the delay in Figure 8.7. The iterations actually converge rather fast; after only about 15 iterations the values converge. Note, that the packet loss is lower for the high priority traffic, but the delay is higher.

8.7.3 Fixed-point equations

The method described can be used to analyze networks as well. The main difference from the single queue in Example 8.7.1 is that the end-to-end performance metrics determine the window size. Assuming a high and a low priority traffic class, the arrival rates offered to each link, given by (8.38)–(8.39), should be determined in each simulation step, where the packet loss rates and delays are determined for each queue as described above.

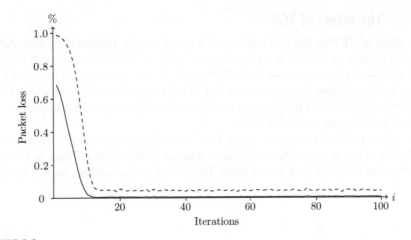

FIGURE 8.6

Convergence of the packet loss probability of the fixed-point method for TCP. Values for high priority are shown with a solid line and for low priority traffic with a dashed line.

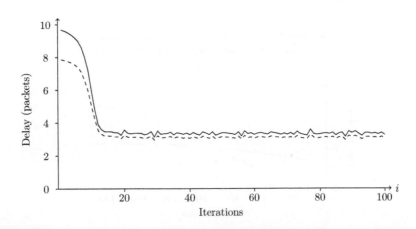

FIGURE 8.7

Convergence of the packet delay of the fixed-point method for TCP. Values for high priority are shown with a solid line and for low priority traffic with a dashed line.

Another complication is that each origin-destination pair may take different routes, and so these loss and delay estimates should be done for alternative routes as well. Even if only a primary route (for example, a shortest path) and an alternative route are specified for each origin-destination pair, the computational effort required to simulate a TCP controlled network increases rapidly with the network size.

8.7.4 The effect of TCP

The effect of TCP on the performance of a queue can be illustrated by calculating the exact queue performance metrics, as in Chapter 6, and compared with a queue with the same parameters, but with TCP active. Figure 8.8 shows the packet loss, delay, and throughput for a queue with finite buffer for an increasing number of jobs (or sessions). The solid lines represent values of the TCP controlled queue and the dashed lines the queue without TCP.

For low arrival rates, traffic to the TCP controlled queue will experience some packet loss even when traffic to the queue without TCP does not. There is a similar situation for delay for low arrival rates. This is due to the fact that packet loss and

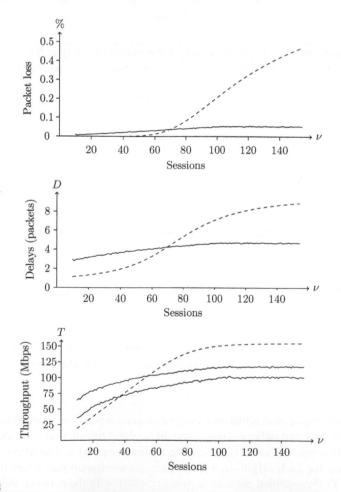

FIGURE 8.8

Comparison between TCP controlled and non-controlled queues.

delay figures used to determine the TCP window size are 'worst-case' rather than mean values. Furthermore, the TCP window size tends to be large for low arrival rates, increasingly allowing traffic as long as the connection does not experience any deterioration.

As the arrival rate increases, however, both packet loss and delay remain almost constant, as opposed to the performance of the queue without TCP. This is the consequence of congestion control that TCP provides. The throughputs of the two queues are shown in the last plot. The two lines of the TCP controlled queue represent the 5 percentile and the median of the throughput. Although, clearly lower than for the queue without TCP, the TCP throughput is somewhat pessimistic.

8.8 Summary

Flow-controlled networks are extensions to the general networks discussed previously. In contrast to general networks, flow-controlled networks are usually modeled by closed Jackson network (Gordon-Newell networks), which require more sophisticated analysis methods than the uncontrolled ones. The chapter also introduces the important concept of decomposition and a fixed-point method for TCP controlled networks.

Some general references on flow-controlled networks are Bertsekas and Gallager (1987) and Gross and Harris (1998). Mean value analysis is discussed in Gross and Harris (1998) and Cremonesi et al. (2002), for example. One of the first discussions on queueing network decomposition was published by Chandy et al. (1975).

Effective Bandwidth

This chapter and the two next discuss networks carrying services of different types, sharing existing network resources, and referred to as multi-service networks. The services typically have different characteristics in terms of requirements on the network so that the service has an acceptable quality. These requirements are collectively known as Quality of Service (QoS) parameters. The services used as representative are voice telephony, streaming video services, and data (for example, file transfer) services. The exact analysis of networks with high capacity, a multitude of services, and various control mechanisms in place is inherently difficult. Some results from queueing theory for general queues can be used. Another approach is the theory of large deviations. The large deviation theory is concerned with events that happen seldom, such as—supposedly—overflow in a buffer.

It appears that, if we allow traffic to have a small but non-zero probability of being lost, the system efficiency can be greatly improved. This effect is referred to as statistical multiplexing. By using large deviation techniques, many important properties of queues serving an inhomogeneous stream of services can be derived.

The technological background of multi-service networks is given in Section 9.1. Some results from queueing theory are presented in Section 9.2. Section 9.3 introduces the large deviation theory, which is put into practice for service modeling in Section 9.4 and 9.5. Some techniques for estimation of complex data types and buffers are discussed in Section 9.6 and 9.7, respectively.

9.1 Broadband services

The two most important technologies for broadband, multi-service networks are Asynchronous Transfer Mode (ATM) and Internet Protocol (IP) based technologies, often using Ethernet as the medium. The main difference is that ATM has a fixed packet length of 53 bytes (called cells), whereas IP-based services allow for variable packet lengths. ATM also uses virtual circuits to facilitate provision for various QoS demands, whereas IP uses technologies such as Transmission Control Protocol (TCP), Multi-Protocol Label Switching (MPLS), and Differentiated Services (DiffServ) to achieve differentiation and robustness in service delivery. TCP is mainly used for lossless data delivery; MPLS provides functionality akin to virtual channels, used particularly for delay sensitive services; and DiffServ is a way of classifying services with different QoS demands.

A broadband multi-service network allows for a variety of services, for example voice calls, streaming video, and data transfer. Each of these services has a set of demands on the network referred to as its Quality of Service (QoS) parameters. The QoS requirements may be limits on delay, packet loss, jitter (the variation in delay), or a combination of these parameters.

Packet loss is related to buffer size in the queues, which also affects the delay. This is a consequence of the fact that buffered jobs may induce delays too long for the QoS demands. Thus, a large buffer may, apart from being expensive, induce longer delays than the QoS requirements planned for.

An ATM switch consists of a buffer and a server, where jobs are served in parallel according to a first-in, first-out fashion by a processor sharing policy. In this way, an ATM server can guarantee a minimum bandwidth while busy. Due to its fixed packet length, an ATM switch is often modeled as a $G/D/1/b$ queue, where traffic is assumed to arrive according to some general process (G denotes general distribution), the service rate is constant, and traffic streams are sharing a buffer space of size b.

An IP router, on the other hand, is rather modeled as a $G/G/1$ queue, where both the arrival process and the service process may be general. Due to the difficulty in analyzing general queues, idealizations and approximations are often necessary. We may, for example, use $M/G/1$ or $G/M/1$ queues as appropriate models in certain situations. In order to analyze $M/G/1$ queues, of which $M/D/1$ is a special case, some results from queueing theory and probability theory are presented.

By using packet network technologies, periods of inactivity of the sources will not occupy any network resources as opposed to circuit-switched networks. As a result of this packetization, traffic becomes bursty, and is delivered in a stream of packets which varies with time. Each traffic source can then be allocated as much bandwidth as required in each instant. This improves the resource utilization, and thereby allows for a larger number of traffic sources sharing these resources.

The statistical leveling of demands is called statistical multiplexing and the gain in utilization, statistical multiplexing gain, described by a number $G \geq 1$. It can be defined as the quotient of the maximum number of sources a system can handle without violating some low packet loss probability p, and the number of sources allowed when the maximum bandwidth demand of each source is allocated. We call the latter quantity the nominal number of sources, which in essence, corresponds to a traditional circuit-switched resource allocation.

The efficiency of a link is dependent on both the characteristics of the traffic sources, the number of sources and the capacity of the link. As a general rule, the more sources and the higher the link capacity of a system, the higher multiplexing gain can be achieved.

Note that a service class may be thought of not only as different service types but also as dependent on the route between different origin and destination pairs. Such classes are often referred to as chains. An example of such a differentiation is when trunk reservation is used to give different priorities to traffic on direct links as compared to transit traffic, as well as for different traffic types.

In this chapter, we will focus on three types of traffic: traditional telephony, video, and data best effort services. These correspond loosely to the paradigm triple play as well as representing different QoS requirements. Traditional telephony remains one of the most important services for which networks have been successfully designed for over a century. The most important QoS demands are limited mean delay and jitter (variation in the delay). Video has similar QoS requirements, but requires more bandwidth and is more bursty. File transfer and secure packet-switched services require low packet loss probability but are less sensitive to delay. We refer to telephony and video services as circuit-switched type services or real-time services. Many data services are classified as best effort, for which any spare capacity in the network is used once traffic with higher QoS demands and/or priority has been served.

These services also map to three classes of DiffServ: expedited forwarding (EF) for circuit-switched type of services, assured forwarding (AF) for secure packet-switched services, and best effort (BE). In ATM, these correspond roughly to constant bit rate (CBR), variable bit rate (VBR), and unspecified bit rate (UBR).

In a multi-service network, traffic streams with greatly varying characteristics may be offered to the network. It is necessary to protect the network and connections in progress from overload and control the amount of assigned resources a new stream might demand.

A common method is a leaky bucket algorithm, based on the analogy of a bucket of water of a certain size with a hole in the bottom. The traffic is represented by water, which can arrive in bursts of large quantity. The size of the hole in the bottom represents the allowed bandwidth and the size of the bucket itself the buffer size. If the arriving work is too bursty to be processed immediately, some of the load is buffered, and the work is processed with maximum allowable rate. This makes the traffic smoother (less bursty) after the leaky bucket as compared to its original characteristics. If large amounts of work are arriving in a short time span, the bucket may overflow, and the data is lost.

A leaky bucket can be described by two parameters, the maximum average arrival rate ρ and the burst size σ, representing the width of the hole and the size of the bucket. A traffic stream passing a leaky bucket will thereby not exceed the burst size σ and the maximum average rate ρ and is said to satisfy a deterministic (σ, ρ) constraint. Based on such a descriptor, the network can reserve an appropriate minimum guaranteed bandwidth and buffer size such that QoS requirements are fulfilled.

When traffic characteristics are not known exactly, such as when the traffic stream exhibits large temporal variation, approximate statistical descriptors can be used. This allows for a variable service rate being allocated, and by statistical multiplexing, resources are utilized more efficiently. In order to guarantee QoS requirements to an individual stream, it may have to be buffered separately. Typically, sources with similar QoS requirements and characteristics are aggregated to share a buffer and output port at a node. The resource allocation in such cases may be difficult, and control mechanisms such as traffic monitoring and policing are often necessary to guard the system against congestion.

The bandwidth and burstiness limits can be specified in a traffic contract. Bandwidth is typically specified in the number packets, bytes, or bits per second. Burstiness can be specified as a limit on allowable jitter or in absolute terms of burst size (in packets, bytes, or bits).

Best effort type of traffic utilizes the remaining capacity after traffic types with higher priority have been allocated resources. Such traffic needs larger buffers than, for example, real-time services. In order to enforce QoS, higher layer protocols are used for providing retransmission and window flow control.

The leaky bucket can be used for traffic shaping or traffic policing. Traffic shaping forces delay of large traffic bursts by buffering so that the stream conforms to allocated bandwidth limits. It is used to prevent congestion by modifying the traffic characteristics. Traffic policing, on the other hand, is a set of rules built on inspection of the traffic stream. Packets that do not conform to bandwidth limits may be discarded or their priority may be reduced by altering the priority marking. Packets marked with low priority are the first to be discarded by traffic management functions in case of congestion.

For variable packet lengths, the amount of work added to the bucket is proportional to the packet length. Then, the packet size as well as the number of packets arriving per time unit influences the statistical properties of the stream. In particular, the probability of overflow is dependent not only on the current content of the bucket and the arrival rate, but on the packet length as well.

9.2 Queues in multi-service networks

In classical communication systems, we have hitherto assumed that traffic arrives according to a Poisson process and/or that the holding times (the duration of a call or the length of a packet) are exponentially distributed. These assumptions simplify the analysis of such systems due to the property of memorylessness. The assumption may be justified for certain network and traffic types. However, in order to reflect different technologies for high-speed communication and an inhomogeneous mix of services, more general types of queues are required for modeling.

This section focuses on single server queues, which are appropriate models when processor sharing is practiced and different job types share a buffer. It is possible to segregate traffic types and handle them separately, but that is neither an efficient nor common way of implementing a multi-service network architecture.

The assumption of having a single server turns out to be crucial, when job arrival and service distributions become more general. Assuming a first-come, first-served policy or a processor sharing policy, the order in which jobs of the same size depart from the queue equals the order in which they arrive. The single server queue is also an assumption for the large deviation results presented in the next section.

9.2.1 The *M/G/*1 queue

Performance metrics for the single-server queue with Poisson arrivals and general service times can be determined exactly, provided the mean $1/\mu$ and variance σ_S^2 of the general service distribution are known. This important result is referred to as the Pollaczek-Khintchine formula. The queue length L is given by

$$L = \rho + \frac{\rho^2 + \lambda^2 \sigma_S^2}{2(1 - \rho)}, \tag{9.1}$$

where $\rho = \lambda/\mu$. The mean waiting time W is given by Little's formula,

$$W = L/\lambda = \frac{1}{\mu} + \frac{\rho + \lambda\mu\sigma_S^2}{2(\mu - \lambda)}.$$

A proof of the Pollaczek-Khintchine formula can be found in most books on queueing theory, see for example (Gross and Harris, 1998).

The presence of the variance in the nominator of Equation (9.1) shows that the smaller variance the service distribution has, the shorter the waiting time. Note that for an exponential service distribution, $\sigma_S^2 = 1/\mu^2$ and for a deterministic service, $\sigma_S^2 = 0$. This is a justification for using ATM in high-speed networks. We can interpret the mean and variance of the service distribution as characteristics of the length of a job at some constant service rate. Then, since data traffic typically has much higher variance than voice traffic, the former creates longer queue lengths and waiting times than the latter. This should also be reflected in the models chosen for these services.

9.2.2 Finite queues

In real systems, the queue length will always be finite due to the fact that the queue can only have a buffer of finite size. Arriving customers are blocked when the buffer is full. There exists a simple relation between the state probabilities $p(i)$, $i = 0, 1, 2, \ldots$ of the infinite system $M/G/1$ and the state probabilities $p_b(i)$, $i = 0, 1, \ldots, b$ of a $M/G/1/b$ queue, where the total number of positions for customers is b (integer), including the customer being served, known as the Kielson formula for finite $M/G/1$ queues (Iversen, 2013)

$$p_b(i) = \frac{p(i)}{(1 - \rho \cdot G_b)}, \quad i = 0, 1, \ldots, b - 1, \tag{9.2}$$

$$p_b(b) = \frac{(1 - \rho) \cdot G_b}{(1 - \rho \cdot G_b)}, \tag{9.3}$$

where $\rho < 1$ is the offered traffic, and

$$G_b = \sum_{j=b}^{\infty} p(j).$$

This is, of course, an approximation if the sum G_b is truncated, but it can be made arbitrarily accurate by including sufficient number of states. We note that the $p(i)$ only exists for $\rho < 1$. In a finite buffer queue, it is also possible to achieve statistical equilibrium for $\rho > 1$. Equations (9.2) and (9.3) cannot be used in this case.

9.2.3 The *M/D*/1 queue

In ATM, where the cell length is fixed, an appropriate model is the $M/D/1$ queue, if the jobs are assumed to arrive according to a Poisson distribution and processed according to a first-come first-served discipline.

When the service distribution is deterministic, the analysis of the queueing properties is actually more complex than when the service distribution is exponential (memoryless). In this case, the progress of ongoing service has to be considered in the analysis at each time instant.

Under the assumption of statistical equilibrium we can derive the state probabilities for $M/D/1$ in a simple way. The queue is assumed to have infinite buffer so that the intensity of offered traffic equals that of the carried traffic. Let the intensity be given by $\lambda \cdot h < 1$, where $h = 1/s$, and s is the deterministic service rate. We then have

$$\rho = \lambda \cdot h = 1 - p(0),$$

as in every state except state zero the carried (serviced) traffic is equal to one.

To study this system, we look at two epochs t and $t + h$ at a distance of $h = 1/s$ apart. Every customer being served at epoch t (at most one) has left the server at epoch $t + h$. Customers arriving during the interval $(t, t + h)$ are still present in the system at epoch $t + h$, either waiting or being served.

The arrival process is a Poisson process, so that we have a Poisson distributed number of arrivals in the time interval $(t, t + h)$

$$p(k, h) = \frac{(\lambda h)^k}{k!} \cdot e^{-\lambda h}, \quad k = 0, 1, 2, \ldots$$

The probability of being in a given state at epoch $t + h$ is obtained from the state at each epoch t by taking account of all arrivals and departures during $(t, t + h)$. We obtain Fry's equations of state for a single server system as

$$p_{t+h}(i) = [p_t(0) + p_t(1)]p(i, h) + \sum_{j=2}^{i+1} p_t(j) \cdot p(i - j + 1, h), \quad i = 0, 1, \ldots,$$

where $p_t(i)$ is the probability of the system being in state i at time t, and $p(k, h)$ is the probability of k arrivals during an epoch of length h. We have

$$p(0) = 1 - \rho,$$

Table 9.1 Truncation of the state probabilities of an infinite queue, and transformation of the these probabilities to those of a finite queue

n	p(n)	np(n)	p(n)	np(n)
0	0.3000	0	0.3002	0
1	0.3041	0.3041	0.3043	0.3043
2	0.1895	0.3791	0.1897	0.3793
3	0.1010	0.3030	0.1011	0.3032
4	0.0517	0.2069	0.0518	0.2070
5	0.0263	0.1316	0.0263	0.1317
6	0.0134	0.0804	0.0134	0.0804
7	0.0068	0.0477	0.0068	0.0477
8	0.0035	0.0278	0.0035	0.0278
9	0.0018	0.0159	0.0018	0.0159
10	0.0009	0.0090	0.0009	0.0090
11	0.0005	0.0050	0.0003	0.0031
Sum	0.9995	1.5105	1.000	1.5095

and under the assumption of statistical equilibrium $p_t(i) = p_{t+h}(i)$, we find successively for $i = 0, 1, \ldots$

$$p(1) = (1 - \rho)\left(e^\rho - 1\right),$$
$$p(2) = (1 - \rho)\left(-e^\rho(1 + \rho) + e^{2\rho}\right),$$
$$\cdots \tag{9.4}$$
$$p(i) = (1 - \rho)\sum_{j=1}^{i}(-1)^{i-j}e^{j\rho}\left(\frac{(j\rho)^{i-j}}{(i - j)! + \frac{(j\rho)^{i-j-1}}{(i-j-1)!}}\right), \quad i = 2, 3, \ldots.$$

The last term corresponding to $j = i$ always equals $e^{i\rho}$, since $(-1)! = \infty$. The method is unsuitable for large queues since it suffers from numerical cancellation due to subtraction of increasingly large terms.

Example 9.2.1.   In order to calculate the state probabilities of an $M/D/1/b$ queue with finite buffer b, we can use one of the two methods. Either we can truncate the state probabilities for an infinite queue at $p(b)$, or we can use the transformation in (9.2) and (9.3) to calculate the exact probabilities for the finite queue (apart from truncating the sum in (9.3)). The result is shown in Table 9.1. Column 2 shows the state probabilities computed using Fry's equations. Column 3 are the terms in the expectation. Column 4 and 5 tabulate the corresponding transformation to a finite queue with $b = 11$. Observe that the truncated probabilities do not sum exactly to unity. The approximate expectation is 1.5105. This can be compared to the $M/G/1$ expectation given by the Pollaczek-Khintchine formula, $\mathbf{E}(n) = 1.5167$. Also, denoting the state probability of the infinite queue by $p_\infty(n)$ and the finite by $p_b(n)$, we note that $p_b(n) < p_\infty(n)$.

The graph of the waiting time distribution (or the queue length) has an irregularity at points in time where the waiting time equals an integral multiple of the constant

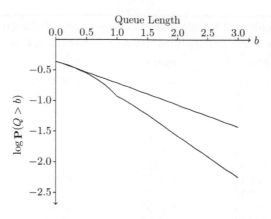

FIGURE 9.1

Comparison between an $M/D/1$ and an $M/M/1$ queue with the same traffic intensity $\rho = 0.7$.

service time. At such points, the graph shows a slight dip. The effect fades quickly with increasing waiting time. The effect is shown in Figure 9.1 in the slope of the logarithmic probability of the $M/D/1$ queue length.

The waiting time can be written in a closed form (Iversen, 2013)

$$p(W \leq t) = (1 - \rho) \sum_{j=0}^{\lfloor t \rfloor} \frac{(\rho(j - t))^j}{j!} e^{-\rho(j-t)}, \tag{9.5}$$

which is a suitable expression for numerical evaluation when waiting times are short. Similarly, the queue length distribution is

$$p(L > q) = 1 - (1 - \rho) \sum_{j=0}^{\lfloor q \rfloor} \frac{(\rho(j - q))^j}{j!} e^{-\rho(j-q)}. \tag{9.6}$$

For larger waiting times we are usually only interested in integral values of t. It can be shown (Iversen, 2013) that for integral t, we have

$$p(W \leq t) = p(0) + p(1) + \cdots + p(t). \tag{9.7}$$

The state probabilities $p(i)$ are calculated accurately by using a recursive formula based on Fry's state equations

$$p(i+1) = \frac{1}{p(0, h)} \left(p(i) - (p(0) + p(1)) p(i, h) - \sum_{j=2}^{i} p(j) \cdot p(i - j + 1, h) \right). \tag{9.8}$$

For non-integral waiting times we are able to express the waiting time distribution in terms of integral waiting times. If we let $h = 1$, then (9.5) may be written in powers of τ, where $t = T + \tau$, T integer and $0 \leq \tau < 1$. We then find

$$p(W \leq T + \tau) = e^{\rho\tau} \sum_{j=0}^{\infty} T \frac{(-\rho\tau)^j}{j!} \cdot p(W \leq T - j), \qquad (9.9)$$

where $p(W \leq T - j)$ is given by (9.7), (9.8), and (9.9).

For finite $M/D/1/b$ queues, the state probabilities $p_b(i)$ of the finite buffer system can be obtained from the state probabilities of the infinite buffer system by (9.2) and (9.3). Waiting times at integral times are then found from the state probabilities and waiting times at non-integral epochs from (9.9).

For an $M/D/1/b$ system with $\rho > 1$ the finite buffer state probabilities can be obtained in the following way. In a system with one server and $(b - 1)$ queueing positions we have $(b + 1)$ states $(0, 1, \ldots, b)$. Fry's balance equations for state probabilities $p_b(i)$, $i = 0, 1, \ldots, b - 2$ yield $b - 1$ linear equations between the states $(p_b(0), p_b(1), \ldots, p_b(b - 1))$. But it is not possible to write down simple time-dependent equation for states $b - 1$ and b. However, the first $(b - 1)$ equations in (9.4) together with the normalization requirement

$$\sum_{j=0}^{b} p_b(j) = 1,$$

and the observation that offered traffic is the sum of carried traffic and rejected traffic

$$\rho = 1 - p_b(0) + \rho \cdot p_b(b)$$

results in $(b + 1)$ independent linear equations, which are easy to solve numerically. The two approaches yield, of course, the same result. The first is only valid when $\rho < 1$, whereas the second is valid for any offered traffic.

9.2.4 **Approximations for general queues**

From Pollaczek-Khintchine's formula, it is easy to see that the more regular the holding time distribution, the shorter becomes the waiting time for jobs. In systems with non-Poisson arrivals, moments of higher order will also influence the mean waiting time.

For $G/G/1$ queues it is possible to give theoretical upper limits for the mean waiting time. Denoting the variance of the inter-arrival times by σ_A^2 and the variance of the service time distribution by σ_S^2, Kingman's inequality gives an upper limit for the mean waiting time

$$W_U \leq \frac{\lambda(\sigma_A^2 + \sigma_S^2)}{2(1 - \rho)}.$$

A lower bound is given by

$$W_L \geq \frac{\lambda^2 \sigma_S^2 + \rho(\rho - 2)}{2\lambda(1 - \rho)},$$

see Gross and Harris (1998). Since this quantity can be negative, we can simply use $\max\{0, W_L\}$ as the lower bound. These formulas show that it is the stochastic variation that results in waiting times, due to the presence of the variances of the inter-arrival and inter-departure processes.

The upper bound can be adjusted (or scaled) to give an approximation for W. Marchal (Gross and Harris (1998)) proposed the quotient

$$\frac{\rho^2 + \lambda^2 \sigma_S^2}{1 + \lambda^2 \sigma_S^2},$$

which multiplied with the upper bound gives the approximation

$$\widehat{W}_q = \frac{\lambda(\sigma_A^2 + \sigma_S^2)}{2(1 - \rho)} \frac{\rho^2 + \lambda^2 \sigma_S^2}{1 + \lambda^2 \sigma_S^2}.$$

Note, that for $\sigma_A^2 = 1/\lambda^2$, that is, a Poisson arrival with rate λ, the expression becomes the Pollaczek-Khintchine formula.

The formula works well for $G/M/1$ queues, but the performance deteriorates for an arbitrary $G/G/1$ queue when the service times or inter-arrival times increasingly deviate from exponentiality. The approximation improves, however, with increasing traffic intensity.

A second approximation proposed by Marchal (Gross and Harris (1998)) is obtained by using a different scaling factor,

$$\frac{\rho^2 \sigma_A^2 + \sigma_S^2}{\sigma_A^2 + \sigma_S^2}.$$

The approximation is then given by

$$\widehat{W}_q = \frac{\rho(\lambda \sigma_A^2 + \mu^2 \sigma_S^2)}{2\mu(1 - \rho)},$$

which is exact for $M/G/1$ as well as $D/D/1$ systems.

9.3 Large deviations

Roughly speaking, large deviations theory is a theory of rare events, that is, events that occur seldom. It is used to analyze the tails of probability distributions in contrast to classical probability which is concerned with events close to the mean of a distribution.

This section is concerned with the input process to a queue, that is the characterization of traffic sources and their aggregation. We can study the behavior of a traffic aggregate with respect to a certain queue length q, far from the mean, and formulate probabilities that the aggregate will exceed that level.

In order to formulate the problem of loss in a queue we formulate two useful congestion probabilities. Suppose we are given a link with capacity C and a stochastic variable describing the aggregate traffic S_n, where n is the number of sources. The system is assumed to be a bufferless $G/G/1$ system in the sense that—similar to a $M/M/n/n$ system—only jobs that can be processed by the server with capacity C can be buffered. A small buffer is necessary since there is only one server available. When the aggregate traffic S_n exceeds the server capacity, some traffic is lost. The first measure, the actual traffic loss probability, p_l, is given by

$$p_l = \frac{\mathbf{E}(\max(0, S_n - C))}{\mathbf{E}(S_n)} = \frac{\int_C^\infty (x - C)p(x)\mathrm{d}x}{\int_0^\infty xp(x)\mathrm{d}x}. \tag{9.10}$$

The loss measure p_l is referred to as a stream-based loss measure since it measures the proportion of traffic that is lost in an offered input stream. The type of blocking p_l measures is sometimes called traffic blocking.

An alternative measure, known as a resource-based congestion measure (or saturation probability), is the probability that a given resource is overloaded. It is defined as

$$p_s = \mathbf{P}(S_n > C). \tag{9.11}$$

The type of blocking measured is also called time blocking, since the probability measures the proportion of time that the resource is overloaded. The saturation measure (9.11) is often larger than (9.10) but might be easier to estimate. The measures are equal in the case of an Erlang link (Poisson arrivals).

In Figure 9.2, the logarithms of the cumulant probability function for a sum S_n of exponential random variables. For large values of the sum, far away from the mean, the logarithm is approximately linear. It turns out that this characteristic is valid for a large number of distributions. This characteristic is formulated in Cramér's theorem.

Theorem 9.3.1.   *Let $X_1, X_2, X_3, \ldots$ be a sequence of bounded, independent, and identically distributed random variables with mean μ, and let S_n be the sum of (the*

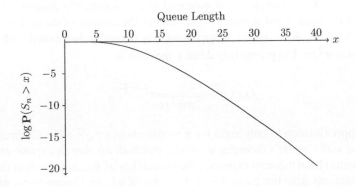

FIGURE 9.2

Logarithm of probability that the sum of $n = 10$ exponentially distributed random variables with $\mu = 1$ exceeds level x.

first) n variables, and

$$M_n = \frac{1}{n}S_n = \frac{1}{n}(X_1 + \cdots + X_n),$$

denote the empirical mean. Then the tails of the probability distribution of S_n decay exponentially with increasing n at a rate given by a convex rate function $I(x)$

$$\mathbf{P}(M_n > x) \asymp e^{-nI(x)} \quad \text{for } x > \mu,$$
$$\mathbf{P}(M_n < x) \asymp e^{-nI(x)} \quad \text{for } x < \mu. \tag{9.12}$$

The sign $\asymp$ (asymptotically equal to) means that the expressions on the left-hand side tend to the right-hand side as n tends to infinity. A proof can be found in Lewis (1996).

Thus, for large systems, the saturation probability $\mathbf{P}(S_n > C)$ in a link of capacity C and a large number of sources n can be approximated by $e^{-I(C/n)}$, where $I(\cdot)$ is the *rate function*. Before turning to the question of how to find the rate function, some reflections on two of the central theorems of mathematical statistics are useful.

The (weak) law of large numbers implies that as the number of sources n increases, the distribution of the sum M_n becomes increasingly concentrated about the mean, which can be written as

$$\lim_{n\to\infty} \mathbf{P}(|M_n - \mu| < \epsilon) = 1,$$

for each $\epsilon > 0$. This statement is equivalent to saying that, as n increases, the tails of the distribution of S_n decrease more sharply, or

$$\lim_{n\to\infty} \mathbf{P}(|M_n - \mu| > \epsilon) = 0, \tag{9.13}$$

for each $\epsilon > 0$. This shows the relation between the weak law of large numbers and large deviations.

Another result from probability theory is the central limit theorem. It states that if $X_1, X_2, \ldots, X_n$ is a sequence of independent and identically distributed random variables with mean μ and variance $\sigma^2 < \infty$, then the average of the first n variables, $M_n = \frac{1}{n}(X_1 + X_2 + \cdots + X_n)$, is approximately normally distributed with mean μ and variance σ^2/n. The probability density function is

$$f(x) = \frac{1}{\sqrt{2\pi\sigma^2/n}} e^{\frac{n}{2}\frac{(x-\mu)^2}{\sigma^2}}, \tag{9.14}$$

and the approximation is only valid for x within about $\sigma/\sqrt{n}$ of μ. Comparing this expression with Cramér's theorem, we notice that both involve an exponential.

The central limit theorem expresses the probability of fluctuations near the mean, where deviations from the mean are of the order of $\sigma/\sqrt{n}$. Fluctuations, which are large in comparison, are called large deviations from the mean. These events happen only rarely, which is the reason why large deviation theory also is called the theory of rare events.

9.3.1 The scaled cumulant generation function

A central representation of a statistical distribution in the context of large deviations theory is the cumulant generation function (CGF), $\Lambda(\theta)$. It is the logarithm of the moment generation function, $M(\theta)$, which is defined as

$$M(\theta) = \mathbf{E}(e^{\theta X}) = \int e^{\theta x} f(x) \mathrm{d}x,$$

where X is a random variable with probability density function $f(x)$. The cumulant generating function is therefore $\Lambda(\theta) = \ln M(\theta)$ and the CGF is sometimes referred to as the logarithmic moment generating function. These functions are convenient to use due to their properties. The values at the origin are

$$M(0) = 1,$$

$$\Lambda(0) = 0,$$

and the mean and variance can be computed as

$$\mu = \mathbf{E}(X) = \frac{\mathrm{d}M(\theta)}{\mathrm{d}\theta}\bigg|_{\theta=0} = \frac{\mathrm{d}\Lambda(\theta)}{\mathrm{d}\theta}\bigg|_{\theta=0},$$

$$\sigma^2 = \mathrm{Var}(X) = \frac{\mathrm{d}^2 M(\theta)}{\mathrm{d}\theta^2}\bigg|_{\theta=0} - \left(\frac{\mathrm{d}M(\theta)}{\mathrm{d}\theta}\bigg|_{\theta=0}\right)^2 = \frac{\mathrm{d}^2\Lambda(\theta)}{\mathrm{d}\theta^2}\bigg|_{\theta=0}.$$

Example 9.3.1.   The moment generating function can be computed from the probability density function. For an exponentially distributed random variable X with mean rate λ^{-1},

$$p(x) = \lambda e^{\lambda x},$$

so that $\mathbf{E}(\exp(\theta X))$ is, by the definition of expectation,

$$\mathbf{E}\left(e^{\theta X}\right) = \int_0^\infty e^{\theta x} p(x) \mathrm{d}x$$

$$= \lambda \int_0^\infty e^{\theta x} e^{-\lambda x} \mathrm{d}x = \lambda \int_0^\infty e^{x(\theta-\lambda)} \mathrm{d}x.$$

The value of the integral is $\lambda/(\lambda - \theta) = (1 + \theta/\lambda)^{-1}$, provided that $\lambda > \theta$. In practice, it is seldom necessary to compute the moment generating function. For common probability distributions, these can be found in reference works.

The additivity property of the cumulant generating function is particularly important. Let $X_1, X_2, \ldots, X_n$ be n independent random variables with cumulant generating functions $\Lambda_1(\theta), \Lambda_2(\theta), \ldots, \Lambda_n(\theta)$ and $S_n = X_1 + X_2 + \cdots + X_n$ the sum of the variables. Then the cumulant generating function $\Lambda(\theta)$ of S_n is given by

$$
\begin{aligned}
\Lambda(\theta) &= \ln\left(\mathbf{E}\left(e^{\theta S_n}\right)\right) \\
&= \ln\left(\mathbf{E}\left(e^{\theta(X_1 + X_2 + \cdots + X_n)}\right)\right) \\
&= \ln\left(\mathbf{E}\left(e^{\theta X_1} e^{\theta X_2} \cdots e^{\theta X_n}\right)\right) \\
&= \ln\left(\mathbf{E}\left(e^{\theta X_1}\right) \cdot \mathbf{E}\left(e^{\theta X_2}\right) \cdots \mathbf{E}\left(e^{\theta X_n}\right)\right) \\
&= \ln\left(\mathbf{E}\left(e^{\theta X_1}\right)\right) \cdot \ln\left(\mathbf{E}\left(e^{\theta X_2}\right)\right) \cdots \ln\left(\mathbf{E}\left(e^{\theta X_n}\right)\right) \\
&= \Lambda_1(\theta) + \Lambda_2(\theta) + \cdots + \Lambda_n(\theta).
\end{aligned}
$$

The moment generating function and the cumulant generating function are defined with a parameter θ, which is used to scale a distribution. Let a random variable X have the probability density function $f(x)$. The scaled distribution is defined as

$$
f_\theta(x) = \frac{e^{\theta x} f(x)}{M(\theta)} = e^{\theta x - \Lambda(\theta)}.
$$

For $\theta x > 0$ the factor is greater than 1. Thus, the probability density function is shifted to larger values of x. The denominator $M(\theta)$ is needed to normalize the probability density function $f_\theta(x)$ so that the area under the function remains unity. The original probability density function can be retrieved by the inverse transformation

$$
f(x) = e^{-\theta x + \Lambda(\theta)} f_\theta(x).
$$

Note that scaling does not require a continuous distribution. We can define scaled variants of the binomial and Poisson distributions, for example.

Formally associated with the scaled distribution is a scaled probability measure

$$
d\mathbf{P}_\theta(x) = e^{\theta x - \Lambda(\theta)} d\mathbf{P}(x).
$$

Using this, we define the expectation $\mathbf{E}_\theta(\cdot)$ of a random variable X with respect to $\mathbf{P}_\theta$. With this, we get the moments of the scaled distribution as

$$
\mu(\theta) = \mathbf{E}_\theta(X) = \frac{1}{M(\hat{\theta})} \left.\frac{dM(\theta)}{d\theta}\right|_{\theta=\hat{\theta}} = \left.\frac{d\Lambda(\theta)}{d\theta}\right|_{\theta=\hat{\theta}},
$$

$$
\sigma^2(\theta) = \mathrm{Var}_\theta(X) = \frac{1}{M(\hat{\theta})} \left.\frac{d^2 M(\theta)}{d\theta^2}\right|_{\theta=\hat{\theta}} - \left(\frac{1}{M(\hat{\theta})} \left.\frac{dM(\theta)}{d\theta}\right|_{\theta=\hat{\theta}}\right)^2 = \left.\frac{d^2\Lambda(\theta)}{d\theta^2}\right|_{\theta=\hat{\theta}},
$$

where $\hat{\theta}$ is the current value of the variable θ.

Again, the scaled cumulant generating functions are additive. Thus, if X_1, $X_2, \ldots, X_n$ are independent random variables with scaled cumulant generating functions $\Lambda_1(\theta), \Lambda_2(\theta), \ldots, \Lambda_n(\theta)$, then for the sum $S_n = X_1 + X_2 + \cdots + X_n$, we have

$$
\mu(\theta) = \mu_1(\theta) + \mu_2(\theta) + \cdots + \mu_n(\theta),
$$

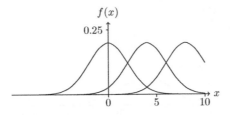

FIGURE 9.3
Scaling of the standard normal distribution showing $\theta = 0, 1, 2$.

$$\sigma^2(\theta) = \sigma_1^2(\theta) + \sigma_2^2(\theta) + \cdots + \sigma_n^2(\theta),$$

where $\mu_i(\theta)$ and $\sigma_i^2(\theta)$ are the expectation and the variance of random variable X_i with respect to the scaled distribution.

As a matter of fact, the scaled distribution of a sum S_n of n independent random variables X_i, $i = 1, 2, \ldots, n$ equals the sum of the scaled distributions of the individual variables X_i.

To see how scaling affects the distribution, we show some examples.

Example 9.3.2. The normal distribution $N(\mu, \sigma^2)$ has moment generating function $M(\theta) = \exp(\mu\theta + \frac{\sigma^2\theta^2}{2})$ and scaled CGF $\Lambda(\theta) = \ln M(\theta)$. Thus, we have, for the scaled distribution,

$$\mu_\theta = \frac{d\Lambda(\theta)}{d\theta} = \mu + \sigma^2\theta,$$

$$\sigma_\theta^2 = \frac{d^2\Lambda(\theta)}{d\theta^2} = \sigma^2.$$

Thus, the scaled distribution is distributed as $N(\mu + \sigma^2\theta, \sigma^2)$. Since only the mean is affected by the scaling, the distribution is simply shifted to the right with increasing θ. This is shown in Figure 9.3, where the standard normal distribution is scaled by parameter $\theta = 0$, 1, and 2.

Example 9.3.3. The Poisson distribution with intensity λ has moment generating function $M(\theta) = \exp(\lambda(e^\theta - 1))$ and cumulant generating function $\Lambda(\theta) = \lambda(e^\theta - 1)$. Differentiation of $\Lambda(\theta)$ gives $\lambda_\theta = \lambda e^\theta$ (which gives both the mean and the variance). Therefore, the scaled distribution is distributed as $Po(\lambda e^\theta)$. Scaling of the $Po(1)$ process is shown in Figure 9.4 with parameter values $\theta = 0$, 1, and 2. As θ grows, the distribution is not only shifted to larger x-values, but is also spread out around its mean.

Example 9.3.4. The binomial distribution $Bin(n, a)$ has moment generating function $M(\theta) = (1 - a + ae^\theta)^n$, so $\Lambda(\theta) = n \ln(1 - a + ae^\theta)$. Differentiation gives

$$\mu_\theta = \frac{nae^\theta}{1 - a + ae^\theta},$$

$$\sigma_\theta^2 = \frac{n(1 - a)ae^\theta}{(1 - a + ae^\theta)^2}.$$

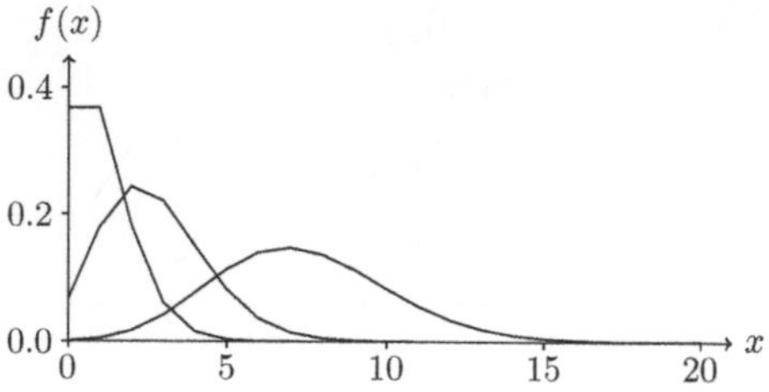

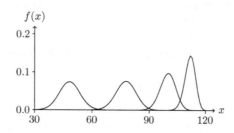

FIGURE 9.4

Scaling of the $Po(1)$ distribution showing $\theta = 0, 1, 2$.

FIGURE 9.5

Scaling of the $Bin(120, 0.4)$ distribution showing $\theta = 0, 1, 2, 3$.

The scaled binomial distribution is $Bin(n, a_\theta)$ with $a_\theta = \frac{ae^\theta}{1-a+ae^\theta}$. Scaling of the $Bin(120, 0.4)$ distribution is shown in Figure 9.5. By the scaling, the distribution is shifted to the right and at the same time, since $a_\theta \to 1$, the probability density function is concentrating around the mean, as θ increases.

9.3.2 The Chernoff formula

In order to compute the rate function in (9.12) for a sequence of independent random variables, we can use Chernoff's formula. We derive this formula by finding an upper bound on the tail probabilities.

Let $X_1, X_2, \ldots, X_n$ be independent and identically distributed random variables with average $M_n = \frac{1}{n}(X_1 + X_2 + \cdots + X_n)$. We can formulate an upper bound on the probability $\mathbf{P}(M_n > x)$ as follows. Define the indicator function $I_A(\cdot)$ of a set $A \subset \mathbb{R}$ as

$$I_A(x) \triangleq \begin{cases} 1 & \text{if } x \in A \\ 0 & \text{otherwise} \end{cases}.$$

On the x-axis, let $A = [a, \infty)$ be an interval. Then, $I_A(x)$ is a step function taking on value one if $x \in [a, \infty)$ and zero otherwise.

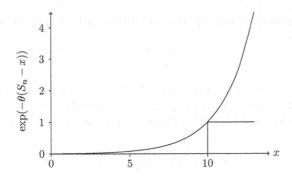

FIGURE 9.6

The Chernoff formula (with $\theta = 0.5$ and $x = 10$).

Figure 9.6 shows graphically that for each number a and each positive number θ, $I_{[a,\infty)} \leq e^{\theta x}/e^{\theta a}$. Now note that $\mathbf{E}(I_{[na,\infty)}(nM_n)) = \mathbf{P}(nM_n > na)$ and so

$$
\begin{aligned}
\mathbf{P}(M_n > a) &= \mathbf{P}(nM_n > na) \\
&= \mathbf{E}(I_{[na,\infty)}(nM_n)) \\
&\leq \mathbf{E}(e^{\theta n M_n}/e^{\theta na}) \\
&= e^{-\theta na}\mathbf{E}(e^{\theta(X_1+\cdots+X_n)}) \\
&= e^{-\theta na}\left(\mathbf{E}(e^{\theta X_i})\right)^n,
\end{aligned}
$$

where the last step follows if the random variables X_i are independent and identically distributed. Denoting the scaled cumulant generation function by $\Lambda(\theta) = \ln(\mathbf{E}(e^{\theta X_1}))$, we have $\mathbf{P}(M_n > a) \leq e^{-n(\theta a - \Lambda(\theta))}$. Since this holds for each θ we can optimize over θ which gives

$$
\mathbf{P}(M_n > a) \leq \min_{\theta > 0} e^{-n(\theta a - \Lambda(\theta))} = \exp\left(-n \max_{\theta > 0}(\theta a - \Lambda(\theta))\right).
$$

If a is greater than the mean m, a lower bound is given by

$$
\mathbf{P}(M_n > a) \asymp \exp\left(-n \max_{\theta > 0}(\theta a - \Lambda(\theta))\right).
$$

Thus, we have

Theorem 9.3.2 (Chernoff's Formula).   *The rate function can be calculated from* $\Lambda(\theta)$, *the cumulant generating function, from*

$$
I(x) = \max_{\theta}\{x\theta - \Lambda(\theta)\}, \tag{9.15}
$$

where $\Lambda(\cdot)$ *is defined by*

$$
\Lambda(\theta) \triangleq \ln(\mathbf{E}(\theta X_i)).
$$

The θ which minimizes the exponent in (9.15) can be found by minimizing the exponent, that is

$$\frac{d\Lambda(\theta)}{d\theta} = x.$$

Example 9.3.5. The Erlang distribution $Erl(\lambda, k)$ is a generalization of the exponential distribution. It describes the probability distribution of a sum of k identical exponential random variables. It has the moment generating function

$$M(\theta) = \left(1 - \frac{\theta}{\lambda}\right)^{-k}, \quad t < \lambda.$$

The (scaled) cumulant generation function is $-k \ln(1 - \theta/\lambda), t < \lambda$. To find $I(x)$, we solve the optimization problem $\max_\theta (x\theta - \Lambda(\theta))$ by setting

$$\frac{d\Lambda(\theta)}{d\theta} = x,$$

and solving for the optimal value θ^*. Thus, we have

$$\frac{k}{\lambda - \theta} = x, \quad \theta^* = \lambda - \frac{k}{x}.$$

Substituting θ^* into $x\theta - \Lambda(\theta)$ gives

$$I(x) = x\lambda - k - k \ln\left(\frac{k}{x\lambda}\right).$$

Example 9.3.6. Figure 9.2 shows a sum S_n of $n = 10$ exponentially distributed variables with mean rate $\lambda = 1$. The exact value is given by the Erlang distribution $Erl(\lambda, n)$ with intensity $\lambda = 1$ and shape $n = 10$. Consider the limit $x = 30$ and suppose we want to know the probability $\mathbf{P}(S_n \geq x)$ that S_n exceeds x. The Chernoff bound gives an estimate for this probability. Its numerical value can be found using the rate function for the exponential distribution and evaluate $e^{nI_{Exp}(x/n)} = e^{-10I_{Exp}(30/10)}$, or using the rate function for the Erlang distribution directly, $e^{I_{Erl}(x)} = e^{-I_{Erl}(30)}$. Both methods give the value $1.22 \cdot 10^{-4}$. The exact value, found from the cumulant generating function, is $5.08 \cdot 10^{-6}$. The Chernoff bound gives in this case a rather pessimistic estimate.

9.3.3 Normal approximation

Suppose we have n independently and identically distributed traffic streams over a link, with $S_n = X_1 + X_2 + \cdots + X_n$. Since they are identical and independent, they can all be described by the mean and variance of one of these streams, X_1 say, so that

$$\mu = \mathbf{E}(X_1),$$

$$\sigma^2 = \text{Var}(X_1).$$

For the aggregate S_n, we have $\mu_{S_n} = n\mu$ and $\sigma^2_{S_n} = n\sigma^2$. When n is large, the aggregate is approximately normally distributed (9.14), so that

$$S_n \sim N(n\mu, n\sigma^2).$$

The saturation probability p_s is then

$$p_s = \mathbf{P}(S_n > C) = 1 - \Phi\left(\frac{n\mu - C}{n\sigma^2}\right), \tag{9.16}$$

where $\Phi(\cdot)$ is the cumulant probability function. The tail probability of the normal distribution $1 - \Phi(x)$ is also known as the Q-function. Thus,

$$Q(x) = 1 - \Phi(x) = \frac{1}{2}\text{erfc}\left(\frac{x}{\sqrt{2}}\right), \tag{9.17}$$

where erfc$(\cdot)$ is the complementary error function. Q(x) is the probability that a normally distributed random variable will assume a value greater than x times the standard deviation above the mean. The complementary error function can be expressed in terms of the upper incomplete gamma function

$$\text{erfc}(x) = \frac{1}{\sqrt{\pi}}\Gamma\left(\frac{1}{2}, x^2\right),$$

$$\Gamma(a, z) = \int_z^\infty e^{-t}t^{a-1}dt.$$

Algorithms for the implementation of erfc$(\cdot)$ and related functions are described in Press et al. (2007).

If the saturation probability p_s is given, the capacity C required in order that $\mathbf{P}(S_n \geq C) \leq p_s$ is simply the inverse of (9.17), that is

$$C \geq n\mu + \eta\sqrt{n}\sigma$$

$$\eta = Q^{-1}(p_s).$$

The inverse of the Q-function (or, the incomplete gamma function) can be found by a series expansion, or by solving $Q(x) = p_s$ for x. The inverse $\Phi(x)$ of the cumulant probability function is also known as the probit function (see Figure 9.7).

Example 9.3.7.   Consider Example 9.3.6. We can construct a normal approximation of the $n = 10$ exponential random variables with $\lambda = 1$ and $\sigma^2 = 1$. The approximate probability of S_n exceeding level x is $\mathbf{P}(S_n \geq x) = 1 - \Phi((x - n\mu)/\sqrt{n}\sigma) = 1.27 \cdot 10^{-10}$. The approximation is an underestimation of the exact value.

The required bandwidth per stream, the effective bandwidth of the stream, is

$$\alpha = \mu + \frac{\eta\sigma}{\sqrt{n}} \tag{9.18}$$

As n grows the effective bandwidth tends to the mean rate. The effective bandwidth is defined in terms of large deviations in the next section.

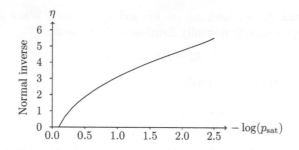

FIGURE 9.7

The inverse of a standard normal distribution—the probit function.

9.3.4 Improved approximation

We can improve the bounds on the congestion probabilities. First, the relation between the stream-based and the resource-based congestion measures follows from the Chernoff bound,

$$
\begin{aligned}
\mathbf{E}(X - C)^+ &= \int_0^\infty \mathbf{P}(X \geq C + x)\mathrm{d}x \\
&\leq \int_0^\infty \exp\left[s\left(\alpha(s) - (C + x)\right)\right]\mathrm{d}x \\
&= \frac{1}{s}\exp\left[s\left(\alpha(s) - C\right)\right].
\end{aligned}
$$

Note that the first line 'accumulates' the probabilities with increasing x, which gives the expected value. The Chernoff bound for a scaled distribution is

$$
\begin{aligned}
\mathbf{P}(X \geq x) &= \int_x^\infty f(y)\mathrm{d}y \\
&= \int_x^\infty e^{-\theta y + \Lambda(\theta)} f_\theta(y)\mathrm{d}y \\
&= e^{-\theta x + \Lambda(\theta)} \int_x^\infty e^{\theta(y-x)} f_\theta(y)\mathrm{d}y,
\end{aligned}
$$

where $e^{-\theta(y-x)} \leq 1$ for $y \in [x, \infty)$, so the value of the integral is less than unity and $\mathbf{P}(X \geq x) \leq e^{-\theta x + \Lambda(\theta)}$.

Since the tightest Chernoff bound is achieved for θ^*, where $m(\theta^*) = x$ is the mean of the scaled distribution $f_{\theta^*}(\cdot)$, we may assume that close to this mean, a normal distribution

$$
f_{\theta^*}(y) \approx \frac{e^{-\frac{1}{2}(y-x)^2/\sigma^2(\theta^*)}}{\sqrt{2\pi}\sigma(\theta^*)},
$$

is a good approximation.

Then,

$$\int_x^\infty e^{-\theta^*(y-x)} f_{\theta^*}(y) dy \approx \frac{1}{\sqrt{2\pi}\sigma(\theta^*)} \int_x^\infty e^{-\theta^*(y-x)} e^{-\frac{1}{2}(y-x)^2/\sigma^2(\theta^*)} dy$$

$$\approx \frac{1}{\sqrt{2\pi}\theta^*\sigma(\theta^*)},$$

and assuming that $\theta^*\sigma(\theta^*) > 1$, we obtain the rather accurate approximation

$$\mathbf{P}(X \geq x) \approx \frac{e^{-I(x)}}{\sqrt{2\pi}\theta^*\sigma(\theta^*)}.$$

By the approximation, we have for the mean

$$p_s = \mathbf{P}(\frac{1}{n}(X_1 + X_2 + \cdots + X_n) \geq x) \approx \frac{e^{-nI(x)}}{\sqrt{2\pi n}\theta^*\sigma(\theta^*)}, \qquad (9.19)$$

where $I(x)$ is the rate function of a single random variable, and $\sigma(\theta^*)$ is its variance. The variance of the scaled distribution of the sum is $\sqrt{n}$ times larger, leading to the presence of n in the denominator.

Similarly, the loss probability can be approximated as

$$p_l = \frac{1}{m}\mathbf{E}(\max(0, X - x)) = \frac{1}{m} e^{-\theta x + \Lambda(\theta)} \int_x^\infty (y - x) e^{-\theta(y-x)} f_\theta(y) dy.$$

Letting $\theta = \theta^*$ gives

$$\int_x^\infty (y - x) e^{-\theta(y-x)} f_\theta(y) dy \approx \frac{1}{\sqrt{2\pi}\sigma(\theta^*)}$$

$$\int_x^\infty (y - x) e^{-\theta^*(y-x)} e^{\frac{1}{2}(y-x)^2/\sigma^2(\theta^*)} dy \approx \frac{1}{\sqrt{2\pi}(\theta^*)^2\sigma(\theta^*)},$$

so that

$$p_l = \frac{1}{m}\mathbf{E}(\max(0, X - x)) \approx \frac{e^{-I(x)}}{\sqrt{2\pi}m(\theta^*)^2\sigma(\theta^*)}. \qquad (9.20)$$

Example 9.3.8.   Consider Example 9.3.6 again. Using (9.19) with $\theta^* = 1 - \frac{1}{10} = 0.9$ and $\sigma^2(\theta^*) = 10^2/1 = 100$, the improved approximation is $e^{-I(x)}/9\sqrt{2\pi} = 5.39 \cdot 10^{-6}$, which is quite close to the exact probability $5.08 \cdot 10^{-6}$.

9.3.5 Varadhan's lemma

So far, the random variables have been assumed as being independent. This is a rather severe restriction for the applicability of large deviations theory. Varadhan's theorem generalizes the theory to cases where the independence assumption between the variables is relaxed. Let M_n be the mean of a sequence of (possibly dependent) random variables X_i, and consider the integral

$$G_n = \int_0^\infty e^{ng(x)} \mathrm{d}\mathbf{P}(M_n \approx x).$$

If M_n is such that $\mathbf{P}(M_n \approx x) \asymp e^{-nI(x)}$, we may set

$$G_n \asymp \int_0^\infty e^{ng(x)} e^{-nI(x)} \mathrm{d}x$$

$$= \int_0^\infty e^{n(g(x) - I(x))} \mathrm{d}x$$

$$\asymp e^{n\max_x\{g(x) - I(x)\}},$$

so that

$$\lim_{n\to\infty} \frac{1}{n} \ln\left(\int_0^{ng(x)} \mathrm{d}\mathbf{P}(S_n \approx x)\right) = \max_x\{g(x) - I(x)\}. \tag{9.21}$$

Equation (9.21) is therefore a generalization of the cumulant generating function for independent random variables. Subject to some rather general technical conditions, the asymptotic formula holds whenever g is a bounded continuous function. Using the expression

$$\mathbf{P}(M_n \approx x) \asymp e^{-nI(x)}, \tag{9.22}$$

we say that the sequence $\{\mathbf{P}(M_n \approx x)\}$ (or more loosely, the sequence $\{M_n\}$) satisfies a *large deviation principle* with rate function I when the expression holds. In this text, we will simply assume that a random sequence satisfies a large deviation principle whenever (9.22) is a justifiable description, for some convex rate function $I(x)$.

Thus, if we can establish the existence of the rate function (or, on good grounds assume that it exists), we can apply Varadhan's lemma. Choosing the function g to be linear

$$g(x) = \theta x \quad \text{for some number } \theta,$$

we have

$$\Lambda(\theta) \triangleq \lim_{n\to\infty} \frac{1}{n} \ln\left(\mathbf{E}(e^{n\theta S_n})\right)$$

$$= \lim_{n\to\infty} \frac{1}{n} \ln\left(\int_0^\infty e^{n\theta x} \mathrm{d}\mathbf{P}(S_n \approx x)\right)$$

$$= \max_x\{\theta x - I(x)\}.$$

The expression $\max_x\{\theta x - I(x)\}$ is known as the Fenchel-Legendre transform of I, and Λ is called the scaled cumulant generating function. This shows that when the rate function $I(\cdot)$ exists, so does the scaled cumulant generating function $\Lambda(\cdot)$, and each can be obtained by performing the Fenchel-Legendre transform of the other. We have, therefore,

$$I(x) = \max_{\theta}\{\theta x - \Lambda(\theta)\}, \tag{9.23}$$

$$\Lambda(\theta) = \max_{x}\{\theta x - I(x)\}. \tag{9.24}$$

9.4 Effective bandwidth

This section is concerned with the output processes from queues. Thus, we have an arrival process and a departure process. The arrival process is the topic of Section 9.3, and the departure process of this section.

Traffic can often be regarded as composed of sources of certain characteristics. Thus, within a traffic stream, there are statistical properties that are different from the actual macroscopic arrival and service processes. A voice call, for example, may be described by an on-off source, switching between a sending state and an idle state. On this level, called the burst level, the arrival process of jobs is often both bursty and the arrivals dependent. By characterizing a source by its effective bandwidth, a parsimonious description of the source on burst level can be obtained. On a call level, the arrival of the call can often still be modeled by a Poisson process. Call level aspects of multi-service traffic are discussed in Chapter 10.

9.4.1 The workload process

Consider a single-server queue in discrete time, and let the queueing discipline be first-come, first-served. There is a single class of jobs, an assumption which is justified if different traffic types are mixing into a single stream of traffic and there are no priorities assigned to any of the different traffic types. We introduce two concepts which are widely used in queueing theory:

Work conservation. A system is said to be work conserving if

(1) No servers are idle when there is at least one job waiting.
(2) The service times are independent of the service disciplines (e.g. queue length).

Load function. Let $U(t)$ denote the time it will require to serve the customers which are in the system at time t. At the time of an arrival, $U(t)$ decreases with a slope depending on the number of working servers until zero, where it stays until next arrival time. This is illustrated in Figure 9.8.

In a $GI/G/1$ queueing system (where GI denotes General Independent arrivals), $U(t)$ will be independent of the queueing discipline, if it is work conserving.

The virtual waiting time is the waiting time a customer experiences if the customer arrives at a random point of time (which leads to time averages). The actual waiting time is the waiting time the real customer experiences (which leads to call averages). When we consider systems with first-come first-served (FCFS) queueing discipline

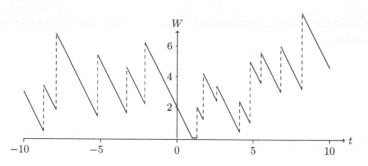

FIGURE 9.8

The workload process for a $G/G/1$ queue.

and Poisson arrival processes, the virtual waiting time will be equal to the actual waiting time, or time averages equal call averages.

The load function U_t denotes the time it will require to serve the customers which are in the system at time t. For FCFS queueing systems the virtual waiting time is equal to the load function at the time of arrival. It is more appropriate to talk about work than traffic, since arrivals may refer to different traffic types with different characteristics.

The waiting time of jobs in a queue (or alternatively, the queue lengths) can be described by Lindley's recursion. Let A_t be the amount of additional waiting time incurred by jobs arriving at time t, and S_t the amount of reduction of waiting time due to the departure of serviced jobs. Then the virtual waiting time in the queue, Q_t, is described by the recursion

$$Q_{t+1} = \max\{0,\, Q_t + A_t - S_t\}. \tag{9.25}$$

Note, that the waiting time (or queue length) never can be negative—it is impossible to 'save' work. Since it may happen that no new jobs arrive at time t, we use the virtual waiting time as the queue is still operating. It is common to refer to the virtual waiting times in this context as workload.

Let $U_t = A_t - S_t$ be the net change in workload. Then, (9.25) can be written

$$Q_{t+1} = \max\{0,\, Q_t + U_t\}. \tag{9.26}$$

It is sometimes convenient to consider the development of the recursion from a past moment in time up to the present time. Consequently, then the time index t is negative. Consider the first step in the recursion from $t = -1$ to $t = 0$,

$$Q_0 = \max\{0,\, Q_{-1} + A_{-1} - S_{-1}\}. \tag{9.27}$$

Let $U_t = A_{-t} - S_{-t}$, observing the change of sign in the index of U for convenience. Then, iterating (9.27) gives

$$Q_0 = \max\{0, U_1 + Q_{-1}\}$$
$$= \max\{0, U_1 + \max\{0, Q_{-2} + U_2\}\}$$
$$= \max\{0, U_1, U_1 + U_2 + Q_{-2}\}.$$

Note that the equation says that the queue length at time t is the work arriving minus the work leaving in one time step, plus the previous queue length. In the second line, the expression for Q is shifted in time. In the third line, the $\max\{\}$ operator is 'moved out' according to $U_1 + \max\{0, U_2 + Q_{-2}\} = \max\{U_1 + 0, U_1 + U_2 + Q_{-2}\}$.

By continuing the iteration, we get a sequence with an increasing number of terms, which are the net change in workload at time t. Let $W_t = U_1 + U_2 + \cdots U_t$, so that

$$Q_0 = \max\{W_0, W_1, \ldots, W_t + Q_{-t}\},$$

where W_0 is defined as $W_0 = 0$. W_t is called the workload process (not to be confused with the symbol W for queue waiting time). If there is some finite time $-\tau$ in the past when the queue was empty, then

$$Q_0 = \max\{W_0, W_1, \ldots, W_\tau\}.$$

Usually, we are interested in the steady-state queue length when the system has been running for a long time. If the queue is stable, that is, $\mathbf{E}(A_t) < \mathbf{E}(S_t)$ (the mean amount of arriving work is less than the mean amount of processed work), then the queue must be empty at some point in time. It follows that the steady-state queue length is independent of the initial queue length. In this case, the steady-state queue length Q can be written

$$Q = \max_{t \geq 0} W_t, \tag{9.28}$$

provided that the stability condition

$$\mathbf{E}(A_t) < \mathbf{E}(S_t), \tag{9.29}$$

is satisfied.

We can now use large deviations to analyze the steady-state queue length Q, and obtain results for the asymptotic probability distribution of $\mathbf{P}(Q \geq q)$ for large q (that is, when q is far away from the mean queue length). This probability can be interpreted as the fraction of time for which the queue length Q exceeds q. Figure 9.1 shows an $M/D/1$ queue and an $M/M/1$ queue. Both queues are stable with $\rho = \mathbf{E}(A_t)/\mathbf{E}(S_t) = 0.7$. Just as in the case of the arrival process in Example 9.2.1, the decay rate of the departure process is also seen to be approximately exponential, or

$$\mathbf{P}(Q > q) \asymp e^{-\delta q},$$

where $-\delta$ is the asymptotic slope of the queue length probability.

Now, we have the following formal result. Suppose the arrival process A_t and the departure process S_t are both stationary and satisfy the stability condition (9.29), and the workload process W_t satisfies a large deviation principle (9.22), so that

$$\mathbf{P}(W_t/t \approx x) \asymp e^{-tI(x)},$$

with rate function I, then

$$\mathbf{P}(Q > q) \asymp e^{-\delta q},$$

and the decay rate δ can be determined from the rate function for the workload process by

$$\delta = \min_x \frac{I(x)}{x}.$$

Here, W_t/t plays the role of M_n in Cramér's theorem 9.3.1.

An informal argument showing this is as follows. Note that

$$\mathbf{P}(Q > q) = \mathbf{P}\left(\max_{t \geq 0} W_t > q\right)$$
$$= \mathbf{P}(\cup_{t \geq 0}\{W_t > q\})$$
$$\leq \sum_{t \geq 0} \mathbf{P}(W_t > q),$$

where the last inequality follows from the Boole inequality for probabilities. A more formal proof can be found in Lewis and Russell (1996).

Since $\mathbf{P}(W_t/t > x) \asymp e^{tI(x)}$, we have

$$\mathbf{P}(W_t > q) = \mathbf{P}(W_t/t > q/t) \asymp e^{-tI(q/t)} = e^{-q\frac{I(q/t)}{q/t}},$$

so that

$$P(Q > q) \asymp e^{-q\frac{I(q)}{q}} + e^{-q\frac{I(q/2)}{q/2}} + \cdots + e^{-q\frac{I(q/t)}{q/t}} + \cdots$$

and the term which dominates when q is large is the one for which $I(q/t)/(q/t)$ is smallest, that is the one for which $I(x)/x$ is a minimum, so that

$$\mathbf{P}(Q > q) \asymp e^{-q\min_x \frac{I(x)}{x}} = e^{-q\delta}.$$

The decay rate δ can also be determined using the scaled cumulant generating function.

$$\theta \leq \min_x \frac{I(x)}{x} \iff \theta \leq \frac{I(x)}{x} \quad \text{for all } x$$
$$\iff \theta x - I(x) \leq 0 \quad \text{for all } x$$
$$\iff \max_x\{\theta x - I(x)\} \leq 0.$$

Thus, $\theta \leq \delta$ if and only if $\Lambda(\theta) \leq 0$ and so

$$\delta = \max\{\theta : \Lambda(\theta) \leq 0\}. \tag{9.30}$$

It is often easier to find δ using the CGF by finding the root to a nonlinear equation, for example by iteration, than by maximizing $I(x)/x$.

In terms of modeling a queue, we can use the net change in workload $U_t = A_t - S_t$. Let $W_t^{(A)} = A_1 + A_2 + \cdots A_t$ and $W_t^{(S)} = S_1 + S_2 + \cdots S_t$, so that $W_t = W_t^{(A)} - W_t^{(S)}$. Assuming that $W_t^{(A)}$ and $W_t^{(S)}$ are independent, we have from the additivity of the CGF,

$$\Lambda(\theta) = \Lambda_A(\theta) + \Lambda_S(-\theta),$$

where $\Lambda_A(\theta)$ and $\Lambda_S(\theta)$ are the asymptotic CGFs for the arrival and departure processes, respectively, that is

$$\Lambda_A(\theta) = \lim_{\tau \to \infty} \frac{1}{\tau} \ln \mathbf{E} \left(\exp \left(\theta \sum_{t=1}^{\tau} A_t \right) \right),$$

$$\Lambda_S(\theta) = \lim_{\tau \to \infty} \frac{1}{\tau} \ln \mathbf{E} \left(\exp \left(\theta \sum_{t=1}^{\tau} S_t \right) \right).$$

This simply means that we add the CGFs for the arrival and departure processes, with the sign of θ indicating the 'direction' of the flow.

Example 9.4.1. Suppose we have an $M/D/1$ queue with a Poisson arrival rate $\lambda = 1$ and a constant service rate $s = 1.43$. The traffic load on the queue is then $\lambda/s = \rho = 0.7$ (just like in Example 9.2.1). Since we have Poisson arrivals, $\Lambda_A(\theta) = \lambda(e^{\theta} - 1)$ and $\Lambda_S(-\theta) = -s\theta$. The queue is therefore described by $\Lambda(\theta) = \lambda(e^{\theta} - 1) - s\theta$. In order to find the root to this equation, the decay rate, we rewrite as

$$\theta = \ln(\theta \frac{s}{\lambda} + 1),$$

which can be iterated until convergence is archived. We can take $\theta^{(0)} = \rho = \lambda/s$ as initial value. (Note that the convergence of the iteration is dependent on the way we rewrite the equation to get an expression for θ.) After a small number of iterations, the equation converges to $\theta = 0.6755$. This is then the value of the decay rate, δ. The rate function for the system can be found by solving

$$I(x) = \max_{\theta}\{x\theta - \Lambda(\theta)\} = \left\{ x\theta - \left[\lambda \left(e^{\theta} - 1 \right) - s\theta \right] \right\}.$$

This gives

$$I(x) = (s + x) \ln \left(\frac{s + x}{\lambda} \right) + \lambda - (s + x).$$

In principle, we can solve

$$\delta = \min_{x>0} \left\{ \frac{I(x)}{x} \right\}.$$

The derivative of $\min\{I(x)/x\}$ is nonlinear with both quadratic and logarithmic terms in x. The equation can be solved numerically, but then we could have used the CGF

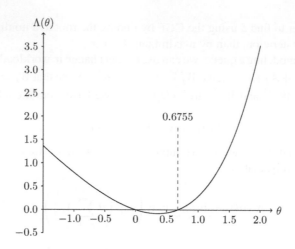

FIGURE 9.9

Cumulant generating function for the $M/D/1$ queue with load $\rho = 0.7$.

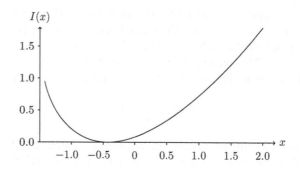

FIGURE 9.10

Rate function for the $M/D/1$ queue with load $\rho = 0.7$.

from the beginning. The CGF and the rate function for this queue are shown in Figures 9.9 and 9.10, respectively. Figure 9.11 shows estimation of δ from the rate function. A comparison between the exact and the large deviations approximation of the queue is shown in Figure 9.15.

Example 9.4.2. For an $M/M/1$, both the arrival process and the departure process are Poisson distributed, as $Po(\lambda)$ and $Po(\mu)$, respectively. Thus, $\Lambda(\theta) = \Lambda_A(\theta) + \Lambda_S(-\theta)$ gives

$$\Lambda(\theta) = \lambda\left(e^{\theta} - 1\right) + \mu\left(e^{-\theta} - 1\right).$$

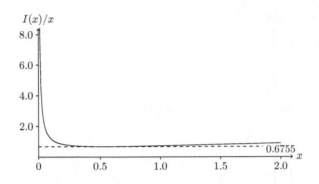

FIGURE 9.11

Estimating δ from the rate function for the $M/D/1$ queue with load $\rho = 0.7$.

Solving $\Lambda(\theta) = 0$ for θ, set $z = e^\theta$ so that

$$\lambda z + \frac{\mu}{z} - (\lambda + \mu) = 0,$$

or

$$z^2 - \frac{\lambda + \mu}{\lambda} z + \frac{\mu}{\lambda} = 0.$$

Thus,

$$z = \frac{\lambda + \mu}{2\lambda} \pm \frac{\sqrt{(\lambda + \mu)^2 - 4\lambda\mu}}{2\lambda}.$$

Finally,

$$\bar\theta = \ln\left(\frac{\lambda + \mu + \sqrt{(\lambda + \mu)^2 - 4\lambda\mu}}{2\lambda}\right),$$

and we have the exact value $\delta = \bar\theta$. The rate function is given by

$$I(x) = \max_\theta \left\{ x\theta - \lambda\left(e^\theta - 1\right) - \mu\left(e^{-\theta} - 1\right) \right\}.$$

This gives

$$\hat\theta = \ln\left(\frac{x + \sqrt{x^2 + 4\lambda\mu}}{2\lambda}\right),$$

and after substituting into the expression for $I(x)$ and some algebra, we have

$$I(x) = x \ln\left(\frac{x + \sqrt{x^2 + 4\lambda\mu}}{2\lambda}\right) + \lambda + \mu - \sqrt{x^2 + 4\lambda\mu}.$$

Figures 9.12 and 9.13 show the CGF and the rate function for this queue. Figure 9.14 shows estimation of δ from the rate function.

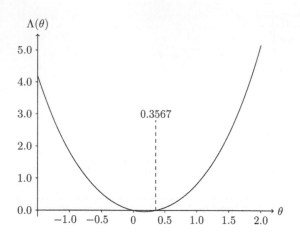

FIGURE 9.12

Cumulant generating function for the $M/M/1$ queue with load $\rho = 0.7$.

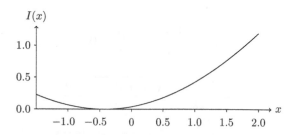

FIGURE 9.13

Rate function for the $M/M/1$ queue with load $\rho = 0.7$.

9.4.2 Buffer allocation

If the queue has only a finite waiting space, then δ gives us an estimate of what that buffer size must be in order to achieve a given probability of overflow. We can use δ to give an estimate $\hat{\mathbf{P}}(b)$ of the probability of a buffer of size b overflowing:

$$\hat{\mathbf{P}}(b) = e^{-\delta b}. \tag{9.31}$$

Example 9.4.3. Suppose we have an $M/D/1$ queue fed by a traffic load of $\rho = 0.7$. In Example 9.4.1, the decay rate is determined to $\delta = 0.6755$. The probability that Q exceeds a buffer size b is then $\mathbf{P}(Q > b) \le e^{-\delta b}$. For $b = 11$, $p_s = \mathbf{P}(Q > 11) \le 5.93 \cdot 10^{-4}$, which can be compared to the value $p_s = 2.79 \cdot 10^{-4}$ computed in Example 9.2.1.

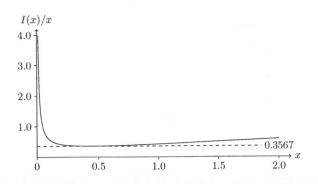

FIGURE 9.14
Estimating δ from the rate function for the $M/M/1$ queue with load $\rho = 0.7$.

9.4.3 Calculating effective bandwidths

Suppose that a queue is fed with an aggregate traffic stream, containing services with different characteristics and that the service rate can be assumed to be constant. This represents a $G/D/1$ queue and is a common model for multi-service ATM networks. The constant is achieved by segmentation of large packets into small ones (cells) and the use of a generalized processor sharing discipline. The scaled CGF for this system can be constructed from the work process, $W_t = X_{-1} + \cdots + X_{-t} - st$, so that

$$
\begin{aligned}
\Lambda(\theta) &= \lim_{t \to \infty} \frac{1}{t} \ln(\mathbf{E}(e^{\theta W_t})) \\
&= \lim_{t \to \infty} \frac{1}{t} \ln(\mathbf{E}(e^{\theta(X_{-1} + \cdots + X_{-t} - st)})) \\
&= \lim_{t \to \infty} \frac{1}{t} \ln(\mathbf{E}(e^{\theta(X_{-1} + \cdots + X_{-t})})) - s\theta \\
&= \Lambda_A(\theta) - s\theta,
\end{aligned}
$$

where Λ_A is the scaled CGF of the arrival process. Given an arrival process A_t, we may wish to know the service rate required in order to guarantee that (see Figure 9.15)

$$
\mathbf{P}(Q \geq b) \leq e^{-\delta b},
$$

for some large b and some given value of δ. The buffer size b may be known, for example, and a saturation probability specified, which then determines δ. (When there are constraints on the maximum delay incurred in a queue, a limit on buffer size can often be determined.) The required capacity can then be expressed as

$$
s = \Lambda_A(\delta)/\delta,
$$

which is called the effective bandwidth of the arrival process A_t. The function $\lambda_A(\theta)/\theta$ is known as the effective bandwidth function and the approximation $\mathbf{P}(Q > b) \approx e^{-\delta(s)b}$ as the effective bandwidth approximation.

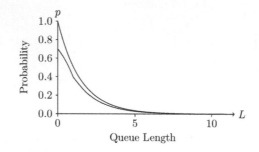

FIGURE 9.15

The exact and the large deviation approximation of the queue length of an $M/D/1$ queue.

Therefore, given the arrivals scaled CGF, we can calculate δ as a function of s as

$$\delta(s) = \max\{\theta : \Lambda_A(\theta) \leq s\theta\}. \tag{9.32}$$

Denoting the saturation probability $\mathbf{P}(Q > b) = p_b(s)$, indicating the dependence on b and s, we have

$$\delta(s) = \frac{-\ln p_b(s)}{b}.$$

Effective bandwidth is therefore a quantity summarizing the statistical properties of the sources, given a quality-service constraint (the exceedance probability of a buffer level b, where b is large), and the capacity of the resource.

Several definitions of effective bandwidth exist, which may be a little confusing. Common to these, however, is an expression for the resources required by a source, given the source and queue characteristics. This may be expressed with or without time dependence. A widely accepted two-parameter version (Kelly, 1996) is (using α to denote the effective bandwidth).

Definition 9.4.1. Let $X(0, t)$ denote the amount of work that arrives from a source in the time interval $(0, t)$, and that $X(0, t)$ has stationary increments (that is, time homogeneous). Then, the effective bandwidth of the source is

$$\alpha(s, t) = \frac{1}{st} \log \mathbf{E}\left(e^{sX(0,t)}\right), \quad 0 < s, t < \infty. \tag{9.33}$$

The effective bandwidth can be shown to have the following properties:

Proposition 9.4.1.

(1) *If $X(0, t)$ has independent increments, then $\alpha(s, t)$ does not depend upon t.*
(2) *If there exists a random variable X such that $X(0, t) = Xt$ for $t > 0$, then $\alpha(s, t) = \alpha(st, 1)$.*
(3) *If $X(0, t) = \sum_i X_i(0, t)$, where $(X_i(0, t))_i$ are independent, then*

$$\alpha(s, t) = \sum_i \alpha_i(s, t). \tag{9.34}$$

(4) *For any fixed value of* t, $\alpha(s,t)$ *is increasing in* s, *and lies between the mean and peak of the arrival rate measured over an interval of length* t, *that is,*

$$\frac{\mathbf{E}(X(0,t))}{t} \leq \alpha(s,t) \leq \frac{\bar{X}(0,t)}{t}, \tag{9.35}$$

where $\bar{X}(0,t)$ *is the maximum value, which may be infinite,*

$$\bar{X}(0,t) = \sup\{x : \mathbf{P}(X(0,t) > x) > 0\}.$$

The value of $\alpha(s,t)$ *near* $s = 0$ *is determined by the mean, variance, and higher moments of* $X(0,t)$, *while the value of* $\alpha(s,t)$ *near* $s = \infty$ *is determined mostly by the distribution of* $X(0,t)$ *near its maximum.*

The class of processes having independent and identically distributed increments is referred to as Lévy processes. Lévy processes can therefore be shown to be time-homogeneous Markov processes, which include the Gaussian, Poisson, and on-off models in this text.

Rather than deriving formulas for the effective bandwidth, we show by examples how this can be determined.

For many models of practical importance, we may use a single parameter definition

$$\alpha(s) = \frac{1}{s} \ln \mathbf{E}\left(e^{s\frac{X(0,t)}{t}}\right). \tag{9.36}$$

The exponent is then the time average of the stochastic process $X(0,t)$. The advantage of this definition is that the effective bandwidth can be written

$$\alpha(s) = \frac{\Lambda(s)}{s},$$

whenever the stochastic process is stationary. We then have

$$\alpha(s) = \lim_{t\to\infty} \alpha(s,t).$$

We determine expressions for the effective bandwidth of three important classes of traffic models.

Fluid sources. Let a stationary source be described by a finite Markov chain with stationary distribution π and matrix Q, where workload is produced at rate h_i while the chain is in state i. Then, from the backward equations for the Markov chain one can deduce that

$$\alpha(s,t) = \frac{1}{st} \log(\pi \cdot \exp[(Q + \mathbf{h}s)t] \cdot \mathbf{1}), \tag{9.37}$$

where $\mathbf{h} = \text{diag}(h_i)_i$, and

$$\lim_{t\to\infty} \alpha(s,t) = \frac{1}{s}\phi(s),$$

where $\phi(s)$ is the largest real eigenvalue of the matrix $Q + \mathbf{h}s$. If $h_1 > h_i, i \neq 1$, then relation (9.37) becomes

$$\alpha(s, t) = h_1 - \frac{1}{s}\left(\mu_1 - \frac{1}{t}\log \pi_1\right) + o\left(\frac{1}{s}\right), \quad \text{as } s \to \infty$$

where μ_1 is the transition rate out of the state with peak rate. A special case of such models is the on-off model, whose effective bandwidth has the shape depicted in Figure 9.16.

Gaussian sources. Suppose that

$$X(0, t) = \mu t + Z(t),$$

where $Z(t)$ is normally distributed with zero mean. The simplicity of calculation with Gaussian variates outweighs the problem of non-conformance of negative increments with the model. Then we have

$$\alpha(s, t) = \mu + \frac{s}{2t}\text{Var}(Z(t)),$$

and so $\alpha(s, t)$ is determined, for all s and t, by the first two terms of the expansion (9.33). By letting $\sigma^2 = \lim_{t\to\infty} \text{Var}(Z(t))/t$, we have

$$\alpha(s) = \mu + \frac{s\sigma^2}{2}.$$

The effective bandwidth for a Gaussian source is shown in Figure 9.17.

Compound poisson sources. If

$$X(0, t) = \sum_{n=1}^{N(t)} Y_n,$$

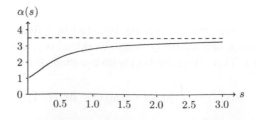

FIGURE 9.16

Effective bandwidth of an on-off source with unit mean, variance 2.5, and maximum rate 3.5.

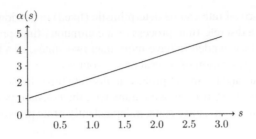

FIGURE 9.17

Effective bandwidth of an $N(1, 2.5)$ source.

where $Y_1, Y_2, \ldots$ are independent identically distributed random variables with distribution function F, and $N(t)$ is an independent Poisson process of rate ν, then

$$\alpha(s) = \frac{1}{s} \int (e^s x - 1) \nu \mathrm{d}F(x).$$

For example, if $Y_1, Y_2, \ldots$ are exponentially distributed with parameter ν, then

$$\alpha(s) = \frac{\nu}{\mu - s} \quad \text{for } s < \mu.$$

9.5 Modeling services

The first important step toward analyzing a multi-service network is the actual modeling of services. In this chapter, traffic streams are modeled on burst level or source level, that is, the dynamics within a traffic stream and how it affects the queue. The purpose of burst-level modeling is to determine the effective bandwidth of the stream, subject to the traffic model and quality-service requirements. Service modeling on a call level is discussed in Chapter 10. Such models rather reflect the arrival process and holding times of a call, and given the effective bandwidth of the source, performance metrics on call level can be determined.

We consider three types of services, which in most contexts well represent the service types of main interest. These are speech telephony, video services, and data transfer. The three service types exhibit different characteristics, both on burst level and on call level.

9.5.1 The basic on-off model

Since a packet technology is used in modern multi-service networks, it is common to model traffic sources as on-off sources. This should be considered a class of traffic models rather than a single model. The main principle is that a source can be in an active state (on), where data is sent, or in a silent state (off), where nothing is sent. There are numerous variations and extensions to this basic model. When in the sending state, the arrival rate can be deterministic (fixed) or random. The model can also be defined as a discrete-time process or a continuous-time process. The on-off model can further be extended to have more than two states, or a traffic stream can be modeled by an aggregation of several on-off sources.

Consider a queue and an arrival process in discrete time. In its simplest form, the on-off source can be described by two parameters; the constant peak rate h, and the probability a of being in an active state, independently of its previous state. Then, its mean rate is

$$m = a \cdot h.$$

If the queue has capacity C, we can define the nominal number of sources, that is, the number of identical sources that the queue can handle simultaneously with zero congestion, as

$$n_0 = C/h.$$

When $a \ll 1$, it is unlikely that all sources are active at the same time. If we allow a small congestion probability, we can allow a much greater number of sources $n > n_0$ onto the queue. Due to statistical multiplexing, more sources make more efficient use of the resources without causing much deterioration in service to a single source. The gain in efficiency is the multiplexing gain.

Since the sources are active independently of previous states and each other, the arrival process can be described by a binomial distribution. Let $p(i)$ denote the probability that i out of n sources are active at any time instant. Then

$$p(i) = \binom{n}{i} a^i (1 - a)^{n-i}. \tag{9.38}$$

The overflow probability is then, using (9.10),

$$p_l(n, n_0, a) = \frac{1}{nm} \sum_{i=\lceil n_0 \rceil}^{n} p(i) \cdot (i - n_0)h = \frac{1}{na} \sum_{i=\lceil n_0 \rceil} p_i \cdot (i - n_0).$$

If the loss probability ε is given, the maximum number of connections $n_\varepsilon(n_0, a)$ that can be accepted by the queue is

$$p_l(n_\varepsilon, n_0, a) \leq \varepsilon.$$

It can be shown that the cumulant generating function $\Lambda(\theta)$ for a single source is

$$\Lambda(\theta) = \ln\left((1 - a) + ae^{h\theta}\right). \tag{9.39}$$

However, in this case, we can find the effective bandwidth directly, as

$$\alpha = \frac{C}{n_\varepsilon} = \frac{n_0}{n_\varepsilon} \cdot h.$$

Note, that $m \leq \alpha \leq h$, where we have $\alpha = h$ for $n_\varepsilon \leq n_0$ and $\alpha \to m$ as n_ε increases.

The multiplexing gain (subject to ε) is given by

$$G = n_\varepsilon/n_0 = h/\alpha.$$

We can also determine the maximum allowable traffic load so that $p_l \leq \varepsilon$ as

$$\rho_\varepsilon = n_\varepsilon m/C = n_\varepsilon ah/C = an_\varepsilon/n_0 = aG.$$

For bursty traffic with small α the multiplexing gain can be great. In order to have a reasonable load in a bufferless system it is required that the peak rate of every single source is a small fraction of the link rate (1% or less).

9.5.2 Telephony

Digitized speech telephony is commonly coded using adaptive multi-rate (AMR), which can be modeled as an on-off source; when one of the parties speaks, the source is active with a constant peak bit rate (henceforth, assumed to be 12.2 kbps per second). When a subscriber is silent, a silence descriptor is sent. We will neglect the influence of this silence descriptor, which is rather short and dependent on the environment. It is used so that periods of silence do not lead to 'ear-deafening' or disturbing absolute silence which may occur during periods of inactivity of one of the parties.

This technique has been adopted to give a feeling of the classical telephony, where it is possible to hear background noise when neither of the participants is talking. However, these periods of inactivity were the inspiration of packet-switched speech—in fact, normally at least up to 50% of transmission cost could be saved.

Using these principles, we can model a traditional speech call as an on-off source. The period of activity is taken to be a constant 0.403, which is suggested by empirical studies. It is less than 0.5, because there might be periods where both parties are silent.

If AMR traffic is modeled as independent on-off sources that are binomially distributed, the resulting traffic streams will be smoother, since the peakness of the binomial distribution is $1 - a < 1$. This can be visualized as follows. Each traffic stream is divided into packets which are interleaved (multiplexed). These packets are typically much shorter than the holding time of the call. Therefore, the system will have faster changes between states as compared to the situation where traffic is served without packetization.

This can be viewed as a sampling problem. Suppose that all possible samples of size N are drawn without replacement from a finite population of size $N_p > N$. If we denote the mean and the standard distribution by μ_X and σ_X and the population mean and the standard deviation by μ and σ, respectively, then

$$\mu_X = \mu,$$

$$\sigma_X = \frac{\sigma}{\sqrt{N}}\sqrt{\frac{N_p - N}{N_p - 1}}.$$

For an infinite population (like in the assumption for Poisson arrivals), the mean and standard deviation of the sampling distribution $\mu_X = \mu$ and $\sigma_X = \sigma/\sqrt{N}$.

Example 9.5.1.   For small examples, the performance can be calculated exactly. Suppose a 500 kbps link is offered a number of AMR sources. We can model an AMR source as an on-off source and investigate the effect of the aggregation of a number of sources offered to the link. Let, therefore, the source rate in the on state be $h = 12.2$ kbps and the activity factor $a = 0.403$. The given required loss probability is $p_l \leq 2.0 \cdot 10^{-3}$. The aggregation of N sources is described by the binomial distribution (9.38).

The loss probability (9.10) for increasing values of N can then be computed as $p_l = \sum_{i=0}^{N} \max\{0, ih - C\} p_i / \sum_{i=0}^{N} i p_i$. Similarly, the saturation probability can be found as $p_s = 1 - \sum_{i=0}^{M} p_i$, where M is the integer number of sources such that $ih \leq C$. Using the loss probability in this case, the number of sources is found to be $N_{max} = 73$. The saturation probability is then $p_s = 1 - \sum_{i=0}^{40} p_i = 4.4 \cdot 10^{-3}$.

The effective bandwidth of an AMR source is found as follows. The nominal number of sources is $N_0 = C/h = 40.98$, so that $\alpha = N_0 h / N_{max} = 6.85$. The multiplexing gain is $G = h/\alpha = 1.78$, and the load on the server is $\rho = aG = 0.72$.

Another approach is to approximate the sources by a normal distribution. We can use the statistics for the number of sources (but the result would be the same if we recalculate the values into bandwidth requirements). The nominal number of sources is $N_0 = 40.98$, the mean number of active sources $\mu_X = aN_{max} = 29.4$, and its variance $\sigma_X^2 = a(N_{max} - \mu_X) = 17.6$. Standardizing the argument, we have $z = \frac{N_0 - \mu_X}{\sigma_X} = 2.76$. The corresponding area under the standard normal probability density function is $1 - \Phi(z) = 2.9 \cdot 10^{-3}$. Note that the tail area under the normal distribution corresponds to the saturation probability, so comparing with this probability obtained earlier, the result is within limits. The effective bandwidth is again found by $\alpha = N_0 h / N_{max} = 6.85$.

Going the other way and using $p_s = 4.4 \cdot 10^{-3}$, we find $\eta = 2.62$ and $\alpha = 6.75$. By evaluating Equation (9.18) for s different number of sources, a maximum of $n = 74$ can be accommodated, using the p_s as limit. In spite of the fact that p_s is more than twice as large as p_l, the results are very close. This shows that the approximation of the loss probability by the saturation probability is rather good.

Finally, the approximate number of sources can be found by using large deviation techniques. By computing the rate function $I(x)$ with $x = N_0/N_{max}$ for increasing number of sources N, we find the mean number of sources m at the point x where $I(x) = 0$. Using the relation $\mathbf{P}(S_N > N_0) \leq e^{-I(x)}$, we can then estimate either p_s or p_l using the improved approximations (9.19) or (9.20), respectively. Trying a different number of sources, we find that $N_{max} = 77$ gives an improved loss probability approximation less than $2.0 \cdot 10^{-3}$. Note, that the results can be obtained either by considering the number of sources, or the actual bandwidth.

Also note, that for $s \to 0$ the mean is obtained, and for $s \to \infty$ the maximum bit rate is indicated (See Figure 9.16).

9.5.3 The Markovian additive process

Many traffic types exhibit burstiness and autocorrelation, and for these types of traffic, of which streaming video and data transfer are used as examples, a process modulated by a Markov process has been proposed (Botvich and Duffield, 1995; Buffet and Duffield, 2012; and Duffield, 1993). We let a jump process be controlled by a Markov chain where the parameters are set so that the probability of remaining in either state (active or silent) is high if the process was in this state immediately before. The resulting arrival stream is also burstier as a result.

Suppose there are N independent sources, controlled by the same Markov chain, which jumps between the active state and the silent state. This is represented by the set $\{0, 1\}$, with zero denoting the silent state and unity the active. The transition probabilities from the silent to the active state and vice versa are defined by

$$a = \mathbf{P}(X_t = 1 | X_{t-1} = 0),$$

$$d = \mathbf{P}(X_t = 0 | X_{t-1} = 1).$$

The controlling Markov chain is denoted

$$T = \begin{pmatrix} 1-a & a \\ d & 1-d \end{pmatrix},$$

and the steady-state probabilities are $\pi_0 = \frac{d}{a+d}$ and $\pi_1 = \frac{a}{a+d}$. The closer the parameters a and d are to zero, the burstier the model is. This is commonly referred to as a Markov additive process (MAP). The model can be used in discrete time, which makes simulation rather straightforward.

The lengths of each source remaining in a state are geometrically distributed. The duration of an active period is, therefore, geometrically distributed with parameter $(1 - d)$ and is $\sum_{n=1}^{\infty} dn(1-d)^{n-1} = 1/d$ time units. Similarly, the mean silence duration is $1/a$ time units.

It is necessary to impose a stability condition to have a stable steady-state queue length. We have:

Proposition 9.5.1.   *Suppose a queue with deterministic service capacity s is fed by N sources, described by a $MAP(a, d)$ process. Then the queue waiting time has a unique stationary distribution provided that*

$$\frac{a}{a+d} < s/N.$$

We can express the maximum number of sources that the system can handle and the multiplexer load as

$$N_{\max} = s(a + d)/a, \tag{9.40}$$

$$\rho = N/N_{\max} = a/s(a + d) \le 1. \tag{9.41}$$

It is only necessary to consider cases where $s/N < 1$, since otherwise traffic from all N sources can be accommodated without causing congestion, and the stationary queue is trivially zero.

It can be shown that the scaled matrix corresponding to T is

$$T_\theta = \begin{pmatrix} 1-a & ae^\theta \\ d & (1-d)e^\theta \end{pmatrix}. \tag{9.42}$$

The scaled CGF is given by the largest eigenvalue of the scaled matrix (9.42). This eigenvalue can be computed from the *characteristic equation* $\det(T - xI)$, where I is the identity matrix of the same order as T. In other words,

$$\begin{vmatrix} 1-a-x & ae^\theta \\ d & (1-d)e^\theta - x \end{vmatrix} = 0.$$

This gives, after some algebra,

$$x = \frac{(1-d)e^\theta + 1 - a}{2} \pm \frac{\sqrt{\left((1-d)e^\theta + 1 - a\right)^2 - 4(1-a-d)e^\theta}}{2}.$$

We arrive at the scaled CGF for the queue $\lambda_A(\theta) = s\theta$

$$\Lambda_A(\theta) = \ln\left(\frac{(1-d)e^\theta + 1 - a + \sqrt{\left((1-d)e^\theta + 1 - a\right)^2 - 4(1-a-d)e^\theta}}{2}\right),$$

or, in a queue with a deterministic service rate s, the equation $\Lambda_A(\theta) - s\theta = 0$, by taking the exponential of both sides,

$$1 = \frac{1}{2}e^{-s\theta}\left((1-d)e^\theta + 1 - a + \sqrt{\left((1-d)e^\theta + 1 - a\right)^2 - 4(1-a-d)e^\theta}\right), \tag{9.43}$$

for a single source and the total service rate is divided by the number of sources. The solution $\hat\theta$ to this equation is the decay rate δ, which can be found numerically.

The basic on-off model described in Subsection 9.5.1 is a special case of a Markov additive process. Thus, and (9.43) reduces to

$$1 = e^{-s\theta}\left(1 - a + ae^\theta\right),$$

since $a + d = 1$.

9.5.4 Streaming video

Video services can be classified into two categories—video conferencing and streaming movies. A video conferencing source has relatively few scene changes, whereas in movie sources, scene changes often occur frequently. Therefore, these two types of video sources are usually encoded in different ways, using different algorithms. Video conferencing allows for higher compression rate due to few changes between frames than a streaming movie source. As a result, the former is also easier to model mathematically.

Two of the most prominent features of a video source are the high burstiness and the autocorrelation of the stream of the arrival process. A video source is sending out frames, comparable to still pictures making up a movie. In video compression, data already sent in earlier frames can be used to reduce the amount of data required for a new frame. As a first approximation, therefore, a video source can be assumed to have a state of high activity, where many scene changes occur and large amounts of data need to be sent, and a state of low activity, where only modifications to a previously sent frame need to be transmitted. A salient feature of most video sources is the autocorrelation between these states, which in general, is a result of filmmaking or teleconferencing behavior. During periods of low activity, the probability that abrupt scene changes will happen is low, and the source remaining in low activity is high. During periods of high activity, on the other hand, the probability of remaining in high activity in the next frame is high, and the probability of the source jumping to low activity state is low.

This means that, in contrast to the service request arrivals, which often are assumed to be memoryless (Poisson arrivals), the sources on a burst level are Markovian, where the state probabilities depend on the state previously in.

A suitable model for video sources are therefore Markov modulated models, such as the Markov additive process presented above (although many other models exist). Such a model can capture the burstiness and the autocorrelation of the source. To achieve flexibility in modeling, another common strategy is to let a source be represented by the superposition of a certain number of identical, independent on-off minisources, each controlled by a two-state Markov chain.

With this model, the arrival rate at a queue is binomially distributed (or, approximately normal). This modeling approach seems to give reasonable results for aggregates of several video sources. When fed to a queue, the resulting queueing model can be analyzed relatively easy.

Since changes in activity levels of the source occur mostly at frame changes, a discrete-time model is appropriate. We therefore consider Markov modulated processes in discrete time.

On the other hand, the rate distribution of a minisource model is always binomial or approximately Gaussian and noticeably different from that of real data which follows a negative binomial or gamma distribution. However, the approach is sufficiently accurate when it is applied to model the aggregate traffic of several sources.

The bit rate produced by a video conferencing can be considered as the aggregate output of M independent minisources, each alternating between on and off states. Each minisource in the on state produces traffic at the constant rate of A bits/s, whereas in the off state, no traffic is generated. The time intervals (in slots) spent in the on and off states are geometrically distributed with parameters $\alpha = 1/p$ and $\beta = 1/q$, where p and q are the mean times spent in on (and off) slots. A minisource in the on (or off) state makes a transition toward the off (or on) state at the end of a slot with probability α (or β). The minisource activity factor is $\Psi_v = p/(p + q)$. We have considered that in a slot at most one minisource can make a transition from on to off or vice versa. Hence, there are no sudden traffic variations for a video conferencing source. The parameters p, q, and A of a minisource can be modeled as follows

$$A = \frac{\mu}{M} + \frac{\sigma^2}{\mu},$$

$$q = \frac{1}{aT_s}\left(1 + \frac{M\sigma^2}{\mu^2}\right),$$

$$p = \frac{1}{aT_s}\left(1 + \frac{\mu^2}{M\sigma^2}\right),$$

where μ is the mean bit rate, σ^2 the variance and the parameter a characterizes the slope of the autocovariance function of the bit rate produced (if a decreases, a more correlated cell generation process is obtained).

The state probability of the modulating process is binomial

$$\mathbf{P}(s = i) = \binom{M}{i}\Psi_v^i(1 - \Psi_v)^{M-i}.$$

In this model, a bit rate iA is produced when the video source is in state i. Hence, the mean bit rate is $\mu = M\Psi_v A$ and the peak cell rate is MA. We can consider $5 \leq M \leq 10$ to model a single video source. We have assumed $\mu = 512$ kbps, $\sigma = 256$ kpbs, $M = 10$ and $a = 1/3.8$. On the basis of these values, the minisource activity factor Ψ_v is 0.28. Also assumed are the values $D_{max} = 90$ ms and $p_l = 10^{-4}$.

Example 9.5.2. Suppose that a video source can be modeled by a Markov additive process with the controlling Markov chain having transition probabilities $a = 0.06$ and $d = 0.04$. Since $a + d = 0.1 < 1$, the process has positive correlation between states. The Markov chain can be described by

$$\begin{pmatrix} 1 - a & a \\ d & 1 - d \end{pmatrix} = \begin{pmatrix} 0.94 & 0.06 \\ 0.04 & 0.96 \end{pmatrix}.$$

Consider a single source ($N_0 = 1$) and a queue service rate of $s = 0.7$. Thus, we have $a/(a + d) < s$, so the queue has stable equilibrium distribution. The decay rate δ can be obtained by solving (9.43) numerically. This gives $\hat{\theta} = \delta = 0.05$. The multiplexing gain and the load are $G = N_{max}/N_0 = s(a + d)/a = 1.17$ and $\rho = a/s(a + d) = 0.86$, respectively. If h is the peak load, the effective bandwidth is $\alpha = \rho h$. (We assume that $h = 1$ in this example.) Congestion starts to occur at $C = N_{max}$, and therefore we may be interested in the probability $\mathbf{P}(X > C) \leq e^{\delta C} = 0.94$, which is a very high figure. In order to find a level for which the queue length probability is, say, 1%, we would need a buffer of $b = -\ln(0.01)/\delta = 92$ jobs. The waiting time to empty a buffer that full is $\mathbf{P}(W > t) = \mathbf{P}(Q > st)$ so that $b = 92/0.7 = 130$ time units. This is probably far too much delay for the service, and we need to change the relation between service rate and buffer size. By changing the server rate to 0.98, we get $\delta = 2.0$ and for a blocking rate of 1%, a buffer size of $b = 2.3$ and an approximate waiting time of 3.3 time units. By reducing the load from $\rho = 86\%$ to 61%, the performance of the queue has been substantially improved. Three simulations of the initial, heavily loaded queue are shown in Figure 9.24 together with the large deviation bound.

Example 9.5.3.   The same method as used in Example 9.5.2 can be used for Example 9.5.1. Assuming a saturation probability of $p_s = 4.4 \cdot 10^{-3}$ and activity rate of $a = 0.403$. By using a service rate of $s = 0.73$, we get a multiplexor load of $\rho = a/s = 0.55$, multiplexing gain $G = 1.81$ and effective bandwidth (with $h = 12.2$) $\alpha = \rho h = 6.74$. This is an approximation to the result in Example 9.5.1, by the use of the saturation probability and its bound in place of the loss probability.

9.5.5 Data services

Data services such as web browsing, are governed by a complex mix of human and machine-to-machine interaction. A web page, for example, usually contains embedded objects which automatically initiate HTTP requests to different servers. Human behavior also influences the intensity of HTTP requests. Typically, when searching for a specific result on the web, the intensity of requests per time unit is high. It is then followed by reading periods when the activity is low (see Figures 9.18 and 9.19).

There are many models proposed, that attempt to capture this complex interactive behavior. One such model is an on-off model extended to several states. Another approach is to aggregate simple on-off minisources, where each source may have different characteristics. (see Figure 9.20).

We will focus on the Markovian additive process on burst level and the negative binomial distribution on call level as models of data services. The first is analytically

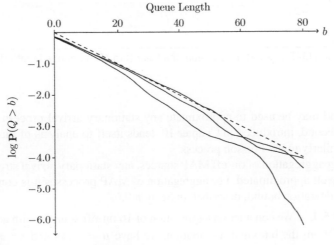

FIGURE 9.18

Three simulations of a single-source MAP in an $M/D/1$ queue.

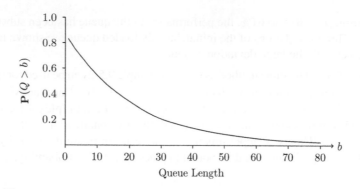

FIGURE 9.19

Simulation of a MAP in an *M/D/*1 queue, queue length distribution.

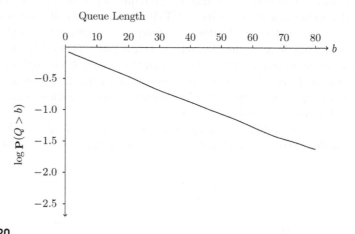

FIGURE 9.20

Simulation of a MAP in an *M/D/*1 queue, the logarithm of the queue length distribution.

tractable and may be used to approximate any stationary arrival process arbitrarily well. The second, introduced in Chapter 10, lends itself to analysis of multi-service queues, similarly to the Poisson process.

By using aggregation of on-off MAP sources, any stationary arrival stream can be arbitrarily well approximated. For aggregation of MAP processes it is convenient to use a large deviation bound, described in Section 9.6.

Example 9.5.4.   We consider an aggregation of 16 on-off sources with activity rate $a = 0.28$. From the binomial distribution, we have $\mu = 4.48$ and $\sigma^2 = 3.23$. In this chapter, we analyze multi-service network where capacity is divided in basic bandwidth units (BBU). If AMR is assumed to have a bandwidth of 12.2 kbps, it is

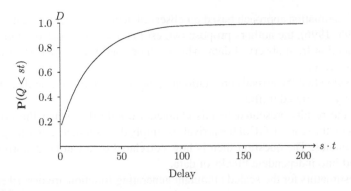

FIGURE 9.21

Simulation of a MAP in an $M/D/1$ queue, delay probability.

natural to use this as BBU. The aggregated sources, denoted by Y, thus have a mean of $\mu_Y = 4.48 \cdot 12.2 = 54.7$ and variance $\sigma^2 = 39.4$. Thus $\mu_Y + 1.5\sigma_Y \approx 64$.

By using the normal approximation, we may chose a number of sources per stream so that there is some saturation probability, say, $p_s = 0.001$. This corresponds (by inversion of the Q-function) to $\eta = 3.1$, or from (9.18), $\alpha = 10.05$. Note, that for this aggregation of on-off sources, $\mu < \alpha < N_{\max} = 16$.

9.5.6 Simulation of Markov Additive Processes

To generate characteristics of a queue fed by a MAP by simulation is rather straigtforward. We consider a $MAP/D/1$ queue and chose a deterministic time step for both the Markov chain modulating the MAP and the service intensity. These time steps need not be of equal length. Usually, to obtain interesting results, the peak rate of the MAP should exceed the service rate s.

In one time step can we therefore simulate state change of the MAP by comparing the probabilities in the Markov chain with a uniform random number. Should the state be in active state, one job, say, is added to the queue, if in idle state, nothing. Next, the amount $s < 1$ of work load is processed and removed from the queue. Simulation of queues with finite buffers are easily accomplished, and various perfomance metrics can be calculated by introduce counters into the simulation process. Figure 9.18 shows the loss probability of three simulated traces of a $MAP/D/1/b$ process. The queue length is illustrated in Figures 9.19 and 9.20, and the delay probability in Figure 9.21.

9.6 Estimation techniques

The scaled cumulant generating function can only be calculated exactly for a few very simple arrival processes. Ideally, any arrival process and related queue performance metrics should be possible to analyze without too much difficulty. Therefore, a numerical estimation approach based on observed data is of great interest. In Crosby et al. (1995, 1996), the authors propose two estimators of the scaled cumulant generating function from observed data, which can be used for estimation of the decay rate δ of a queue.

We assume that the arrivals are stationary, that is, there is no time dependence of the intensity of offered traffic.

From the results presented in this chapter, we note that when the arrivals are independent, the scaled CGF of the arrivals is simply the cumulant generating function of the estimates. The question then is how correlated data can be decorrelated, or, partitioned into independent blocks of data.

Two estimators for the scaled cumulant generating function are described, which can be used to determine the decay rate δ in a queue based on an observed (or simulated) arrival process. The main idea is to form blocks of arrivals, defined by the aggregation of arrivals occurring within a certain time window, and formulate empirical estimators based on these blocks. However, the choice of these block sizes is crucial for the accuracy of the estimates. The blocks should be long enough to decorrelate the arrivals as much as possible, but not too long to smooth out significant bursts of data.

9.6.1   Constant time-window estimator

The main idea is to form blocks of data from the arrival (or queue output) process. By appropriately choosing a time-window size T, this partition process of the data decorrelates the arrivals, so that the arrivals in each block can be considered to be approximately independent. By the assumption of stationarity, the sequence is also identically distributed. The stronger the dependence in the data is, the larger the window size needs to be to decorrelate the sequence. It follows that the data sequence needs to be longer for strongly correlated data in order to obtain reasonable accuracy in the estimator. On the other hand, large bursts of data, which are contributing the most to the queue length should be captured, require a small window size.

For a large class of arrival processes it is possible to find a window size T such that the aggregated arrivals A_T are approximately independent and identically distributed. For any T such that $A_T = \sum_{t=1}^{T} X_t$, where X_t is the number of arrivals in each discrete time instant t, we have

$$\hat{\Lambda}_A(\theta) \approx \Lambda_A^{(T)}(\theta) = \frac{1}{T} \ln \mathbf{E}\left(e^{\theta A_T}\right).$$

If the data is approximately independent and identically distributed, an empirical estimator $\hat{\Lambda}_T(\cdot)$ of the cumulant generating function is given by

$$\hat{\Lambda}_T(\theta) \frac{1}{T} \ln \frac{1}{K}\left(\sum_{k=1}^{K}(\theta A_T^{(k)})\right), \tag{9.44}$$

where K is the number of blocks of arrivals. The estimator (9.44) is easy to use and is based on minimal assumptions about the traffic. It is, however, necessary to choose a window size T. The size of T should be sufficiently large so that successive blocks of arrivals are approximately independent, but not so large that characteristic burstiness is smoothed out.

Since the block size T is dependent on the source data, a trial-and-error approach is often the only way to determine this parameter. Once the estimated scaled cumulant generating function $\hat{\Lambda}_A(\theta)$ has been determined, the scaled CGF of the workload process, $\hat{\Lambda}(\theta)$, is then calculated through $\hat{\Lambda}(\theta) = \hat{\Lambda}_A(\theta) - s\theta$ and the estimated decay rate $\hat{\delta}$ is given by

$$\hat{\delta} \triangleq \max\{\theta : \hat{\Lambda}(\theta) \leq 0\}. \tag{9.45}$$

For any set of data, we must typically determine both T and θ. As mentioned, the block size T depends on the data, so a systematic approach other than trial-and-error is difficult to devise. For a given T, however, θ also needs to be determined from the data. A simple and robust approach to do so is by using the bisection method (see for example Dahlqvist and Björk, 1974).

Let $f(\theta) = \hat{\Lambda}(\theta) - s\theta$, which clearly is continuous. We want to find a θ so that $f(\theta) = 0$. First, we create an interval (a_0, b_1), such that $f(a_0) \cdot f(b_0) < 0$. We can determine a sequence of intervals $(a_1, b_1) \supset (a_2, b_2) \supset (a_3, b_3) \supset \ldots$, containing the root θ of $f(\theta) = 0$ as follows. Suppose that $f(a_0) < 0$ and $f(b_0) > 0$ and let $I_0 = (a_0, b_0)$ be the first interval containing θ. For each interval $I_{k-1}, k = 1, 2, \ldots$, calculate its midpoint

$$m_k = \frac{1}{2}(a_{k-1} + b_{k-1}).$$

Compute $f(m_k)$, which we may assume to be non-zero (since otherwise we have found a root of the equation), and let the new interval be

$$I_k = (a_k, b_k) = \begin{cases} (m_k, b_{k-1}), & \text{if } f(m_k) < 0 \\ (a_{k-1}, m_k), & \text{if } f(m_k) > 0 \end{cases}.$$

From the construction of (a_k, b_k) we have that $f(a_k) < 0$ and $f(b_k) > 0$, and that each interval I_k contains a root of $f(\theta) = 0$. After n steps, we know that the θ lies in the interval (a_k, b_k) having

$$(b_n - a_n) = 2^{-1}(b_{n-1} - a_{n-1}) = \ldots = 2^{-n}(b_0 - a_0).$$

After n iterations, we have the estimate of the root

$$\theta = m_{n+1} \pm \epsilon, \quad \epsilon = 2^{-n-1}(b_0 - a_0).$$

Example 9.6.1.   Estimator (9.44) used on three simulated two-state MAP arrival processes with $a = 0.06, d = 0.04$, each run for 10,000 cycles, gave decay rate estimates shown in Figure 9.22. For each T, bisection was used to determine $\hat{\delta}$ with an error $\epsilon \leq 10^{-4}$. The initially high estimates drop to a fairly stable value roughly in the range $50 \leq T \leq 200$. For larger T, the estimates start to oscillate between high and low values, indicating that this range of T is useless for estimation. A time window $T = 125$ gives $\hat{\delta} = 0.052$ for one of the simulated data sets, which is comparable

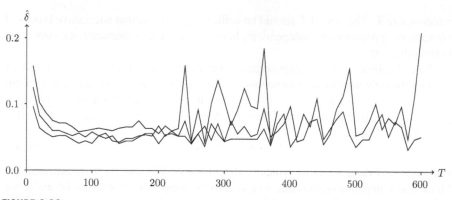

FIGURE 9.22

Estimation of the decay rate δ for a simulated MAP process using a constant time window T.

to the exact value $\delta = 0.0501$. Unfortunately, correlation between arrivals makes simulation difficult in that much longer simulation runs need to be made to obtain accurate results. An additional complication is the choice of time window size, which usually cannot be ascertained beforehand. Therefore, different window sizes should be tested. In the same simulation set up, using a window size $T = 10$, the decay rate was estimated to $\hat{\delta} = 0.12$, a rather bad estimate. With this choice of T, the data is not decorrelated enough, which leads to high variance in the samples. In the case of MAP, a window size of a magnitude of at least $T \approx 1/\min\{a, d\}$ seems to be appropriate.

It does not matter whether we want to estimate the CGF of the arrivals $\Lambda_A(\cdot)$ based on data generated by the arrival process itself, or the output from the queue $\Lambda_S(\cdot)$. The principle is the same. If the arrival and departure processes are independent, we can construct the estimator

$$\hat{\Lambda}(\theta) = \hat{\Lambda}_A(\theta) + \hat{\Lambda}_S(-\theta).$$

9.6.2 Variable time-window estimator

Not only can we easily calculate the scaled CGF for independent sequences, but we can also do so for Markov sequences. Thus, if we use an estimator based on a Markov structure, we only need T to be large enough that the blocks A_T are approximately Markov; such an estimator is tailor-made for the case of Markovian arrivals.

An alternative estimator to the estimator (9.44) is using a variable time-window size T_n rather than a fixed T. A variable window can better capture burstiness in the arrivals by letting the window size being short in periods of high activity and long in periods of low activity. Such blocks of arrivals will then approximately capture the Markovian structure of the arrival process.

An estimator based on variable time-window sizes is based on a result for random time changes, which is only coarsely outlined here. Suppose a stochastic process S_t satisfies a large deviation principle

$$\mathbf{P}(S_t \approx x) \asymp e^{-tI(x)}. \tag{9.46}$$

This remains true for a deterministic sequence t_n of times, with t replaced by t_n in (9.46). It can then be shown that for a sequence of random times T_n such that $T_n/n \to \tau$, and the corresponding sequence of arrivals S_{T_n}, we have the joint large deviation principle

$$\mathbf{P}(S_{T_n} \approx s, T_n \approx \tau) \asymp e^{-nJ(s,\tau)}$$

for some rate function $J(s, \tau)$. By introducing the counting function N_t associated with T_n, we have

$$N_t = \sup\{n : T_n \leq t\},$$

where N_t is the number of random times T_n which have occurred up to time t. Now, if S_t and N_t satisfy a joint large deviation principle

$$\mathbf{P}(S_t \approx x, N_t/t \approx y) \asymp e^{-tI(x,y)},$$

then the rate function $J(\cdot)$ can be expressed in terms of the rate function for S_τ, $I(\cdot)$ so that

$$J(s, \tau) = \frac{I(s, 1/\tau)}{\tau}.$$

The scaled cumulant generating function is now given by

$$\bar{\Lambda}(\theta) = \max_{x>0,\tau>0} \tau\{\theta x - I(x, 1/\tau)\}.$$

It follows that

$$\delta = \max_{\theta}\{\theta : \bar{\Lambda}(\theta) \leq 0\},$$

as before. The definition of $\bar{\Lambda}(\theta)$ is

$$\bar{\Lambda}(\theta) = \lim_{n\to\infty} \frac{1}{n} \ln \mathbf{E}\left(e^{\theta(X_{-1}+\cdots+X_{-T_n}-sT_n)}\right), \tag{9.47}$$

where s is the service rate of the queue and T_n is any increasing sequence of random times. Details of this derivation can be found in Lewis (1996) and O'Connell (1999), for example.

Expression (9.47) suggests the form of the estimator. We can choose any sequence $\{T_n\}$ to define the window size applied to the arrivals.

A straightforward way of constructing a variable window size is by letting the size T_N be the time it takes for N cells to arrive at the queue. If the arrival intensity is high, the window size will be short and larger window sizes follow at lower activity.

Let W_N be the modified workload process, where W_N is the workload after N arrivals and t_i denotes the i th inter-arrival time. Then

$$W_N = N - (st_1 + st_2 + \cdots + st_N) = N - sT_N,$$

valid for both discrete and continuous times t_i. The modified workload process follows immediately from the definition (9.47) of $\tilde{\Lambda}(\cdot)$. The scaled cumulant generating function for the modified workload process is now given by

$$\Lambda(\vartheta) = \lim_{N \to \infty} \frac{1}{N} \ln \mathbf{E}(e^{\vartheta} W_N)$$

$$= \lim_{N \to \infty} \frac{1}{N} \ln \mathbf{E}(e^{\vartheta} (N - sT_N)),$$

where s is the service rate of the queue.

The estimator $\tilde{\Lambda}(\vartheta)$ can now be constructed by computing the scaled CGF from the empirical distribution of arrival block lengths T_k. If the arrivals are split into K time intervals each containing N arrivals, then $\tilde{\Lambda}(\vartheta)$ is given by

$$\tilde{\Lambda}_N(\vartheta) = \frac{1}{N} \ln \frac{1}{K} \sum_{k=1}^{K} \exp\left(\vartheta N - \vartheta T_N^{(k)}\right), \tag{9.48}$$

where $T_N^{(k)}$ is the length of the k th time interval.

The decay rate can be obtained from the scaled cumulant generating function by

$$\tilde{\delta} = \max\{\vartheta > 0 : \tilde{\Lambda}(\vartheta) \leq 0\},$$

The partition of the data according to variable time-window sizes better reflects the burstiness of the arrivals and may give estimators with lower bias and variance, as compared to (9.44).

Example 9.6.2.   Using the estimator (9.48) on the same data as in Example 9.6.1 and proceeding in the same way, successively trying different arrival window sizes N which determine the time window sizes T_N, we get the result shown in Figure 9.23. The estimator was applied to three sets of simulated data. For large N, there may not exist a solution θ since

$$\tilde{\Lambda}(\vartheta) < s\vartheta,$$

for all ϑ. The estimated decay rate $\tilde{\delta}$ is fairly stable in the range $N \in (20, 150)$. With $N = 85$, the estimate is $\tilde{\delta} = 0.053$ for one of the simulated data sets. For each N, bisection was used to find $\tilde{\delta}$ with an error $\epsilon \leq 10^{-4}$.

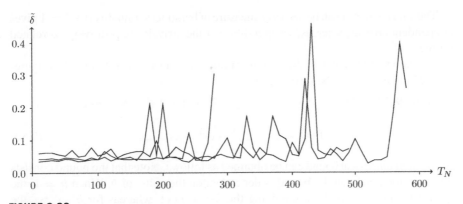

FIGURE 9.23

Estimation of the decay rate δ for a simulated MAP process using a variable time window T_N.

9.7 Finite buffers

The Markov additive process is a flexible model for bursty traffic. Buffet and Duffield (2012), Duffield (1993) and Duffield et al. (1995) construct a large deviation bound which is tighter than the bound obtained by the methods presented in Section 9.5. A detailed derivation of this bound can be found in these references.

Let the discrete-time Markov additive process be defined by the Markov chain

$$T = \begin{pmatrix} 1-a & a \\ d & 1-d \end{pmatrix},$$

and let the process generate a unit of work (packet) when in the on state. The steady-state probabilities are found to be $\pi_0 = \frac{d}{a+d}$ and $\pi_1 = \frac{a}{a+d}$. The generated work is offered to a queue with service rate s. Note that we consider the virtual waiting time, and since packets are assumed to be of constant length, the virtual waiting time is identical to the queue length at each epoch.

The bound can be formulated in terms of three functions

$$
\begin{aligned}
y(x) &:= \frac{x\left((1-a)+ax\right)}{(1-d)x+d}, \\
f(x) &:= \frac{ax+d}{(a+d)x^{s/M}}, \\
g(x) &:= \frac{y(x)^{s/M}}{ax+1-a},
\end{aligned}
\tag{9.49}
$$

where M is the number of sources.

The sum $a+d \le 1$ can be seen as a measure of burstiness. Equality $a+d = 1$ gives independent arrivals, whereas, when $a + d < 1$ the arrivals are positively correlated in time.

An exponential bound on the virtual waiting time caused by a stream of M aggregated MAP arrivals can be formulated as:

Theorem 9.7.1. *Let $a + d \le 1$ and $x \ge 1$. If $g(x) \ge 1$ then for any $b \ge 1$*

$$\mathbf{P}(Q \ge b) \le x^{-1} f(x)^M y(x)^{1-b}. \tag{9.50}$$

The desired numerical bound can be obtained as follows. Note that the bound consists of two factors, and that the bound is dependent on the value of b. When $b = 1$, the tightest bound is found by minimizing the factor $f(x)$, whereas for $b \to \infty$, the bound becomes better with increasing values of $y(x)$. The value of x minimizing $f(x)$ can easily be found explicitly. The maximum value of $y(x)$ can be found by iteration, as shown in Section 9.5.

The procedure is justified by the following:

Theorem 9.7.2. *Let $a + d \le 1$, $s/M > a/(a + d)$, and $x > 1$. Then, it holds that*

(1) *if $f'(x) = 0$, then $g(x) > 1$;*
(2) *if $g(x) \ge 1$, then $f(x) \le 1$.*

By differentiating f, we get the value x^* such that $f'(x^*) = 0$,

$$x^* = d(s/M)/a(1 - s/M).$$

By Theorem 9.7.2, we have that $g(x_\sigma) > 1$ so that $f(x^*) \le 1$.

Inserting $x = x^*$ in (9.50) gives the explicit bound for the asymptotic queue length distribution

$$\mathbf{P}(Q \ge b) \le \frac{1}{y(x^*)^{b-1}x^*} \left(\frac{ax^* + d}{(a + d)(x^*)^{s/M}} \right)^M$$

$$= \frac{a \left(1 - (a + (s/M)(1 - a - d))\right)}{d \left(a + (s/M)(1 - a - d)\right)} \left[\frac{(1 - (s/M)) \left(a + (s/M)(1 - a - d)\right)}{(s/M) \left(1 - (a + (s/M)(1 - a - d))\right)} \right]^b$$

$$\times \left[\frac{1}{a + d} \left(\frac{d}{1 - (s/M)} \right)^{1-(s/M)} \left(\frac{a}{s/M} \right)^{s/M} \right]^M, \tag{9.51}$$

for any $b \ge 1$, provided that the conditions $a + d \le 1$ and $s/M > a/(a + d)$ are satisfied. It is worth noting that $f(x^*)^M$ is in fact the estimate of the overflow probability in a bufferless model, that is $\mathbf{P}(Q \ge 1)$.

It is convenient to express the first factor in (9.50) in exponential form, so that

$$f(x^*)^M = e^{-\mu M}, \tag{9.52}$$

where μ is determined by a queue fed by a single source and served at rate s/M. It is then possible to immediately extend the bounds obtained for the single-source case to a queue fed with M independent sources.

Equation (9.52) also illustrates the statistical multiplexing gain. If $\mu > 0$, then by adding another source and increasing the service rate so that constant load is maintained, the probability of the queue exceeding any limit b is reduced by a factor e^{μ}.

Using (9.52), the general bound for M sources can be written as

$$\mathbf{P}(Q > b) \leq e^{-\mu M - \delta b}. \tag{9.53}$$

The bound is therefore determined by the two constants δ and μ. The decay rate δ is found by solving Equation (9.30).

Rather than using $f(x^*)$ in (9.49), Duffield et al. (1995) suggest the approximation $\tilde{\mu}$ to μ

$$\tilde{\mu} = (s/M)\ln\left(\frac{s/M}{\pi_1}\right) + (1 - s/M)\ln\left(\frac{1 - s/M}{1 - \pi_1}\right) \tag{9.54}$$

to be used in (9.53).

These bounds are valid for queues with infinite buffers, but the can also be useful for queues with finite buffers. Denote by Q_{∞} a queue with infinite buffer and by Q_b a queue with a finite buffer of size b. Then,

$$\mathbf{P}(Q_b > b) \leq \mathbf{P}(Q_{\infty} > b),$$

so any upper bound for Q_{∞} is also an upper bound for Q_b. Thus,

$$\mathbf{P}(Q_b > b) \leq e^{-\mu L - \delta b}.$$

As the buffer size b is increasing, the bound becomes better. For small buffers, however, it may be too pessimistic.

The upper bounds on the queue length distributions can be used to derive upper bounds for important QoS parameters. The probability that a queue with a finite buffer of size b overflows is given by

$$\mathbf{P}(Q_b = b) \leq \mathbf{P}(Q_{\infty} > b) \leq e^{-\mu L - \delta b},$$

the probability that the length of a queue with infinite buffer exceeds b. The cell loss ratio in a finite buffer queue is the number of cells arriving while $\mathbf{P}(Q_b = b)$, or approximately

$$p_s \approx \mathbf{P}(Q_b = b) \leq e^{-\mu M - \delta b}.$$

The mean cell delay can be found with the following argument. The delay of a cell in a queue is approximately the queue length at its arrival divided by the service rate. Thus, we get

$$\mathbf{P}(W > t) \approx \mathbf{P}(Q > st) \leq e^{-\mu M - \delta st}.$$

In a queue with finite buffer size b, an improved bound is given by

$$\mathbf{P}(W > t) \leq \frac{e^{-\mu M}(1 - e^{-\delta b})}{s(1 - e^{-\delta})}.$$

Example 9.7.1. In Example 9.5.2, a video source is modeled by a Markov additive process. Using the same assumptions as in the example, the Buffet-Duffield bound is computed as follows. We know from Example 9.5.2 that $\delta = 0.501$. The prefactor is

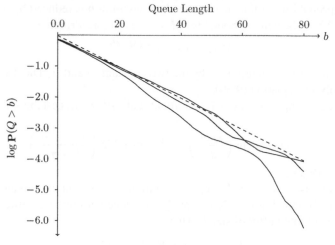

FIGURE 9.24

Three simulations of a single-source MAP in an $M/D/1$ queue with the Buffet-Duffield bound.

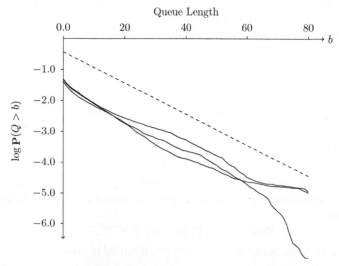

FIGURE 9.25

Three simulations of 20 aggregated MAP sources in an $M/D/1$ queue with the Buffet-Duffield bound.

computed using (9.54). Since the sources are assumed independent, we can use the parameters for a single source. With $s = 0.7$, $M = 1$ and $\pi_1 = 0.6$, we get

$$\tilde{\mu} = 0.0216, \quad \text{or } e^{-\tilde{\mu}} = 0.9786.$$

The bound for a single source is then given by

$$\mathbf{P}(Q > b) \le e^{-\tilde{\mu}-\delta b},$$

shown in Figure 9.24, together with three simulated traces. By aggregating M sources, the bound is

$$\mathbf{P}(Q > b) \le e^{-M\tilde{\mu}-\delta b},$$

with $M = 20$ and $\tilde{\mu}$ as before. The result is shown in Figure 9.25 with three simulated traces.

9.8 Summary

The concept of effective bandwidth, a statistically asymptotic quantity is of profound importance in modern networks, where services and network components are highly dependent and traffic exhibit irregular patterns. The theory, however, is not easily accessible. Good introductions can be found in Lewis and Russell(1996), O'Connell (1999) and Kelly (1996), just to mention a few.

The effective bandwidth can be expressed analytically for a rather wide selection of traffic models and used in queueing models.

For some models, analytic expressions appear to be non-existing. In other cases, the characteristics of the traffic is available only in traffic measurement data. Numerical estimation and simulation techniques are here the only approaches available for analysis.

Multi-Service Systems

A system carrying traffic with different characteristics is termed a *multi-service system* in this text. Such a system may be a queue with multiple service types having different requirements, either of loss or delay type. Similarly, a network consisting of such systems is called a *multi-service network*. Typically, such a network is designated to carry both circuit-switched type of services like voice calls, as well as best effort type of services.

Since modern network technologies are using packet-switched technologies, services are characterized by their effective bandwidth. However, from a service point of view, service-specific requirements persist.

For single service systems, the Erlang formulas are convenient for analysis of loss networks or delay networks. In this chapter, various extensions to the single service case are presented. The extensions are made in several directions. We define a broader class of traffic models, known as the *binomial-poisson-pascal* (BPP) class of models, for which performance metrics of a system can be found analytically.

The Erlang B-formula can be extended to multiple Poisson traffic streams, leading to the algorithm known as the Fortet-Grandjean or the Kaufman-Roberts algorithm, which efficiently calculates the blocking probabilities in multi-service systems.

When the different traffic classes are subject to trunk reservation, the algorithms become more complex. Two approaches are explored in the text; convolution and an algorithm based on discrete-event simulation.

Finally, algorithms for queues with processor sharing disciplines are discussed. Such systems form a sort of hybrid between loss and delay systems, and they turn out to be some of the most least restrictive queueing models in a network context. The presented models can conveniently be used in analytic or fixed-point methods for network analysis.

10.1 The acceptance region

The acceptance region is defining the combination of the number of sources that a resource of a given capacity can accept without violating the QoS parameters of the services. Due to the effect of statistical multiplexing, the border of the admission region is typically a upwards concave curve. In order to be on the safe side, a linear approximation, lying within the region, is used.

This defines a linear relationship between the demands of the sources, which is used in, for example, the multi-dimensional Erlang formula. By this procedure, classical design methods can be used for multiple circuit-switched type services.

The admission region is—for two different service types—an area limited by a curve having each end point at the maximum number of sources that can be supported by a given capacity. Thus, the abscissa is representing one of the services, and the ordinate the other. Since the statistical multiplexing gain is greater for a larger number of sources, and if the sources are considered to be independent, the limiting curve is downwards convex. The more different types of services, the less gain and the less serving capacity. This region defines a feasible traffic mix.

The relationship between the total bandwidth requirement and available capacity of the system can be expressed as

$$\sum_{k=1}^{K} n_k \alpha_k(s^*) \leq C^*, \tag{10.1}$$

where k is the number of traffic types. This relationship depends on how the linear restriction of the acceptance region is defined, that is, how the hyperplane touching the acceptance region is chosen. We will refer to this as the linearized acceptance region. The hyperplane will be different for different values of (s^*, C^*). The pair is chosen so that given performace limits are guaranteed for all combinations of $\mathbf{n} = (n_1, n_2, \ldots, n_K)$ within the linearized acceptance region, where n_k is the maximum number of sources of type k allowed.

We denote by X the aggregate traffic

$$X = \sum_{k=1}^{K} \sum_{j=1}^{n_k} X_{kj}, \tag{10.2}$$

that is, we add the contributions from all n_k sources of all service types k. Assuming that X_{kj} are independent stochastic variables with effective bandwidths

$$\alpha_k(s) = \frac{1}{s} \ln(\mathbf{E}(e^{sX_{kj}})), \tag{10.3}$$

we can regard X_{kj} as the instantaneous arrival rate of work from a source type k at a bufferless resource of capacity C.

Chernoff's bound then gives for the aggregate traffic X

$$\ln(\mathbf{P}(X \geq C)) \leq \ln(\mathbf{E}(e^{s(X-C)})) = s(\alpha(s) - C), \tag{10.4}$$

where $\alpha(s) = \sum_k n_k \alpha_k(s)$. We can formally define the acceptance region as follows. If we require a loss constraint $\ln(\mathbf{P}(X \geq C)) \leq -\delta$, we see that the constraint will be satisfied whenever the vector $\mathbf{n} = (n_1, n_2, \ldots, n_K)$ lies within the set

$$A = \left\{ n : \inf_s \left(s \left(\sum_{k=1}^{K} n_k \alpha_k(s) - C \right) \right) \leq -\delta \right\}, \tag{10.5}$$

where $n_k \geq 0$. We construct the linear restriction by letting a hyperplane touching the boundary of the set A at a chosen point $\mathbf{n}^*$, that is

$$\sum_{k=1}^{K} n_k \alpha_k(s^*) \leq C - \frac{\delta}{s^*}, \qquad (10.6)$$

where s^* attains the infimum in (10.5) with $\mathbf{n}$ replaced by $\mathbf{n}^*$. Note that $\mathbf{n}^*$ is the vector of simultaneously active sources that brings the system closest to the required performance limit $\ln(\mathbf{P}(X \geq C)) \leq -\delta$. Thus (10.6) provides a conservative bound on the loss, and in this region we have the relation (10.1). If $\mathbf{n}$ satisfies (10.6), then the performance requirement $\ln(\mathbf{P}(X \geq C)) \leq -\delta$ is assured.

The problem is how to find $\mathbf{n}^*$. In principle, it amounts to investigating all possible combinations of sources $\mathbf{n}$ such that the performance bound just is satisfied.

Let us say that we have an acceptance region for two services given (two dimensions) with the maximum number of sources $(N^1_{\max}, N^2_{\max})$. These extremes define a line which usually lies outside the acceptance region. Geometrically, we may find $\mathbf{n}^*$ by displacing this line towards the origin, until it becomes a tangent to A.

This may give us an idea where to look for $\mathbf{n}^*$: decrease the distance to this line until we reach values (n_1, n_2) such that the performance bound is satisfied. First, we need to find the distance to the original line. This can be done with a simple geometric calculation.

When there are more than two services, a geometric approach is rather useless. It is possible to analyze services pairwise to get an idea of the region where to look for $\mathbf{n}^*$, but this is rather crude.

A better approach is to use a randomized method, sampling the space of possible vectors $\mathbf{n}$ to find a reasonable traffic mix that just satifies the performance bound.

Example 10.1.1. Suppose we have a resource of capacity $C = 500$ and two services, AMR speech telephony (from Example 9.5.1) and a simple video service. We have determined $N^1_{\max} = 73$ and $N^2_{\max} = 16$. In this simple case, we can choose a bandwidth step Δ and for each $j = 0, 1, \ldots, C/\lceil \Delta \rceil$, divide the capacity into two bins with capacity $j\Delta$ and $C - j\Delta$, respectively, and determine the number of sources and the effective bandwidth for each j. The result is shown in Figure 10.1.

10.2 The Binomial-Poisson-Pascal models

In order to model services on a call level, there are two basic approaches. The first approach is based on the source characteristics using the effective bandwidth of the source and letting the call arrival and holding time processes be Poisson and exponentially distributed, respectively. An alternative is to use analytical traffic distributions that give tractable models directly.

A class of such traffic models is the class of *Binomial-Poisson-Pascal* (BPP) models, for which congestion measures can be found for many queueing systems. The

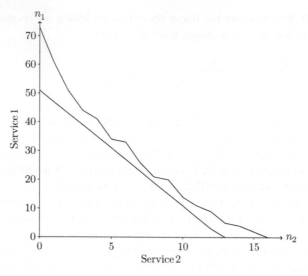

FIGURE 10.1

Acceptance region 1.

Pascal distribution is often considered to be a special case of the *negative binomial* distribution, so rather the latter term is used throughout the text. These distributions correspond to different values of peakedness in the traffic pattern. Recalling that peakedness $Z = \sigma^2/\mu$ is defined as the variance divided by the mean of a traffic stream, the BPP models represent traffic which exhibit peakedness in the ranges of $0 < Z < 1, Z = 1$, and $Z > 1$, respectively.

Traffic with $0 < Z < 1$, called smooth traffic, corresponds to, for example, aggregation of AMR coded speech (see Section 9.5). Most data sources, on the other hand, are rather bursty, with peakedness $Z > 1$. Thus, rather than using a Poisson arrival process having an effective bandwidth α as traffic model, a binomial or negative binomial distribution can be adopted, with parameters estimated from the mean and variance of the traffic.

Both the binomial and the negative binomial distribution have a limited number of sources S, which can be in the states active (busy) or silent (idle).

10.2.1 The binomial distribution

Consider a loss system with n channels, fed by a binomially distributed source with S number of sources. Any source can be in either an idle state, when no data is transmitted, or in a busy state, when data is tranmsitted at a constant rate. A source is idle and busy for time intervals which are exponentially distributed with intensity γ and μ, respectively. This type of source is called an on-off source.

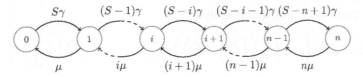

FIGURE 10.2
State transitions for the binomial distribution.

The state diagram for binomially distributed traffic offered to a loss system is shown in Figure 10.2. Initially, we let the number of channels be greater than the number of sources, $n \geq S$, so that no calls are lost.

Let $p(i)$ denote the probability of being in state i. By using the state diagram 10.2, we obtain the detailed balance equations

$$S\gamma \cdot p(0) = \mu \cdot p(1),$$
$$(S - 1)\gamma \cdot p(1) = 2\mu \cdot p(2),$$
$$\cdots$$
$$(S - (i - 1))\gamma \cdot p(i - 1) = i\mu \cdot p(i),$$
$$(S - i)\gamma \cdot p(i) = (i + 1)\mu \cdot p(i + 1),$$
$$\cdots$$
$$\gamma \cdot p(S - 1) = S\mu \cdot p(S).$$

Rearranging these equations gives

$$p(1) = \frac{S\gamma}{\mu} p(0) = p(0) \binom{S}{1} \left(\frac{\gamma}{\mu}\right)^{1},$$

$$p(2) = \frac{(S - 1)\gamma}{2\mu} p(1) = p(0) \binom{S}{2} \left(\frac{\gamma}{\mu}\right)^{2},$$

$$\cdots$$

$$p(i) = \frac{(S - (i - 1))\gamma}{i\mu} p(i - 1) = p(0) \binom{S}{i} \left(\frac{\gamma}{\mu}\right)^{i},$$

$$p(i + 1) = \frac{(S - i)\gamma}{(i + 1)\mu} p(i) = p(0) \binom{S}{i + 1} \left(\frac{\gamma}{\mu}\right)^{i+1},$$

$$\cdots$$

$$p(S) = \frac{\gamma}{S\mu} p(S - 1) = p(0) \binom{S}{S} \left(\frac{\gamma}{\mu}\right)^{S}.$$

Normalization of the sum of probabilities gives $p(0)$

$$1 = p(0) \left(1 + \binom{S}{1}\left(\frac{\gamma}{\mu}\right) + \binom{S}{2}\left(\frac{\gamma}{\mu}\right)^2 + \cdots + \binom{S}{S}\left(\frac{\gamma}{\mu}\right)^S\right)$$

$$= p(0) \left(1 + \frac{\gamma}{\mu}\right)^S,$$

where we have used the identity

$$(1 + x)^y = \sum_{i=1}^{\infty} \binom{y}{i} x^i.$$

By letting $\beta = \gamma/\mu$ we get

$$p(0) = \frac{1}{(1 + \beta)^S}.$$

The parameter β is the offered traffic per idle source (the number of call attempts per time unit for an idle source—the offered traffic from a busy source is zero, since it is already "carried"). The state probabilities $p(i)$ expressed in β are

$$p(i) = \binom{S}{i} \beta^i \frac{1}{(1+\beta)^S}$$

$$= \binom{S}{i} \left(\frac{\beta}{1+\beta}\right)^i \left(\frac{1}{1+\beta}\right)^{S-i}, \quad i = 0, 1, \ldots, S, \quad 0 \le S \le n,$$

which is the binomial distribution. We can also express the probabilities in the offered traffic per source a, to get:

$$a = \frac{1}{1+\beta} = \frac{\gamma}{\mu + \gamma},$$

$$p(i) = \binom{S}{i} a^i (1 - a)^{S-i}, \quad i = 0, 1, \ldots, S, \quad 0 \le S \le n.$$

In a system with no blocking, we have the following parameters characterizing the binomial distribution,

$$a = \frac{\beta}{1+\beta}, \quad \text{offered traffic per source,}$$

$$\rho = Sa = \frac{S\beta}{1+\beta}, \quad \text{total offered traffic,}$$

α, carried traffic per source,

$Y = S\alpha$, total carried traffic,

$y = Y/n$, carried traffic per channel with random channel selection.

Note that the quantity offered traffic per source a depends on the level of congestion is the system; the higher congestion, the higher probability of a source being idle, and the more call attempts.

It is possible to define three different congestion measures: time congestion, call congestion and traffic congestion. For Poisson traffic, these three quantities take the same value. For binomial and the negative binomial traffic, however, these measures have different values. The time congestion is

$$E = 0, \quad S < n,$$

$$E = p(n) = an, \quad S = n.$$

The carried traffic is the traffic that has been admitted into the system. It equals the expectation

$$Y = S\alpha = \sum_{i=0}^{S} ip(i) = Sa = \rho,$$

that is, the mean value of the binomial distribution. In the no blocking case have $a = \alpha$ and

$$C = \frac{Ay}{A} = 0.$$

The number of call attempts per time unit are

$$\Lambda = \sum_{i=0}^{S} p(i)(S - i)\gamma$$

$$= \gamma S - \gamma \sum_{i=0}^{S} ip(i) = \gamma S - \gamma Sa$$

$$= S(1 - \alpha)\gamma,$$

where $S(1 - \alpha)$ is the number of idle sources. The peakedness of the binomial distribution is

$$Z = \frac{\sigma^2}{m} = \frac{Sa(1 - a)}{Sa} = 1 - a = \frac{1}{1 + \beta} < 1.$$

We observe that the peakedness is independent of the number of sources and always less than unity, corresponding to smooth traffic. The duration $q(i)$ of state i is exponentially distributed with rate

$$q(i) = (S - i)\gamma + i\mu, \quad 0 \le i \le S \le n. \tag{10.7}$$

Observing that the mean and variance of the binomial distribution are

$$\mu = S\frac{\beta}{1 + \beta} = Sa,$$

$$\sigma^2 = Sa(1 - a),$$

the traffic model parameters can be estimated from the empirical mean $\hat{\mu}$ and variance $\hat{\sigma}^2$ by

$$a = \frac{\hat{\mu} - \hat{\sigma}^2}{\hat{\mu}},$$

$$S = \frac{\hat{\mu}^2}{\hat{\mu} - \hat{\sigma}^2}.$$

10.2.2 Engset distribution

Now, consider a system where $S \geq n$, so that there is a non-zero probability of blocking. The transition diagram is shown in Figure 10.2, from which we get the detailed balance equations and the state probabilities

$$p(i) = \frac{\binom{S}{i} \beta^i}{\sum_{j=0}^{n} \binom{S}{j} \beta^j}, \quad 0 \leq i \leq n. \tag{10.8}$$

or, equivalently,

$$p(i) = \frac{\binom{S}{i} a^i (1-a)^{S-i}}{\sum_{j=0}^{n} \binom{S}{j} a^j (1-a)^{S-j}}, \quad 0 \leq i \leq n.$$

The truncated binomial distribution is called the Engset distribution.

It is characterized by the number of sources S, the offered traffic per idle source $\beta = \gamma/\mu$, and the number of channels n.

Since the number of sources exceeds the number of channels by assumption, the congestion measures evaluate to non-zero values. The time blocking is

$$E(\beta, S, n) = p(n) \frac{\binom{S}{n} \beta^n}{\sum_{j=0}^{n} \binom{S}{j} \beta^j}, \quad S \geq n. \tag{10.9}$$

In order to determine the call congestion, we need to determine the proportion of call attempts that are lost of the total number of offered call attempts. A call attempt can only be blocked when the system is saturated, that is, in state n. Thus, we have

$$B(\beta, S, n) = \frac{p(n)(S - n)\gamma}{\sum_{j=0}^{n} p(j)(S - j)\gamma}, \tag{10.10}$$

We also have the relation $B(\beta, S, n) = E(\beta, S - 1, n) < E(\beta, S, n)$, $S \geq n$, which is to say that the call blocking probability from an idle source is equal to the probability that the remaining $(S - 1)$ sources occupy all n channels in the system. This result is known as the *arrival theorem*, valid for both loss and delay systems.

Theorem 10.2.1 (The Arrival Theorem).   *For a fully accessible system with a limited number of sources, a random source upon arrival will observe the state of the system as if the source itself does not belong to the system.*

In order to calculate the traffic congestion, we need to know the amount of carried traffic $Y(\beta, S, n)$, which is determined from

$$Y(\beta, S, n) = \sum_{i=1}^{n} i p(i) = \sum_{i=0}^{n} \beta(S - i) p(i) - \beta(S - n) p(n), \qquad (10.11)$$

or

$$Y(\beta, S, n) = \frac{\beta}{1 + \beta}(S - (S - n)E(\beta, S, n)).$$

The traffic congestion is $C = C(\beta, S, n)$ is given by

$$C(\beta, S, n) = \frac{\rho - Y(\beta, S, n)}{\rho} = \frac{S - n}{S} E(\beta, S, n). \qquad (10.12)$$

The duration of state i is exponentially distributed with intensity

$$q(i) = (S - i)\gamma + i\mu, \quad 0 \le i < n,$$

$$q(n) = n\mu, \quad i = n.$$

For the Engset system, we have $C(\beta, S, n) < B(\beta, S, n) < E(\beta, S, n)$. We can formulate the following functional relations between E, B and C (dropping the arguments):

$$E = \frac{S}{S - n}\frac{B}{1 + \beta(1 - B)},$$

$$B = \frac{(S - n)E(1 - \beta)}{S + (S - n)E\beta},$$

$$C = \frac{S - n}{S}E,$$

$$E = \frac{S}{S - n}C,$$

$$C = \frac{B}{1 + \beta(1 - B)},$$

$$B = \frac{(1 + \beta)C}{1 + \beta C}.$$

When the time congestion E is known, it is easy to obtain the call congestion B and the traffic congestion C by using the relations above. Similarly to the Erlang formulas, the Engset formula (10.9) itself is not stable for numerical evaluation. More suitable for practical use is the recursive formula (where n and S can be non-integral)

$$E(\beta, S, x) = \frac{(S - x + 1)\beta E(\beta, S, x - 1)}{x + (S - x + 1)\beta E(\beta, S, x - 1)}, \quad E(\beta, S, 0) = 1. \quad (10.13)$$

where $\beta = \gamma/\mu$. Letting $I(\beta, S, n) = 1/E(\beta, S, n)$, we obtain the alternative formula

$$I(\beta, S, x) = 1 + \frac{x}{(S - x + 1)\beta} I(\beta, S, x - 1), \quad I(\beta, S, 0) = 1. \quad (10.14)$$

Any of the formulas can be iterated for increasing x until the desired system capacity n is reached. Both formulas are numerically stable and accurate. It can be shown that the Engset distribution is independent of service time distribution. Only the mean service time μ^{-1} enter into the formulas above.

10.2.3 The negative binomial distribution

In a binomial model, the arrival intensity decreases with an increasing number of busy sources. This effect follows from that traffic from busy sources is carried by the system and only the remaining idle sources offer new arrivals. It is also possible to construct a model where the arrival intensity increases with the number of busy sources. Formally, the arrival intensity in state i is

$$\lambda_i = \gamma(S + i), \quad 0 \le i \le n,$$

where γ and S are positive constants. We assume that the holding times are exponentially distributed with intensity μ.

First, consider a system with an infinite number of channels. Some of the state probabilities are shown in Figure 10.3. From these, we can formulate the detailed balance equations

$$S\gamma \cdot p(0) = \mu \cdot p(1),$$
$$(S + 1)\gamma \cdot p(1) = 2\mu \cdot p(2),$$
$$\dots$$
$$(S + i - 1)\gamma \cdot p(i - 1) = i\mu \cdot p(i),$$
$$(S + i)\gamma \cdot p(1) = (i + 1)\mu \cdot p(i + 1),$$
$$\dots$$

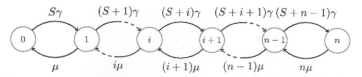

FIGURE 10.3

State transitions for the negative binomial (Pascal) distribution.

Denote by β the offered traffic per idle source

$$\beta = \gamma/\mu < 1$$

The state probabilities can be expressed as

$$p(i) = \binom{S+i-1}{i}(\beta)^i(1-\beta)^S, \quad 0 \le i < \infty, \beta < 1.$$

For an infinite system, the mean and peakedness of the traffic (which is offered and carried) are

$$\rho = \frac{S\beta}{1-\beta},$$

$$Z = \frac{1}{1-\beta}.$$

The mean and variance of the negative binomial distribution are

$$\mu = S\frac{\beta}{1-\beta},$$

$$\sigma^2 = S\frac{\beta}{(1-\beta)^2}.$$

Solving for the mean μ and variance σ^2 and substituting for the empirical mean $\hat{\mu}$ and variance $\hat{\sigma}^2$ gives

$$\beta = \frac{\hat{\sigma}^2 - \hat{\mu}}{\hat{\sigma}^2},$$

$$S = \frac{\hat{\mu}^2}{\hat{\sigma}^2 - \hat{\mu}},$$

which can be used as estimators for the traffic model parameters.

10.2.4 The truncated negative binomial distribution

We now consider the same traffic process offered to a system with a finite number n of channels. The resulting state probability distribution is referred to as the truncated negative binomial distribution, and the state probabilities are found from

$$p(i) = \frac{\binom{-S}{i}(-\beta)^i}{\sum_{j=0}^{n}\binom{-S}{j}(-\beta)^j}, \quad 0 \le i \le n. \tag{10.15}$$

The truncated negative binomial distribution can be obtained from the Engset case by substituting S for $-S$ and γ for $-\gamma$. With these substitutions, all formulas of the Engset cases are valid for the truncated Pascal distribution.

For the negative binomial distribution, we have

$$C(\beta, S, n) > B(\beta, S, n) > E(\beta, S, n). \tag{10.16}$$

Just like for the Erlang and Engset distributions, the truncated negative binomial distribution is independent of service time distribution.

Under the assumption of exponentially distributed holding times, the negative binomial distribution has the same state probabilities as a system with a Poisson arrival process with a random intensity distributed according to a gamma distribution. Inter-arrival times are then Pareto distributed, which is a heavy-tailed distribution.

The negative binomial distribution can therefore be seen as an extension of the Poisson distribution that allows for greater variance than the mean intensity. We illustrate this by explicitly deriving the state probabilities $\mathbf{P}(X = x)$.

Suppose $X|\Lambda$ is a Poisson random variable and Λ is a Gamma(k, θ) random variable, having the probability density function

$$f(y; k, \theta) = \frac{1}{\Gamma(k)\theta^k} y^{k-1} e^{-y/\theta}, \tag{10.17}$$

where $\Gamma(k)$ is the gamma function evaluated at k. We can construct a new random variable based on a Poisson distributed variable, but allowing the mean parameter to be random as well by letting $y = \lambda$. Thus,

$$
\begin{aligned}
\mathbf{P}(X = x) &= \frac{1}{\Gamma(k)\theta^k} \int_0^\infty \frac{e^{-\lambda}\lambda^x}{x!} \lambda^{k-1} e^{-\lambda/\theta} d\lambda \\
&= \frac{1}{x!\Gamma(k)\theta^k} \int_0^\infty \lambda^{k+x-1} e^{-\lambda(1+1/\theta)} d\lambda \\
&= \frac{1}{\Gamma(x+1)\Gamma(k)\theta^k} \Gamma(k+x) \left(\frac{\theta}{\theta+1}\right)^{k+x} \\
&= \binom{k+x-1}{x} \left(\frac{1}{\theta+1}\right)^k \left(1 - \frac{1}{1+\theta}\right)^x .
\end{aligned}
\tag{10.18}
$$

This shows that the marginal distribution of X is negative binomial with $S = k$ and $\beta = 1/(1 + \beta)$.

The mean of Λ is $\lambda = k\theta$ and its variance is $k\theta^2$. Letting $k \to \infty$ while keeping $\theta = \lambda/k$, so that the mean is constant, the variance of Λ approaches zero. The distribution of X then converges to a Poisson distribution. Due to this result, the negative binomial distribution is also called the Gamma-Poisson (mixture) distribution.

10.3 Loss systems with multiple services

Modern networks are designed to handle different types of traffic. Such networks provide statistical multiplexing gain, improving on utilization efficiency, but, on the other hand, the service requirements remain similar to older network types in terms of performance.

When analyzing networks carrying circuit-switched type of services, where blocking is the crucial performance metric, an extension of the Erlang B-formula (5.39) to multiple concurrent services can be used to compute the link blocking experienced.

On a service level, the effective bandwidth can be used to determine the resource requirement of each service in number of channels (or basic bandwidth units), and the corresponding blocking rates can then be determined by the extensions of the Erlang formula.

10.3.1   Systems in product form

The important concept of a system solution having product form leads to the possibility of constructing efficient algorithms for analysis. The product-form property is defined as

Definition 10.3.1 (Product-Form Solution).   Whenever the state probabilities $p(\cdot)$ of a system with J components and state space Ω can be written

$$p(x_1, x_2, \ldots, x_J) = G^{-1}(\Omega) \prod_{j=1}^{J} p(x_j), \tag{10.19}$$

where $G(\Omega)$ is a normalization constant, the system is said to have a *product-form solution*, and the components are mutually independent.

First, we investigate the conditions for a product form to exist. Next, we describe three approaches for analyzing systems in product form; convolution, state-space recursion, and Monte Carlo simulation.

As shown in Sections 5.2 and 5.3, the state space of a system can often be modeled by a (continuous-time) Markov chain. The dynamics of the system can then be described by a set of equations referred to as the global balance equations.

If a Markov chain has state space S and transition rates q_{ij} from state i to j, then the equilibrium distribution π is given by

$$\sum_{j \in S \setminus \{i\}} \pi_i q_{ij} = \sum_{j \in S \setminus \{i\}} \pi_j q_{ji}, \tag{10.20}$$

where $\pi_i q_{ij}$ represents the probability flux from state i to state j. In most cases, this system of equations is impossible to solve analytically. Equation (10.20) can also be expressed

$$\pi Q = 0,$$

where Q is the infinitesimal generator of the chain.

If a continuous-time Markov chain has the infinitesimal generator Q, and equilibrium probabilities π_i can be found so that for every pair of states (i, j)

$$\pi_i q_{ij} = \pi_j q_{ji}, \tag{10.21}$$

holds, then the global balance equations are satisfied and π is the equilibrium distribution of the Markov chain. Equation (10.21) is called the detailed balance equations of the Markov chain.

For a Markov continuous-time chain we define

$$q(j,k) = \lim_{\tau \to 0} \frac{\mathbf{P}(X(t+\tau) = k | X(t) = j)}{\tau}, \quad j \neq k.$$

assuming that $q(j,j) = 0$. Then,

$$\pi(j) \sum_{k \in S} q(j,k) = \sum_{k \in S} \pi(k) q(k,j).$$

The Markov process remains in state j for a length of time which is exponentially distributed with parameter

$$q(j) = \sum_{k \in S} q(j,k).$$

When it leaves state j it moves to state k with probability

$$p(j,k) = \frac{q(j,k)}{q(j)}.$$

The possibility to find efficient algorithms for solving for the equilibrium probabilities of a Markov chain is closely related to the concept of *reversibility*. Reversibility means that the dynamics of a system "looks the same" if the time is reversed so that output becomes input and vice versa. Such a process has the same equilibrium distribution in forward time and in reversed time.

Theorem 10.3.1. *A stationary Markov process is reversible if and only if there exists a collection of positive numbers $\pi(j)$, $j \in S$, $\sum_j \pi(j) = 1$ that satisfy the detailed balance conditions*

$$\pi(j)q(j,k) = \pi(k)q(k,j), \quad j,k \in S.$$

When $\pi(j)$ exists, it is the equilibrium distribution of the process.

For the sake of clarity, we will often write q_{ij} as $q(i,j)$ for models with more complex state spaces. These numbers should still be interpreted as elements of the infinitesimal generator Q. In particular, for a birth-death process (stationary) we have

$$\pi(j)q(j,j-1) = \pi(j-1)q(j-1,j),$$

so that

$$\pi(j) = \pi(0) \prod_{i=1}^{j} \frac{q(i-1,i)}{q(i,i-1)}.$$

A certain class of models lend themselves to analytical treatment in a multi-dimensional queueing context. These are the binomial, the Poisson, and the negative binomial distributions. The Poisson distribution is characterized by a single parameter, and has the same mean and variance, whereas the binomial and the negative binomial distribution have two parameters—a finite number of sources and a traffic intensity. The number of sources can be a real number and should be considered a modeling parameter rather than the actual number of sources.

We assume that the capacity C of any resource can be discretized into n units of basic bandwidth units (BBU). This unit is chosen such that the bandwidth requirement of any source can be expressed in integer numbers of this unit. Fractions of this unit cannot be used. We let services have different bandwidth requirements in terms of number of channels, which are determined by their respective effective bandwidths.

In Section 5.5, the peakedness factor is used to characterize the burstiness of traffic. The peakedness Z is defined as the ratio of the variance σ^2 to the mean μ of the traffic,

$$Z = \frac{\sigma^2}{\mu}.$$

Since the mean and variance of the Poisson process is both λ, its peakedness is 1. The binomial distribution has peakedness

$$Z = \frac{na(1-a)}{na} = 1 - a < 1,$$

where a is the offered traffic per source (or, the *activity factor*). In contrast, the negative binomial distribution has peakedness

$$Z = \frac{S\beta/(1-\beta)^2}{S\beta/(1-\beta)} = \frac{1}{1-\beta},$$

which lies in the range $1 < Z < \infty$. The three models—Poisson, binomial and negative binomial, can therefore be used to model any traffic that can be characterized by their mean and variance. The models also have the advantage of being relatively easy to analyze.

For example, most data services are bursty and can be modeled by the negative binomial distribution. This argument can roughly be justified by the following observation (see Choudhury et al., 1994). Many data sources have requests that arrive in batches, which make the traffic arrival process more bursty than the ordinary Poisson process. When the sizes of the batches are geometrically distributed, arrive according to a Poisson process and the service times are exponentially distributed, the job arrival process is negative binomial. In other words, this batch-Poisson arrival process is equivalent to a linear state-dependent arrival process without batches—the negative binomial arrival process.

To begin the analysis of a multi-service loss system, we consider a system with n channels (BBU) offered two independent Poisson traffic streams characterized by (λ_1, μ_1) and (λ_2, μ_2), repectively, each with bandwidth demand $d = 1$. The offered traffic by each stream is $\rho_1 = \lambda_1/\mu_1$ and $\rho_2 = \lambda_2/\mu_2$, and since the streams are independent, the total offered traffic is $\rho = \rho_1 + \rho_2$. Denoting the state of the system

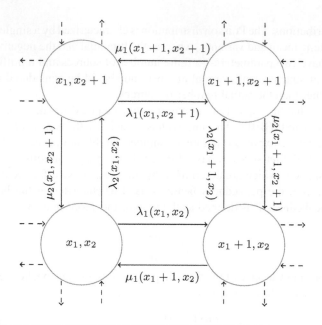

FIGURE 10.4

The Kolmogorov cycle condition for a two-dimensional system.

by (x_1, x_2), where x_i is the contribution from stream i, $i = 1, 2$ (that is, the number of channels used), we must have

$$
\begin{aligned}
0 &\le x_1 \le n, \\
0 &\le x_2 \le n, \\
0 &\le x_1 + x_2 \le n.
\end{aligned}
\tag{10.22}
$$

The system can be modeled by a continuous time Markov chain and the equilibrium probabilities can therefore (in principle) be found by solving the global balance equations (10.20). The state space S has size $(n + 1)(n + 2)$ and therefore we have $(n + 1)(n + 2)/2$ equations. It turns out that this system is reversible so that it is sufficient to solve the detailed balance equations (10.21).

With two traffic streams we can depict a two-dimensional state transition diagram shown in Figure 10.4. The diagram shows state transitions between neighboring states (recall that with Poisson traffic, the Markov process only can make one "jump" at any time instant, and the transitions must be to any of the neighboring states). Then we have the condition for reversibility by the following theorem.

Theorem 10.3.2 (Kolmogorov cycle condition). *A necessary and sufficient condition for reversibility is that the probability flux around any cycle of states in one direction equals the flow in the reverse direction.*

The condition is illustrated in Figure 10.4, where we must have,

$$\lambda_2(x_1, x_2) \cdot \lambda_1(x_1, x_2 + 1) \cdot \mu_2(x_1 + 1, x_2 + 1) \cdot \mu_1(x_1 + 1, x_2)$$
$$= \lambda_1(x_1, x_2) \cdot \lambda_2(x_1 + 1, x_2) \cdot \mu_1(x_1 + 1, x_2 + 1) \cdot \mu_2(x_1, x_2 + 1).$$

If the two fluxes are equal, then there is detailed balance. As a consequence we must have that if there is a flux (shown by an arrow in the diagram) from state x_2 to state x_1, these must be equal.

We can now formulate the detailed equations by looking at the cuts between neighboring states. Any state probability $p(x_1, x_2)$ can be expressed by considering a path, and be formulated in terms of the state probability $p(0, 0)$. We may, for example, choose the path

$$(0, 0), (1, 0), \ldots, (x_1, 0), (x_1, 1), \ldots, (x_1, x_2),$$

then we have the following balance equation:

$$p(x_1, x_2) = \frac{\lambda_1(0, 0)}{\mu(1, 0)} \frac{\lambda_1(1, 0)}{\mu_1(2, 0)} \cdots \frac{\lambda_1(x_1 - 1, 0)}{\mu_1(x_1, 0)} \cdot$$
$$\cdot \frac{\lambda_2(x_1, 0)}{\mu_2(x_1, 1)} \frac{\lambda_2(x_1, 1)}{\mu_2(x_1, 2)} \cdots \frac{\lambda_2(x_1, x_2 - 1)}{\mu_2(x_1, x_2)} \cdot p(0, 0).$$

State probability $p(0, 0)$ is obtained by normalization of the total probability mass.

The reversibility condition is fulfilled for many state-dependent transition probabilities, such as when

$$\lambda_1(x_1, x_2) = \lambda_1(x_1), \quad \mu_1(x_1, x_2) = x_1\mu_1, \tag{10.23}$$

$$\lambda_2(x_1, x_2) = \lambda_2(x_2), \quad \mu_2(x_1, x_2) = x_2\mu_2, \tag{10.24}$$

which includes the Poisson, binomial and negative binomial traffic models.

10.3.2 Multi-rate systems

For systems with multiple traffic streams conditions for reversibility are obtained similarly to the two-dimensional case. The Kolmogorov cycle condition must still be fulfilled for all possible cycles in the state transition diagram. This generalization is straightforward. The BPP models are insensitive to the holding time distribution, that is, the system corresponds to $BPP/G/n$ systems with multiple inputs, and general service time G, which means that the state probabilities only depend on the mean service time. Now, returning to our system with two Poisson traffic streams with parameters (λ_1, μ_1) and (λ_2, μ_2), requiring that the constraints (10.22) are fulfilled, we can write the state probability as

$$p(x_1, x_2) = G \cdot p_1(x_1) \cdot p_2(x_2) = G \frac{\rho_1^{x_1}}{x_1!} \frac{\rho_2^{x_2}}{x_2!}, \tag{10.25}$$

where $\rho_1 = \lambda_1/\mu_1$, $\rho_2 = \lambda_2/\mu_2$, and $p_1(x_1)$ and $p_2(x_2)$ are the one-dimensional truncated Poisson distributions for traffic streams one and two, respectively. The constant G is a normalization constant which has to be determined.

By convolution of two Poisson distributions, we can find the marginal state probabilities for any state $y \leq n$,

$$
\begin{aligned}
p(x_1 + x_2 = y) &= G \sum_{x_1=0}^{y} p_1(x_1) \cdot p_2(y - x_1) \\
&= G \sum_{x_1=0}^{y} \frac{\rho_1^{x_1}}{x_1!} \cdot \frac{\rho_2^{y-x_1}}{(y-x_1)!} \\
&= G \frac{1}{y!} \sum_{x_1=0}^{y} \binom{y}{x_1} \rho_1^{x_1} \cdot \rho_2^{y-x_1} \\
&= G \frac{1}{y!} (\rho_1 + \rho_2)^y = G \frac{\rho^y}{y!},
\end{aligned}
\tag{10.26}
$$

where $\rho = \rho_1 + \rho_2$ and the normalization constant is given by

$$
G^{-1} = \sum_{y=0}^{n} \frac{\rho^y}{y!}.
\tag{10.27}
$$

We recognize this as the truncated Poisson distribution.

Thus, we may also interpret this model as an Erlang loss system with a single Poisson arrival process with rate $\lambda = \lambda_1 + \lambda_2$, since the superposition of two Poisson processes is another Poisson process,

$$
\lambda = \lambda_1 + \lambda_2.
$$

The holding time distribution for the composite process is obtained by weighting the two exponential distributions by the relative number of calls per time unit which results in a hyper-exponential distribution with

$$
f(t) = \frac{\lambda_1}{\lambda_1 + \lambda_2} \mu e^{-\mu_1 t} + \frac{\lambda_2}{\lambda_1 + \lambda_2} \mu_2 e^{-\mu_2 t}.
\tag{10.28}
$$

The mean service time is

$$
m = \frac{\lambda_1}{\lambda_1 + \lambda_2} \frac{1}{\mu_1} + \frac{\lambda_2}{\lambda_1 + \lambda_2} \frac{1}{\mu_2} = \frac{\rho_1 + \rho_2}{\lambda_1 + \lambda_2} = \frac{\rho}{\lambda}.
\tag{10.29}
$$

This procedure is easily generalized to the case where there are K independent Poisson traffic streams. We have

$$
\begin{aligned}
p(x_1, x_2, \ldots, x_K) &= G p_1(x_1) p_2(x_2) \cdots p_K(x_K) \\
&= G \frac{\rho_1^{x_1}}{x_1!} \frac{\rho_2^{x_2}}{x_2!} \cdots \frac{\rho_K^{x_K}}{x_K!}, \quad 0 \leq x_k \leq n, \sum_{k=1}^{K} x_k \leq n,
\end{aligned}
$$

which is the general multi-dimensional Erlang B-formula. Applying the multinomial theorem, this can be written

$$p(x_1 + x_2 + \cdots + x_K = y) = G \frac{(\rho_1 + \rho_2 + \cdots + \rho_K)^y}{y!} = G \frac{\rho^y}{y!},$$

where $\rho = \sum_{k=1}^{K} \rho_k$.

A multi-service system has more potential parameters than a single service system; services may have different bandwidth requirements and the system may have admission control (of trunk reservation type) to guarantee accessibility of each service.

Usually, the requested bandwidth vary from service to service. In a system of total capacity C, the capacity is divided into n channels of bandwidth equal to the basic bandwidth unit. For example, a voice call may require a single channel, whereas a video conference call requires several channels. Denoting the bandwidth requirement of service k by d_k and let K be the total number of services, we have the system restrictions

$$0 \leq x_k = m_k d_k \leq n, \quad k = 1, 2, \ldots, K, \tag{10.30}$$

$$0 \leq \sum_{k=1}^{K} m_k d_k \leq n, \tag{10.31}$$

where m_k is the integer number of type k calls and x_k is the number of channels occupied by service k. It turns out that the corresponding state transition diagram still is reversible and its solution is in product form.

Example 10.3.1. Consider a system with six channels and two traffic classes. Traffic of the first class arrives according to a Poisson process with intensity $\lambda_1 = 1$, holding time $\mu_1 = 1$, and effective bandwidth $d_1 = 1$. The second traffic class is binomially distributed with four sources, arrival rate $\gamma_2 = 1/2$, holding time $\mu_2 = 1$, and effective bandwidth $d_2 = 2$. The traffic parameters are summarized in Table 10.1. The state space is shown in Figure 10.5. We can set $p(0, 0) = 1$ and use the detailed balance

Table 10.1 Traffic parameters for Example 10.3.1

Description	Parameter	Value	Parameter	Value
Sources	S_1	∞	S_2	4
Arrival rate	λ_1	1	γ_2	1/2
Holding time	μ_1	1	μ_2	1
Offered traffic/idle source			β_2	1/2
Peakedness	Z_1	1	Z_2	2/3
Bandwidth	e_1	1	e_2	2
Traffic load	ρ_1	2	ρ_2	4/3
Resources	n_1	6	n_2	6

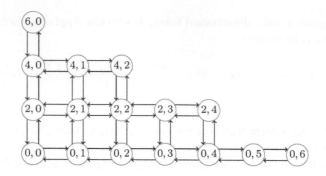

FIGURE 10.5

State space for Example 10.3.1.

Table 10.2	Normalized state probabilities for Example 10.3.1						
x_2	**0**	**1**	**2**	**3**	**4**	**5**	**6**	$\sum_i p(i, j)$
3	0.0170	–	–	–	–	–	–	0.0170
2	0.0511	0.1022	0.1022	–	–	–	–	0.2555
1	0.0681	0.1363	0.1363	0.0908	0.0454	–	–	0.4769
0	0.0341	0.0681	0.0681	0.0454	0.0227	0.0091	0.0030	0.2506
$\sum_j p(i, j)$	0.1703	0.3066	0.3066	0.1363	0.0681	0.0091	0.0030	1.000

equations to determine the relative probabilities. For service one, we have

$$\lambda_1 p(x_1 - 1, 0) = x_1 \mu_1 p(x_1, 0).$$

For service two, the balance equation reads

$$p(0, x_2 - 1)[S_2 - (x_2 - 1)]\gamma_2 = x_2 \mu_2 p(0, x_2).$$

The services are independent by assumption, so that the joint distribution is given by

$$p(x_1, x_2) = p(x_1, 0) \cdot p(0, x_2).$$

By summing all relative probabilities and dividing all relative probabilities with this quantity, the probabilities are normalized. These probabilities are shown in Table 10.2. Next, we can calculate performance measures for the queue based on these probabilities. The time congestion values are

$$E_1 = p(0, 3) + p(2, 2) + p(4, 1) + p(6, 0) = 0.1677, \tag{10.32}$$
$$E_2 = p(0, 3) + p(1, 2) + p(2, 2) + p(3, 1) + p(4, 1)$$
$$+ p(5, 0) + p(6, 0) = 0.3698. \tag{10.33}$$

These are simply the sum of the right-most probabilities on each row, and the sum of the top-most probabilities in each column, respectively. To determine the call congestion, we need to determine the number of blocked call attempts to the number of offered call attempts. For service class 1, the offered call attempts equal $\lambda_1 = 2$ and the number of blocked call attempts $\lambda_1(p(0, 3) + p(2, 2) + p(4, 1) + p(6, 0)) = 0.3354$. For the second traffic stream, the total number of call attempts is $\sum_{x_2=0}^{N_2} \gamma(S_2 - x_2)p_2(x_2) = 1.4805$, where $N_2 = 3$ is the maximum simultaneously active sources possible in the system, and $p_2(x_2)$ is the marginal probability distribution for service two. The blocked call attempts are given by $\gamma(S_2 - 0)(p(5, 0) + p(6, 0)) + \gamma(S_2 - 1)(p(3, 1) + p(4, 1)) + \gamma(S_2 - 2)(p(1, 2) + p(2, 2)) + \gamma(S_2 - 3)p(0, 3) = 0.4415$. Thus, we have

$$B_1 = \frac{0.3354}{2} = 0.1677, \tag{10.34}$$

$$B_2 = \frac{0.4415}{1.4805} = 0.3698. \tag{10.35}$$

Finally, the traffic congestion is the ratio between overflowing traffic to offered traffic. For service one, the carried traffic is $\sum_{x_1=1}^{N_1} x_1 d_1 p_1(x_1) = 1.6646$, where N_1 is the maximum number of simultaneous calls of type one in the system. The offered traffic is $\lambda_1 = 2$. Similarly, the carried traffic of type two is $\sum_{x_2=1}^{N_2} x_2 d_2 p_2(x_2) = 2.0780$. The class-two offered traffic is $\rho_2 d_2 = 2$. Thus,

$$C_1 = \frac{2.000 - 1.6646}{2} = 0.1677, \tag{10.36}$$

$$C_2 = \frac{2.6667 - 2.0780}{2} = 0.2208. \tag{10.37}$$

10.3.3 Convolution

Consider a system with n channels offered K different service classes. A call of type k requires d_k channels for the duration of the connection, so that d_k channels are seized simultaneously at call set-up and relased at the end of the connection. If there are less than d_k available at the time of call request, the call is blocked and lost. The state of the system is described by the vector $(x_1, x_2, \ldots, x_K)$, where x_k is the number of channels occupied by service k. The system restrictions (10.30) and (10.31) are assumed to be satisfied. We require that x_k is an integral multiple of d_k, so that $x_k/d_k = m_k$.

Since the system is reversible and the equilibrium probability distribution has product form, we have

$$p(x_1, x_2, \ldots, x_K) = p_1(x_1)p_2(x_2) \cdots p_N(x_K), \tag{10.38}$$

subject to the system restrictions (10.30) and (10.31). Since the equilibrium probability distribution is in product form, that is, the state probabilities of the individual traffic streams are independent, we can use convolution to find the global state probabilities. We may express the state probabilities of traffic stream k as a k-dimensional vector,

$$\mathbf{p}_k = (p_k(0), p_k(1), \ldots, p_k(x_k)), \tag{10.39}$$

where $p_k(j) = 0$ when $j \neq id_k$, $i = 0, 1, \ldots, \lfloor x_k/d_k \rfloor$.

The global state probabilities can be found by the following convolution algorithm described in Iversen (2013). The algorithm is described in three steps: finding individual one-dimensional state probabilities for each service type, convolution of the one-dimensional state probabilities, and computation of system performance measures. The method is essentially a generalization of the one-dimensional convolution algorithm used in Section 8.3.

Let $q_k(x)$ be the relative (unnormalized) state probabilities for service type k, and $p_k(x)$ the corresponding normalized state probabilities, where $p_k(x) = q_k(x)/G_k$ for some normalization constant G_k. The normalization is often performed in several steps. The one-dimensional state probabilities (10.39) for service type k are found by considering each service k separately in the system. By letting $q_k(0) = 1$, successive relative state probabilities $q_k(n_k)$ are found by solving the detailed balance equations like in Example 10.3.1, or by using the product form solution directly for the truncated distributions, (5.21), (10.8) or (10.15).

Having thus obtained the one-dimensional state probabilities, the probabilities for aggregated traffic streams can be found by convolution. Denoting the convolution operator by $*$, we convolve step by step, including one traffic type at a time. The convolution of two probability vectors $\mathbf{p}_i$ and $\mathbf{p}_j$ subject to system capacity n is

$$\mathbf{p}_i * \mathbf{p}_j = \left(p_i(0)p_j(0), \sum_{s=0}^{1} p_i(s)p_j(1-s), \ldots, \sum_{s=0}^{N} p_i(s) \cdot p_j(N-s) \right), \tag{10.40}$$

where $N = \min\{N_i + N_j, n\}$ is the maximum number of simultaneous connections possible in the system of type i and j. We may first convolve $\mathbf{p}_1$ with $\mathbf{p}_2$ to obtain $\mathbf{p}_{12}$, then $\mathbf{p}_{12}$ with $\mathbf{p}_3$ to obtain $\mathbf{p}_{123}$, etc. It should be noted that even if $\mathbf{p}_i$ and $\mathbf{p}_j$ are normalized, the result of a convolution is in general not normalized due to truncation. Normalization should be performed after each convolution to avoid numerical overflow.

In order to compute performance measures for service k, we need the convolution of all other traffic streams (denoting the set by K as well as the number of streams, identifying the streams by their index in the set) $\mathbf{p}_{K\backslash k}$ which together with the probability vector $\mathbf{p}_k$ is used to determine time, call and traffic blocking rates. The convolution can be formally written

$$\begin{aligned} \mathbf{q}_{K\backslash k} &= (q_{K\backslash k}(0), q_{K\backslash k}(1), \ldots, q_{K\backslash k}(x)) = \\ &= \mathbf{p}_1 * \mathbf{p}_2 * \cdots * \mathbf{p}_{k-1} * \mathbf{p}_{k+1} * \cdots * \mathbf{p}_K, \end{aligned} \tag{10.41}$$

which can be seen as a decomposition of the state space into traffic stream k, and the aggregate of all other streams. We note that we have

$$\mathbf{p}_K = \mathbf{q}_{K \setminus k} * \mathbf{p}_k. \tag{10.42}$$

Performing this convolution yields the vector components

$$q_K(s) = \sum_{x_j=0}^{s} q_{K \setminus k}(s - x_k) \cdot p_k(x_k) = \sum_{x_k=0}^{s} p_k(x_k | s), \tag{10.43}$$

where $p_k(x_k | s)$ is the probability of x_k channels occupied by service k out of a total of s busy channels in the system. The time congestion E_k for service k is

$$E_k = \frac{1}{Q} \sum_{s \in \mathcal{B}_k} p_k(x_k | s). \tag{10.44}$$

where the set of blocking states $\mathcal{B}_k$ is defined as

$$\mathcal{B}_k = \{(x_k, s) : x_k \le s \le n \cap (\{x_k > N_k - d_k\} \cup \{s > n - d_k\})\}, \tag{10.45}$$

that is, the state where either the restriction N_k on traffic stream k is limiting access, or the total system capacity is exhausted. The sum (10.44) is taken over all blocking states. The quantity Q is the normalization constant

$$Q = \sum_{m=0}^{n} q_K(m) \tag{10.46}$$

for the total state probabilities.

Call congestion B_k for traffic stream k is the ratio between the number of blocked call attempts and the total number of offered call attempts. We have

$$B_k = \frac{\sum_{s \in \mathcal{B}_k} \lambda_k(x_k) \cdot p_k(x_k | s)}{\sum_{s=0}^{n} \sum_{x_k=0}^{s} \lambda_k(x_k) \cdot p_k(x_k | s)}.$$

The traffic congestion C_k is the ratio of blocked traffic, that is, the difference between offered traffic and carried traffic, to the offered traffic. The carried traffic is

$$Y_k = \sum_{s=0}^{n} \sum_{x_k=0}^{m} x_k \cdot p_k(x_k | s), \tag{10.47}$$

so that

$$C_k = \frac{\rho_k - Y_k}{\rho_k}. \tag{10.48}$$

Example 10.3.2. Consider a queue with six channels and three traffic classes. Traffic of the first class arrives according to a Poisson process. The second traffic class is binomially distributed and the third traffic stream is negatively binomially distributed. The traffic parameters are summarized in Table 10.3. The state space is shown in Figure 10.5. For the first two streams, we can set $p(0, 0) = 1$ and use the detailed

Table 10.3 Traffic parameters for Example 10.3.2

Parameter	Service 1		Service 2		Service 3	
Sources	S_1	∞	S_2	4	S_3	-2
Arrival rate	λ_1	2	γ_2	1/2	γ_3	$-1/4$
Holding time	μ_1	1	μ_2	1	μ_3	1
Offered traffic/idle source			β_2	1/2	β_3	$-1/4$
Peakedness	Z_1	1	Z_2	2/3	Z_3	4/3
Bandwidth	e_1	1	e_2	2	e_3	1
Traffic load	ρ_1	2	ρ_2	4/3	ρ_3	2/3
Resources	n_1	6	n_2	6	n_4	4

Table 10.4 Convolution of probabilities for Example 10.3.2

x	$p_1(x)$	$p_2(x)$	$(p_1 * p_2)(x)$	Norm.	$p_3(x)$	Norm.	$(p_{12} * p_3)(x)$	Norm.
0	0.1360	0.2000	0.0272	0.0341	1.0000	0.5651	0.0193	0.0218
1	0.2719	0.0000	0.0544	0.0681	0.5000	0.2826	0.0481	0.0544
2	0.2719	0.4000	0.1088	0.1363	0.1875	0.1060	0.0999	0.1129
3	0.1813	0.0000	0.1450	0.1817	0.0625	0.0353	0.1496	0.1691
4	0.0906	0.3000	0.1677	0.2101	0.0195	0.0110	0.1873	0.2117
5	0.0363	0.0000	0.1613	0.2021	0.0000	0.0000	0.1984	0.2243
6	0.0121	0.1000	0.1338	0.1677	0.0000	0.0000	0.1820	0.2058

balance equations to determine the relative probabilities as in Example 10.3.1, that is

$$p_{12}(x) = (p_1 * p_2)(x) = (0.0341, 0.0681, 0.1363, 0.1817, 0.2101, 0.2021, 0.1677).$$

We proceed to find the total state probabilities of all three streams by convolution of $p_{12}(x)$ and $p_3(x)$, shown in Table 10.4.

For the third (modeled by negative binomial distribution) service, we have the time blocking rate

$$\sum_{s \in B_3} (p_{12} * p_3)(s) = p_{123}(6) + p_3(4)(p_{12}(0) + p_{12}(1)) = 0.2071.$$

Notice that the first term corresponds to a saturated system and the second term to the restriction of service three to four channels.

The call blocking is found from the blocked call attempts

$$\sum_{s=0}^{4} (S_3 - s)\gamma_3 p_3(s) p_{12}(6 - s)$$

divided by the offered call attempts

$$\sum_{s=0}^{4}(S_3 - s)\gamma_3 p_3(s)\sum_{j=0}^{6-j} p_{12}(j)$$

and equals $B_3 = 0.2270$. The traffic congestion is found from the carried traffic

$$\sum_{s=1}^{4} sp_3(s)\sum_{j=0}^{6-s} p_{12}(j) = 0.4791,$$

and $\rho_3 = 2/3$ and equals $C_3 = 0.2814$.

If we would like to know the blocking rates for service one, we compute the convolution $\mathbf{p}_{123}$. The time blocking is given by $p_{123}(6) = 0.2058$. The call blocking and traffic congestion calculated in the same way as for service three turn out to have the same value as the time blocking. This is just what we expect since the service arrival is Poisson distributed.

10.3.4 **State-space recursion**

When all offered services are Poisson and there is no individual limitations on how many channels a particular service can occupy, a very efficient algorithm to compute blocking rates is by state-space recursion, first proposed by Fortet and Grandjean Fortet and Grandjean (1964) but widely known as the *Kaufman-Roberts algorithm*.

$$\lambda_k \cdot p(x - d_k) = \frac{x}{d_k}p_k(x)\mu_k, \quad x = d_k, d_k + 1, \dots, n. \tag{10.49}$$

Suppose a system is offered K services, arriving according to Poisson processes with intensity λ_k, service rate μ_k and bandwidth demand of d_k channels. The detailed balance equations are

$$\lambda_k \cdot p(m - d_k) = \frac{m}{d_k}p_k(m)\mu_k, \quad m = d_k, d_k + 1, \dots, n. \tag{10.50}$$

In Equation (10.50), $p(m)$ are the global state probabilities and $p_k(m)$ are the invidual state probabilities for each service k. The left-hand side is the flux from a global state $m - d_k$ to state m due to an arrival of type k call. The right-hand side is the flux from state m to state $m - d_k$ due to a departure of a type k call.

Solving for $p_k(m)$ and expressing the global state probabilities $p(m)$ as the sum of all individual traffic streams give

$$p_k(m) = \frac{1}{m}d_k\rho_k \cdot p(m - d_k),$$

$$p(m) = \frac{1}{m}\sum_{k=1}^{K} d_k\rho_k \cdot p(m - d_k), \quad p(m) = 0 \text{ for } m < 0. \tag{10.51}$$

Notice that in a state diagram such as Figure 10.5 of Example 10.3.1, we can draw parallel lines through some states which make up a global state where all channels

Table 10.5 Blocking probabilities for Example 10.3.3

m	$p(m)$
0	$1.1216 \cdot 10^{-19}$
$\vdots$	$\vdots$
61	0.0450
62	0.0464
63	0.0475
64	0.0434
65	0.0416

are occupied by some service. Such states in the figure are (0.0), $(2, 0)$, $(0, 2)$, $(4, 0)$, $(2, 2)$, $(0, 4)$ and $(6, 0)$, $(4, 2)$, $(2, 4)$, $(0, 6)$, corresponding to the global states $p(0)$, $p(2)$, $p(4)$ and $p(6)$. These states are equal from a global point of view. In two dimensions, a transition due to an arrival can either take the global state upwards in the diagram by an arrival of a call of type two, or to the right in the diagram by an arrival of type one (when there is capacity available). Transitions due to departures of calls, however, are state dependent, where $mp_k(m)$ is the average number of calls of type k in the system. This is the reason for the presence of $p(m - d_k)$ on the left-hand side and $mp_k(m)$ on the right-hand side in (10.50).

Example 10.3.3. To illustrate the Kaufman-Roberts algorithm, consider a system with $n = 65$ channels offered two services, a voice telephone service arriving at rate $\lambda_1 = 40$ with service rate $\mu_1 = 1$ and bandwidth requirement $d_1 = 1$ and a video conference service arriving at rate $\lambda_2 = 4$ with service rate $\mu_2 = 1$ and bandwidth requirement $d_2 = 5$. The algorithm only takes as arguments the capacity (in channels) n, the offered traffic $\rho_k = \lambda_k/\mu_k$ and the bandwidth requirements d_k, and is used as follows. For each state, we successively form a vector $\mathbf{q}$ of unnormalized global state probabilities $q(x)$, with $q(0) = 1$. The individual unnormalized state probabilities are $q_k(x) = 0$ if $x - d_k < 0$, and $q_k(x) = q(x - d_k)$ otherwise. The global state probabilities for $x > 0$ are $q(x) = \sum_{k=1}^{K} \rho_k d_k q_k(x)/x$. In order to normalize $q(x)$, we compute the normalization constant $G = \sum_{x=0}^{n} q(x)$ and form the normalized global state probabilities $p(x) = q(x)/G$. Using the algorithm on the example, we get the global state probabilities some of which are shown in Table 10.5. The blocking probabilities for service one equals the probability of state $x = n$, that is 0.0416, and the blocking probability for service two the sum of probabilities $\sum_{x=61}^{65} p(x) = 0.2238$.

10.3.5 A generalized algorithm

The state-space recursion can be generalized to handle BPP traffic (Iversen (2005), Iversen (2007)). The generalization of (10.50) to state-dependent arrivals is

$$p(x - d_k) \cdot S_k \gamma_k - p_k(x - d_k)\frac{x - d_k}{d_k}\gamma_k = \frac{x p_k(x)}{d_k}\mu_j. \tag{10.52}$$

On the left-hand side the first term assumes that all sources of type k are idle for the time to a transition. Since

$$\frac{x - d_k}{d_k} p_k(x - d_k)$$

sources of type k are busy on average in global state $x - d_k$, we reduce the first term with the second term to get the value of offered traffic. Thus, we have the global state probabilities

$$p(x) = \begin{cases} 0, & x < 0 \\ p(0), & x = 0 \\ \sum_{k=1}^{N} p_k(x), & x = 1, 2, \ldots, n, \end{cases}$$

where

$$p_k(x) = \frac{d_k}{x} \frac{S_k \gamma_k}{\mu_k} p(x - d_k) - \frac{x - d_k}{x} \frac{\gamma_k}{\mu_k} p_k(x - d_k), \tag{10.53}$$

$$p_k(x) = 0, \quad x < d_k. \tag{10.54}$$

The state probabilities are expressed in the terms of $p(0)$, which is found by normalization such that

$$\sum_{i=0}^{n} p(i) = p(0) + \sum_{i=1}^{n} \sum_{k=1}^{K} p_k(i) = 1, \tag{10.55}$$

where $p_k(0) = 0$, and $p(0) \neq 0$. Alternatively, expression (10.53) can be written

$$p_k(x) = \frac{d_k}{x} \frac{\rho_k}{Z_k} p(x - d_k) - \frac{x - d_k}{x} \frac{1 - Z_k}{Z_k} p_k(x - d_k), \tag{10.56}$$

where Z_k is the peakedness of service k.

$$p_k(x) = \frac{d_k}{x} \frac{\rho_k}{Z_k} p(x - d_k) - \frac{x - d_k}{x} \frac{1 - Z_k}{Z_k} p_k(x - d_k), \tag{10.57}$$

where Z_k is the peakedness of service k. In case of Poisson arrivals, Equation (10.53) reduces to (10.51), as expected. In evaluation of the recursive formula, it is often necessary to perform normalization in each step to avoid numerical overflow.

Since we have BPP traffic, time blocking, call blocking and traffic blocking are different for the non-Poissonian traffic streams. Since stream k requires d_k free channels, the time blocking is

$$E_k = \sum_{x=n-d_k+1}^{n} p(x).$$

The total carried traffic of stream k becomes

$$Y_k = \sum_{x=1}^{n} x \cdot p_k(x),$$

and the traffic congestion

$$C_k = \frac{\rho_k d_k - Y_k}{\rho_k d_k}.$$

The total carried traffic is given by the sum

$$Y = \sum_{k=1}^{K} Y_k,$$

so that the total traffic congestion becomes

$$C = \frac{\rho - Y}{\rho},$$

where ρ is the total offered traffic in channels,

$$\rho = \sum_{k=1}^{K} \rho_k d_k.$$

The call congestion for traffic stream k is given by

$$B_k = \frac{(1 + \beta_k)C_k}{1 + \beta_k C_k}.$$

There is no such formula for the total call congestion, since we do not have a global value of β. However, from individual values of carried traffic and call congestion, we can find the total number of offered calls and admitted calls for each stream, and from this calculate the total call congestion.

Example 10.3.4. By using the recursive algorithm on Example 10.3.1, we get the partial results shown in Table 10.6. Next, by normalization, the marginal state probabilities are found as in Table 10.7. The blocking rates are found to be

$$E_1 = 0.1677 \tag{10.58}$$
$$E_2 = 0.3698 \tag{10.59}$$
$$C_1 = 0.1677 \tag{10.60}$$

Table 10.6 Recursion of un-normalized state probabilities for Example 10.3.4

x	$q_1(x)$	$q_2(x)$	$q(x) = q_1(x) + q_2(x)$
0	0.000	0.000	1.000
1	2.000	0.000	2.000
2	2.000	2.000	4.000
3	2.667	2.667	5.333
4	2.667	3.500	6.167
5	2.467	3.467	5.933
6	1.978	2.944	4.922

Table 10.7 Normalized probabilities and expectations for Example 10.3.4

x	$p_1(x)$	$xp_1(x)$	$p_2(x)$	$xp_2(x)$	$p(x)$	$xp(x)$
0	0.0000	0.000	0.0000	0.000	0.0341	0.000
1	0.0681	0.068	0.0000	0.000	0.0681	0.068
2	0.0681	0.136	0.0681	0.136	0.1363	0.272
3	0.0908	0.272	0.0908	0.272	0.1817	0.545
4	0.0908	0.363	0.1192	0.477	0.2101	0.840
5	0.0840	0.420	0.1181	0.590	0.2021	1.012
6	0.0674	0.404	0.1003	0.602	0.1677	1.006
Sum	1	1.665	1	2.078	1	3.7426

$$C_2 = 0.2982 \tag{10.61}$$
$$B_1 = 0.1677 \tag{10.62}$$
$$B_2 = 0.2208, \tag{10.63}$$

just as expected.

10.3.6 Monte Carlo simulation

Multi-dimensional state probabilities can also be found using Monte Carlo simulation. Suppose that K traffic types with effective bandwidths d_k arrive at a resource with capacity n according to Poisson processes with intensities $\lambda_k, k = 1, 2, \ldots, K$. The method is based on the simulation of arrivals, the number of which is denoted by N_k. The service distribution is assumed to have mean rates μ_k^{-1}.

In each unit time step, the number of arrivals of each traffic type is simulated with their respective intensities, resulting in N_k arrivals, and the total bandwidth requirement is calculated as

$$d = \sum_{k=1}^{K} N_k d_k / \mu_k,$$

that is, the state as measured in the total number of occupied channels. By using a counter for the number of total feasible states G_0 and the states G_k where one additional call of type k can be accepted, the blocking rates B_k can be estimated. This value is the time blocking, but since we have Poisson arrivals all blocking measures are equal and the notation using "B" is may be more intuitive.

The counter G_0 is stepped up whenever a simulation result gives $n \geq \sum_k^K N_k d_k$, corresponding to a feasible state. Clearly, whenever the simulated bandwidth sum exceeds n, the bandwidth sum cannot represent a feasible state, and so the simulation result is discarded and a new bandwidth sum is generated.

The counter G_k is stepped up whenever the system is a non-blocking state with respect to traffic type k. The condition is that $n - \sum_{i=1}^{K} N_i d_i \geq d_k$.

Finally, the blocking probability of traffic class k is given by

$$B_k = 1 - \frac{G_k}{G_0}. \tag{10.64}$$

Note, that no assumptions have been made regarding the service processes apart from their mean rates. The algorithm can be summarized as follows:

Algorithm 10.3.3 (Monte Carlo Simulation). Given a system with capacity n and K traffic classes, each with Poisson arrival rate λ_k, general holding time of mean μ_k and bandwidth requirement d_k. Let G_0 be the number of feasible states, and G_k the number of non-blocking states for service k.

STEP 1:
Simulate K Poisson variables $(N_1, N_2, \ldots, N_K)$.

STEP 2:
if $\sum_k^K N_k d_k \leq n$ **then**
The state is feasible, set $G_0 \leftarrow G_0 + 1$,
else Go to STEP 1.

STEP 3:
for $k = 1, 2, \ldots, K$ **do**
if $\sum_{i=1}^K N_i d_i \leq n - d_k$ **then**
The system is in a non-blocking state for class k,
Set $G_k \leftarrow G_k + 1$.

STEP 4:
Repeat until a total number of N iterations has been performed.

Output blocking rates $B_k = 1 - G_k/G_0$.

Example 10.3.5. Using Monte Carlo simulation with 1,000,000 iterations on the same problem as in Example 10.3.3 gave the blocking probability 0.0413 for service one and 0.2231 for service two (compared to 0.0416 and 0.2238, respectively). A possible reason for the slight inaccuracy in the simulated results may be due to insufficient randomness of the generated Poisson variates. It is therefore a good idea to use high quality pseudo random number generators and try different seeds to compensate for slight dependencies that may exist.

10.4 Admission control

Trunk reservation is a conceptually simple and very efficient mechanism to prioritize between different traffic classes.

Consider a resource with capacity C offered a total number of K traffic streams. A traffic stream $k = 1, 2, \ldots, K$ is assumed to arrive according to a Poisson process with intensity λ_k and have exponential holding times with mean rate μ_k^{-1}, all of which are (mutually) independent.

The system capacity C is assumed to be channelized into basic bandwidth units (BBU), so that it can be interpreted as consisting of n BBUs in total. Each traffic stream requires an integer value $d_k \geq 1$ of BBUs for a connection. The resource requirement d_k of traffic class k is related to its effective bandwidth α_k by $d_k = \alpha_k c$, where c is the capacity of the BBU.

So far, the system is identical to the one in Section 10.3. In addition to this, however, we now introduce an integer trunk reservation parameter $t_k \geq 0$ for each service. The trunk reservation controls the rate at which traffic stream is admitted to the system. A connection of class k is accepted only if at least $d_k + t_k$ BBUs are available at the time of its arrival; otherwise it is lost. The condition for call acceptance is therefore

$$n_k \geq n - \sum_{j=1}^{K} N_j d_j - t_k,$$

where the left-hand side is the free capacity—the total system capacity minus the resources already occupied—less the trunk reservation parameter.

When $t_k = 0$ for all k, the equilibrium distribution of the number of calls in progress of each traffic class is independent of the holding time distribution. When some $t_k > 0$, this is no longer true, in general. Therefore, we assume exponentially distributed holding times in this section.

Trunk reservation provide a logically simple and a very robust mechanism for controlling the behavior of a system. The strength of the strategy lies in that a flexible prioritization of traffic streams can be achieved by assigning different trunk reservation parameters to traffic classes, while still utilizing the full capacity of the system. It also has a stabilizing effect as discussed in Section 6.6. An illustration of a state space with trunk reservation is shown in Figure 10.6.

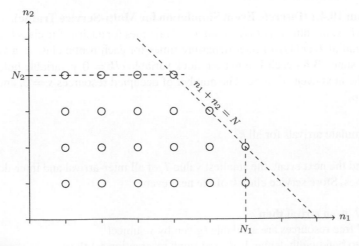

FIGURE 10.6

State space truncated by trunk reservation.

Bean et al. (1995) formulate the global balance equations for the system described above and propose a fixed-point method to determine the equilibrium probabilities. The method is, however, rather technical as it involves finding the roots of a polynomial in each iteration.

We describe a conceptually simple method for calculating blocking rates in systems subject to trunk reservation. The main idea is to simulate a queue, like in Section 6.5, with K traffic sources and trunk reservation. The input to such a procedure is the traffic the arrival rates λ_k, the holding times with mean μ_k^{-1}, the bandwidth requirements of the traffic streams d_k and the trunk reservation parameters t_k.

The method is based on discrete-event simulation, where a vector and a matrix of exponential variables determine the succession of events, that is, specifying the inter-arrival times and inter-departure times, respectively. A vector is also defined to keep track of the states of the queue and to keep track of the number of rejected calls of each type. In each iteration, the simulation jumps to the closest coming event by inspection of the vectors containing inter-arrival times and inter-departure times. Bookkeeping must be performed after each step to reduce the time variables with the current closest future event, as well as updating the total time elapsed and the total time the queue is in a blocked state for service k.

A new exponential random variable replaces the one for the event that just occurred. Testing for available bandwidth is straightforward with this approach: we simply test whether $n - \sum_{j=1}^{K} d_j N_j - t_k > d_k$ for each service k; if the expression evaluates true, the new call is accepted, otherwise it is rejected. If rejected, the counter for the number of blocked calls of type k is increased by one. At the end of the simulation, the time blocking rate is given by the total time the system spent in a congested state, divided by the total simulation time, for each service k. The call blocking rate is the number of rejected calls divided by the total number of call attempts. The procedure can be summarized as follows.

Algorithm 10.4.1 (Discrete Event Simulation for Multi-Service Traffic). Inputs: a vector of exponential $Exp(1/\lambda_k)$] inter-arrival times for each traffic class k, a matrix of exponential $Exp(1/\mu_k)$ inter-departure times for each traffic class k, a vector of blocking states **B** for each k, a master clock variable $clk = 0$, a variable holding the time of the next event $T = \infty$, The number of occupied resources $s = 0$, and vector of states **p**.

STEP 0:
  Simulate arrivals for all k.

STEP 1:
  Find the next event, the smallest value T of all inter-arrival and inter-departure times. Store service class k of the next event.

STEP 2:
  if T is an arrival **then**
    if free resources are available (given by s subject to bandwidth demands d_k and trunk reservation t_k) **then**
    (Accept call)

> Update all inter-arrival times and inter-departure times;
> Generate an inter-departure time for the new call;
> Generate a new inter-arrival time for a future arrival;
> **else**
>> Update blocking state vector **B** by B_k;
>> Update all inter-arrival times and inter-departure times;
>> Generate a new inter-arrival time for a future arrival;
>> Update clk, s and **p**
> **elseif** T is a departure **then**
>> Remove the call.
>> Update all inter-arrival times and inter-departure times.
>> Update clk, s and **p**

Output normalized (with respect to clk) vectors **p** and **B**.

A natural benchmark is to test the method on a system without trunk reservation, for which exact results are obtainable. In order to verify the accuracy of this approximation, we can compare the figures for some traffic without trunk reservation with the exact blocking rates that can be calculated using the Kaufman-Roberts recursion. Using the values from Example 10.3.3 for an AMR coded voice call (service 1) and a video service (service 2), we let $d_1 = 1$ and $d_2 = 5$ capacity units determined from the acceptance region. The capacity is $n = 65$ units and calls arrive according to Poisson processes with $\lambda_1 = 80\kappa$ and $\lambda_2 = 8\kappa$, where $0 < \kappa \leq 1$ is a scaling parameter. Each simulation is run for 100,000 events. The comparison is shown in Figure 10.7. The exact values are shown with dashed lines and the simulated values with solid lines. The method shows very good results, as they are almost completely overlapping. The relative errors are depicted in Figure 10.8. The method becomes more accurate in relative terms with higher load. This is a result of the simulation 'small number inaccuracy.' On the other hand, small errors in small blockings are small in absolute value. For loads higher than 50%, the relative error in blocking is

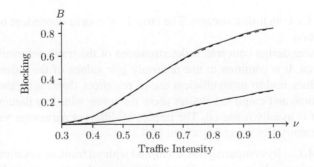

FIGURE 10.7

Comparison of the discrete-event simulation results with the exact blocking rates for a link without trunk reservation.

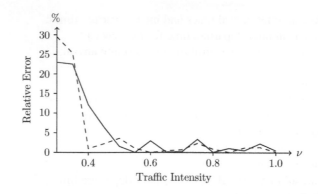

FIGURE 10.8

Comparison of the relative error between discrete-event simulation and exact blocking rates for a link without trunk reservation.

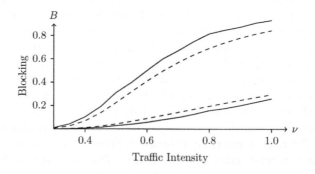

FIGURE 10.9

Comparison of blocking rates for two services on a link with trunk reservation $t_2 = 2$.

less than 4% for both traffic streams. The error is of course dependent on the length of the simulation.

Among other design concerns is determination of the trunk reservation parameters themselves. It is common to use relatively low values of the parameters. Low parameter values lead to more efficient use of resources than high values. Furthermore, theoretical and empirical studies show that even with low parameter values, desired effect is usually achieved. The impact of different parameter values can be investigated empirically using Algorithm 10.4.1.

Example 10.4.1. By comparing a link with and without trunk reservation, the effects on traffic streams can be estimated. Figure 10.9 shows the effect of trunk reservation parameters $t_1 = 0, t_2 = 2$. The dashed lines represent blocking rates without trunk reservation. As expected, the blocking rate of service 1 drops and the blocking rate

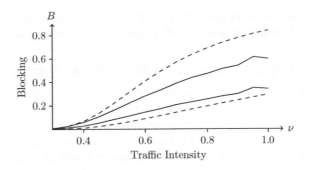

FIGURE 10.10

Comparison of blocking rates for two services on a link with trunk reservation $t_1 = 5$.

of service 2 increases rather drastically. In figure 10.10 the situation is reversed. Here $t_1 = 5$, and the blocking rate for service 1 increases whereas it drops for service 2.

By using discrete-event simulation, we are not limited to exponential inter-arrival and inter-departure times. In principle, these can be specified arbitrarily. It may, however, be difficult to model these distributions in general. The exponential assumption is usually a rather good approximation in most situations. As with almost all methods, exact as well as approximate, it is increasingly difficult to obtain accurate results as the system size increases and the state space quickly becomes enormous. This also requires longer simulation runs in order to obtain reasonable accuracy. The memory requirement for storage of system vectors also becomes an issue.

10.5 Processor load sharing

An ATM switch can be considered bufferless; it only has a relatively small buffer for handling burst level variations. Its constant service rate and guaranteed QoS make multichannel systems suitable models. In contrast, IP multiservice nodes are often better described by a single server queue with exponential service times, and which, therefore, will require larger buffer space. In a multi-service network based on IP, the service discipline is processor sharing (PS) (or generalized processor sharing), which provides a fair and efficient use of resources. The model can be extended to multiserver systems corresponding to an IP router with several queues.

In the queueing systems discussed so far, there is non-sharing of processing resources between traffic streams. This means that a job in a queue occupies the server all by itself until service has been completed. A busy server does not necessarily imply traffic loss, since a job may be waiting in a buffer. Thus, a customer is either waiting or being served.

This section discusses systems with processor sharing disciplines, where all jobs share the available capacity so that no job is strictly waiting for service. All jobs are therefore serviced with some rate, which may be lower than the requested rate. When this is the case the node will need to buffer some of the job contents.

10.5.1   The single-server queue with processor sharing

A job of type k is characterized by the offered load ρ_k. This is equal to the resources required from the system. If jobs of type k are served at a lower rate than requested, this results in an increase of the sojourn time. Since a job is served all the time at some rate, the sojourn time cannot be divided into waiting time and time being serviced.

Let $\overline{W}$ be the total sojourn time and $\overline{L}$ the queue length in a processor sharing system. We define the virtual waiting time

$$W_k = \overline{W}_k - \rho_k,$$

which describes the increase of waiting time due to insufficient capacity. Similarly, we define the mean virtual queue length as

$$L_k = \bar{L}_k - \rho_k.$$

To begin with, consider a single server queue with $K = 2$ streams—or chains—of jobs. Jobs from chain k arrive to the node according to a Poisson process with intensity λ_k for $k = 1, 2$. We may describe the state of the system by (x_1, x_2), where x_1 is the number of jobs of chain 1 and x_2 the number of jobs of chain 2. Let the global state be

$$x = \sum_{k=1}^{2} x_k.$$

Note that these assumptions are not very practical as any service usually has a demand which is much less than the capacity. The model will be extended below to reflect a situation similar to a loss system with multiple servers representing basic bandwidth units (BBU).

We introduce the notation $\sum_{k=1}^{K} M_k/M_k/1$ for a single server system that is offered K different chains with possibly different arrival rates and mean service times.

If we let the capacity be described by the number of servers-initially a single server-all service rates are reduced whenever the global state $(x > 1)$. To do so, we introduce the reduction factors for state (x_1, x_2) as

$$0 < g_1(x_1, x_2) \leq 1,$$

$$0 < g_2(x_1, x_2) \leq 1.$$

Jobs of chain k receive capacity $x_k g_k(x_1, x_2)$, $k = 1, 2$. The service rate with which chain k jobs are processed is $x_k \mu_k g_k(x_1, x_2)$. By using the Kolmogorov cycle condition for reversibility, we get the state diagram shown in Figure 10.11.

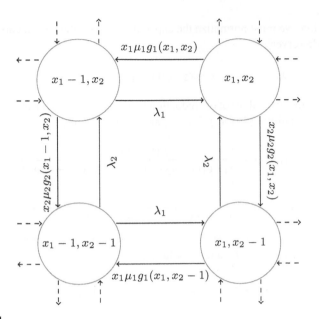

FIGURE 10.11

State diagram for a $\sum_k M_k/M_k/1$ server showing transition rates modified by the reduction factors.

We note that, for any multiserver system with n servers, the relative state probabilities of the states where $x_1 + x_2 \leq n$ can be found by the methods described in Section 10.3.

In order to construct a reversible system, we need to specify the reduction factors so that the flow balance equations are satisfied. Since the reduction factors are state dependent, we characterize the states as follows:

(1) The states $x_k < 0$ are infeasible, and so $g_k(x_1, x_2) = 0$.
(2) For states $x_k \geq 0$ and $0 < x_1 + x_2 \leq 1$, we set $g_k(x_1, x_2) = 1$. In these states, any call gets the capacity it requires and there no reduction of service rates is necessary.
(3) For states with only one service, for example $x_1 \geq 1$ and $x_2 = 0$, we set $g_1(x_1, x_2) = 1/x_1$ (and similarly when $x_2 \geq 1$ and $x_1 = 0$). These cases correspond to single job class $M/M/1 - PS$ systems, and all jobs are therefore sharing capacity equally, since they are identical.
(4) The states where the sum of demands exceeds capacity, both demands being non-zero, that is, $x_k > 0$ and $x_1 + x_2 > 1$, require capacity scaling, and these factors are determined as follows. In order for the system to be reversible, we require that the flow balance according to the Kolmogorov cycle theorem holds, that is

$$g_2(x_1, x_2) \cdot g_1(x_1, x_2 - 1) = g_1(x_1, x_2) \cdot g_2(x_1 - 1, x_2).$$

In addition, we must normalize the capacities, so that the partial capacities equal the single server

$$x_1 \cdot g_1(x_1, x_2) + x_2 \cdot g_2(x_1, x_2) = 1, \quad x_1 + x_2 \geq 1.$$

If we know the two values of the reduction factors $g_1(x_1, x_2-1)$ and $g_2(x_1-1, x_2)$, we may solve condition (4) for $g_1(x_1, x_2)$ and $g_2(x_1, x_2)$ and arrive at

$$g_1(x_1, x_2) = \frac{g_1(x_1, x_2 - 1)}{x_1 \cdot g_1(x_1, x_2 - 1) + x_2 \cdot g_2(x_1 - 1, x_2)},$$

$$g_2(x_1, x_2) = \frac{g_2(x_1 - 1, x_2)}{x_1 \cdot g_1(x_1, x_2 - 1) + x_2 \cdot g_2(x_1 - 1, x_2)}.$$

It can be shown, Iversen (2013), that by using these recursion formulas with initial conditions (1)–(3), we get the unique solution

$$g_k(x_1, x_2) = \frac{1}{x_1 + x_2}, \quad x_1 + x_2 \geq 1, k = 1, 2.$$

Note that the service of the two chains is reduced by the same factor $1/(x_1 + x_2)$. The total service rate $\mu(x_1, x_2)$ in state (x_1, x_2) is accordingly

$$\mu(x_1, x_2) = \frac{x_1 \mu_1}{x_1 + x_2} + \frac{x_2 \mu_2}{x_1 + x_2} = \frac{x_1 \mu_1 + x_2 \mu_2}{x_1 + x_2}.$$

These results can easily be extended to a system accepting K traffic classes. In this case, the state vector is

$$\mathbf{x} = (x_1, x_2, \dots, x_K),$$

where x_k is the channel demand of chain k. For the states $\sum_{k=1}^{K} x_k > 1, x_k \geq 0$ we have the reduction factors

$$g_k(\mathbf{x}) \frac{1}{\sum_{k=1}^{K} x_k}, \quad k = 1, 2, \dots, K.$$

Similarly, for a system with K traffic streams the total service intensity in state $\mathbf{x}$ is

$$\mu(\mathbf{x}) = \frac{\sum_{k=1}^{K} x_k \mu_k}{\sum_{k=1}^{K} x_k}, \quad k = 1, 2, \dots, K.$$

The model is reversible and independent of service time distributions, where each chain may have different mean service time. We state the result as a theorem.

Theorem 10.5.1. *The $\sum_j M_j/G_j/1 - PS$ single server system with processor sharing is reversible and independent of the service time distributions. Each service class is allowed to have different mean service times.*

The processor sharing discipline is a fundamental assumption for the properties stated in Theorem 10.5.1 of a system to hold. Non-sharing here means that a single job is served at a time and any job to be processed is selected with equal probability. The system is only reversible when the service times are exponentially distributed, since the inter-departure times equal the service times. In other words, in order to have a Poisson distributed departure process, the service time distribution must be exponential. Thus, given the processor sharing discipline, the queueing model is more general—in terms of reversibility—than a non-sharing system. In the latter, the service time distributions for the K chains have to be exponentially distributed and all K chains have to have the same mean service time.

We will extend this model in two steps. First we divide a single server into n (logical) servers, and finally allow services to require an integer number of servers, corresponding to basic bandwidth units (BBU). This models the situation where different services require an integer number of channels for service.

The state probabilities for the $\sum_k M_k/G_k/1 - PS$ model can be found by using the detailed balance equations, which follow from reversibility. We have, for example,

$$\lambda_1 p(x_1 - 1, x_2) = \mu_1 x_1 p(x_1, x_2)/(x_1 + x_2),$$

where the right-hand side has to be divided by the reduction factor. Rearranging gives

$$p(x_1, x_2) = \frac{\rho_1}{x_1}(x_1 + x_2)p(x_1 - 1, x_2). \tag{10.65}$$

Expressing $p(x_1 - 1, x_2)$ in the same way, we have

$$p(x_1, x_2) = \frac{\rho_1^2}{x_1(x_1 - 1)}(x_1 + x_2)(x_1 - 1 + x_2)p(x_1 - 2, x_2). \tag{10.66}$$

For two chains, we obtain

$$p(x_1, x_2) = p(0, 0)\frac{\rho_1^{x_1}}{x_1!}\frac{\rho_2^{x_2}}{x_2!}(x_1 + x_2)! \tag{10.67}$$

as usual expressed in terms of the zero state probability $p(0, 0)$. The presence of the factor $(x_1 + x_2)!$ destroys the product form, which we have in the multi-dimensional Erlang formula. As a result,

$$p(x_1, x_2) \neq p_1(x_1)p_2(x_2). \tag{10.68}$$

Losing the product form makes analysis of the system more complicated as the state space grows, due to the normalization step, in which we need to compute

$$\sum_{x_1=0}^{\infty} \sum_{x_2=0}^{\infty} p(x_1, x_2) = 1.$$

We may use the binomial expansion to arrive at an expression for the global state probabilities, as

$$p(x_1 + x_2 = x) = p(0, 0)(\rho_1 + \rho_2)^x$$
$$= (1 - \rho)\rho^x,$$

where $\rho = \rho_1 + \rho_2$. The state probability $p(0,0) = 1 - \rho$ is known explicitly, since it is identical with the state probabilities of an $M/M/1$ system with offered traffic $\rho = \rho_1 + \rho_2$. We will see, however, that it is possible to formulate an algorithm which finds the state probabilities, performing normalization in each iteration.

Extending to K chains, the state probabilities become

$$p(x_1, x_2, \ldots, x_K) = p(0) \frac{\rho_1^{x_1}}{x_1!} \frac{\rho_2^{x_2}}{x_2!} \cdots \frac{\rho_K^{x_K}}{x_K!} (x_1 + x_2 + \cdots x_K)!$$

$$= p(0) \left(\prod_{j=1}^{K} \rho_k^{x_k} \right) \frac{\left(\sum_{k=1}^{K} x_k \right)!}{\left(\prod_{k=1}^{K} x_k! \right)}. \tag{10.69}$$

Similarly to the two-chain case, this can be expanded using the multinomial distribution, giving

$$p(\mathbf{x}) = p(0) \left(\prod_{k=1}^{K} \rho_k^{x_k} \right) \binom{x_1 + x_2 + \cdots + x_K}{x_1, x_2, \ldots, x_K}.$$

For an infinite system, the global state probabilities are identical with the state probabilities of the $M/M/1$ queue, so that

$$p(x) = (1 - \rho)\rho^x,$$

where $\rho = \rho_1 + \rho_2 + \cdots + \rho_K$.

The global state probabilities for K chains can be found by the following algorithm. Assuming K chains, we first determine the relative non-normalized state probabilities, which are denoted $q(x)$, letting $q(0) = 1$. In the single server case, the normalization is straightforward as we know that $p(0) = 1 - \rho$, with $\rho = \rho_1 + \rho_2 + \cdots + \rho_K$. We have the relations

$$p(x) = \begin{cases} 0 & \text{for } x < 0, \\ 1 - \rho & \text{for } x = 0, \\ \sum_{k=1}^{K} p_k(x) & \text{for } x = 1, 2, \ldots \end{cases}$$

where

$$p_k(x) = \begin{cases} 0 & \text{for } x < 1, \\ \rho_k p(x - 1) & \text{for } x = 1, 2, \ldots \end{cases}$$

This algorithm is extended to multiple servers and chains with different bandwidth requirements below.

With K chains, the mean queue length $\bar{L}$, which includes all jobs in the system, are the same as for an $M/M/1$ queue. The queue length is geometrically distributed with mean

$$\bar{L} = \frac{\rho}{1 - \rho},$$

where total offered traffic is $\rho = \rho_1 + \rho_2 + \cdots \rho_K$. In state x the average number of type k calls is $x p_k(x)$. The mean queue length for stream k is then

$$\bar{L}_k = \sum_{x=0}^{\infty} x p_k(x) = \frac{\rho_k}{\rho} \bar{L},$$

$$\frac{\bar{L}_k}{\rho_k} = \frac{\bar{L}}{\rho},$$

where $\bar{L} = \sum_{k=1}^{K} \bar{L}_k$. Since there is no blocking, the carried traffic equals the offered traffic. Therefore, increase in $\bar{L}$ due to limited capacity is $L = \bar{L} - \rho$, and for chain k the increase in queue length is $L_k = \bar{L}_k - \rho_k$.

The mean sojourn time for type k jobs is found by Little's formula to be

$$\bar{W}_k = \frac{\bar{L}_k}{\lambda_k},$$

The mean sojourn time for all chains is consequently

$$W = \sum_{k=1}^{K} \frac{\lambda_k}{\lambda} W_k.$$

10.5.2   **Extensions to the model**

These results can be extended to the multiserver case. Rather than that the system is having n servers, we interpret the total capacity as being divided in channels corresponding to some basic bandwidth unit (BBU). The derivation of the result is similar to the single server case, and we will omit the details and only state the results. In the first step of the extension, each chain is assumed to require exactly one channel.

The state of the system is now given by:

$$\mathbf{x} = (x_1, x_2, \ldots, x_K),$$

$$x = \sum_{k=1}^{K} x_k,$$

where x_k is the number of jobs of type k in the system and x is the global state. Jobs of chain k are assumed to arrive according to a Poisson process with intensity λ_k, and the service time exponentially distributed with intensity μ_k. When the system capacity is limited to n channels, the service rates are reduced in states where the demand exceeds n channels. We assume that there are K traffic types offered to the system.

The reduction factors are found in a way analogous to the single server case. By classifying the states, formulating initial values, and solving the flow balance equations, we can derive the unique solution

$$g_k(\mathbf{x}) = \begin{cases} 1 & \text{for } 0 \le x \le n, \\ \frac{n}{x} & \text{for } n \le x, \end{cases} \quad k = 1, 2, \ldots, K,$$

where $x = \sum_{k=1}^{K} x_k, x_k \ge 0$. Again, all chains are reduced by the same factor and jobs share the capacity equally.

In the n-server context, the queueing discipline is called the generalized processor sharing (GPS) discipline. The system is independent of service time distribution and each chain may have different mean service times. We can thus formulate an analog to Theorem 10.5.1.

Theorem 10.5.2.   *A $\sum_{k=1}^{K} M_k/G_k/n - GPS$ system with generalized processor sharing is reversible and independent of the service time distributions. In addition, each service type may have different mean service time.*

We consider an $M/M/n$ non-sharing system. A customer being served always has a server by itself. A customer is either waiting or being served. To maintain reversibility for $x_1 + x_2 > n$ we have to require that all services have the same mean service rate $\mu_j = \mu$, which furthermore has to be exponentially distributed.

This corresponds to an $M/M/n$ system with total arrival rate $\lambda = \sum_j \lambda_j$ and service rate μ. The state probabilities are identical to an $M/M/n$ system with arrival rate λ and service rate μ.

The state probabilities for K chains and n servers are given by

$$
\frac{p(x_1, x_2, \ldots, x_K)}{p(0, 0, \ldots, 0)}
$$
$$
= \begin{cases} \rho_1^{x_1}/x_1!\rho_2^{x_2}/x_2! \cdots \rho_K^{x_K}/x_K! & \text{if } \sum_{k=1}^{K} x_k \leq n \\ \rho_1^{x_1}/x_1!\rho_2^{x_2}/x_2! \cdots \rho_K^{x_K}/x_K!(x_1 + x_2 + \cdots x_K)!/n! \cdot n^{x_1+x_2+\cdots x_K)-n} & \text{if } \sum_{k=1}^{K} x_k \geq n \end{cases}.
$$

$$(10.70)$$

The state probability $p(0, 0, \ldots, 0)$ is obtained by normalization. The global state probability of state x is given by

$$
p(x) = \sum_{\sum_i x_i = x} p(x_1, x_2, \ldots, x_K).
$$

Multinomial expansion gives

$$
\frac{p(x)}{p(0)} = \begin{cases} \rho^x/x! & \text{if } 0 \leq x \leq n \\ \rho^x/n! \cdot n^{x-n} & \text{if } x \geq n \end{cases}.
$$

For a system with n servers and K chains, we have the relative global state probabilities

$$
q(x) = \begin{cases} 0 & \text{if } x < 0, \\ 1 & \text{if } x = 0, \\ \sum_{k=1}^{K} q_k(x) & \text{if } x = 1, 2, \ldots, \infty \end{cases},
$$

where

$$
q_k(x) = \begin{cases} \rho_k/x \, q(x-1) & \text{if } x \leq n, \\ \rho_k/n \, q(x-1) & \text{if } x > n \end{cases}.
$$

The quantity $q_k(x)$ is the contribution of chain k to the relative global state probability $q(x)$, where

$$
q_k(x) = \sum_{\sum_i x_i = x} \frac{x_k}{x} q(x_1, x_2, \ldots, x_K).
$$

In this case, the state probability $p(0)$ is obtained by computing the normalization constant G, according to

$$G = \sum_{i=0}^{\infty} q(i) = \sum_{i=0}^{\infty} \sum_{k=1}^{K} q_k(i).$$

Note that the state, that is, the total channel requirement, can be far greater than the number of available channels n. It can be infinite, in fact, since the queue is infinite. That is why the state x sums from zero to infinity. The actual state probabilities $p_k(x)$ and $p(x)$ are found by normalizing the relative state probabilities $q_k(x)$ and $q(x)$ by the constant G. To maintain stability of the algorithm, the normalization should be performed in each step.

The performance measures for the n-server system can be derived similarly as for the single server case. The total mean queue length $\bar{L}$ is

$$\bar{L} = \sum_{x=0}^{\infty} x p(x). \tag{10.71}$$

Chain k contributes to the mean number of jobs in system state x by $x p_j(x)$. The mean queue length consisting of chain k is denoted by $\bar{L}_k$ and is given by

$$\bar{L}_k = \sum_{x=0}^{\infty} x p_k(x) = \frac{\rho_k}{\rho} \bar{L},$$

where $\bar{L} = \sum_{k=1}^{K} \bar{L}_k$ and $\rho = \sum_{k=1}^{K} \rho_k$. The mean sojourn time for jobs of type k can be found by Little's formula,

$$\bar{W}_k = \frac{\bar{L}_k}{\lambda_k}. \tag{10.72}$$

and total mean sojourn time

$$\bar{W} = \lambda \bar{L}. \tag{10.73}$$

10.5.3 A recursive algorithm

We can now formulate a recursive algorithm for the state probabilities of a $\sum_k M_k/G_k/n$ system. Suppose that we have an n-server system offered K chains with Poisson arrivals of intensity λ_k and service requirement $d_k \mu_k$.

Denote the relative state probabilities for state x by $q(x)$. After normalization, we obtain the actual state probabilities, denoted by $p(x)$. Each traffic type k contributes to $q(x)$ by $q_k(x)$, so that

$$q(x) = \begin{cases} 0 & \text{for } x < 0 \\ 1 & \text{for } x = 0 \\ \sum_{k=1}^{K} q_k(x) & \text{for } x = 1, 2, \ldots, b, \end{cases} \tag{10.74}$$

We allow to truncate the system at b, some finite buffer size. The contribution of stream k $q_k(x)$ is found by using the recursion

$$q_k(x) = \frac{1}{\min\{x, n\}} \left(\frac{d_k}{x} \lambda_k \cdot q(x - d_k) + \sum_{i=1}^{K} \left(\frac{x - d_i}{x} \lambda_i \cdot q_k(x - d_i) \right) \right). \quad (10.75)$$

In order to find performance measures for a traffic type k, we decompose the contribution of stream k to system state x into two parts:

$$q_k(x) = q_{k,y}(x) + q_{k,l}(x), \quad k = 1, 2, \dots, K.$$

The component $q_{k,y}(x)$ corresponds to the proportion of servers used by service k in the global state x, and this reflects the average number of servers allocated to a type k job. Being in state x, the mean number of channels serving type k jobs is

$$n_{k,y}(x) = \frac{p_{k,y}(x)}{p_k(x)} x.$$

The total of allocated servers is

$$\sum_{k=1}^{K} n_{k,y}(x) = n, \quad x \geq n,$$

with the total contributions to the relative state probabilities in state x

$$q_y(x) = \sum_{k=1}^{K} q_{k,y}(x).$$

The second component $q_{k,l}(x)$ corresponds to the proportion of buffer position used by traffic type k. In the global state x, type k buffer positions are given by

$$n_{k,l}(x) = \frac{q_{k,l}(x)}{q_j k(x)} x,$$

with the total contributions to the relative state probabilities in state x

$$q_l(x) = \sum_{k=1}^{K} q_{k,l}(x).$$

For the global states, we must have

$$\sum_{k=1}^{K} (n_{k,y}(x) + n_{k,l}(x)) = x,$$

and for all streams

$$q(x) = q_y(x) + q_l(x).$$

The quantities $q_{k,y}(x)$ can be found as follows. The local balance equations for type k jobs between the states $x - d_k$ and x give in equilibrium

$$\lambda_k q(x - d_k) = x q_{k,y}(x) \mu_k.$$

The bandwidth requested by a type k job is $d_k \mu_k$, and the bandwidth granted to stream k is $x q_{k,y}(x)$. Therefore,

$$q_{k,y}(x) = \frac{\lambda_k}{x \mu_k} q(x - d_k),$$

$$q_{k,l}(x) = q_k(x) - q_{k,y}(x),$$

and $q(0) = 1$. Also note that $q_{j,y}(x) = q_j(x)$ and $q_{j,l}(x) = 0$ whenever $x \leq n$, and then there is no delay incurred by the system. The initial values of the algorithm for the relative state probabilities are

$$q_k(x) = q_{k,y}(x) = 0, \quad \text{when } x < d_k.$$

The algorithm is used as follows. Letting the state probability $q(0) = 1$ for a system with $x = 0$ occupied channels, we proceed by calculating the relative state probabilities $q_k(1)$ using (10.75). From (10.74), the relative global state is found. Notice that the state probability $q_k(x)$ only depends on the global state $q(x - d_k)$, the local state probability $q_k(x - d_k)$, and the traffic parameters of all the chains.

Next, we increase the number of channels by one, calling this state x and repeat the calculations, first the value $q(x)$ (from which we may compute $q_{k,y}(x)$ and $q_{k,l}(x)$ as well), and then the $q_k(x)$ for all streams k.

To ensure stability of the algorithm, the state probabilities up to some state x should be normalized in each step (apart from the first, where $q(0) = 1$). This will produce a successive convergence of the state probabilities, which in each step sum to unity. Suppose we have the normalized probabilities up to state $x - 1$. The relative probabilities $q(x)$ calculated based on previously normalized probabilities is in general not normalized. Let $q^{(x)}(i)$ denote the state probabilities calculated up to state x. Then

$$q^{(x)}(k) = \begin{cases} p^{(x-1)}(k) & k = 0, 1, \ldots, x - 1 \\ q(x) & k = x. \end{cases} \tag{10.76}$$

The normalization constant G_x after each new state x is

$$G_x = \sum_{k=0}^{x} q^{(x)}(k) = 1 + q(x), \tag{10.77}$$

since $\sum_k p^{(x-1)}(k) = 1$.

Thus, in each step, compute $q_j(x)$ and $q(x)$ based on $p_j^{(x-1)}(k)$ and $p^{(x-1)}(k), k = 0, 1, \ldots, x-1$. Set $G_x = 1 + q(x)$ and calculate the new normalized state probabilities $p_j^{(x)}(k) = p_j^{(x-1)}(k)/G_x$ and $p^{(x)}(k) = p^{(x-1)}(k)/G_x$ for $k = 0, 1, \ldots, x - 1$, and set $p^{(x)}(x) = q(x)/G_x$. Note that normalization usually should be applied to the recursive algorithms in Section 10.3 as well.

A number of system performance measures are of interest. We assume that the system consists of n channels and b buffer positions. Each chain k contributes in state m to the carried traffic by an amount

$$y_k(m) = m p_{k,y}(m),\tag{10.78}$$

so that the total carried traffic of type k is

$$Y_k = \sum_{m=0}^{n+b} y_k(m) = \sum_{m=0}^{n+b} m p_{k,y}(m).\tag{10.79}$$

When there is a finite buffer, the carried traffic is less than the offered traffic, or $Y_k < \rho_k$, and the traffic congestion for type k traffic is

$$C_k = \frac{\rho_k - Y_k}{\rho_k}.\tag{10.80}$$

Since the arrivals are assumed to be Poisson, where all congestion types have the same value, the congestion can also be read immediately from the state probabilities.

The contribution from chain k to the virtual queue length in state m is

$$l_k(m) = m p_{k,l}(m),\tag{10.81}$$

so the total queue length of type k jobs is

$$L_k = \sum_{m=0}^{n+b} l_k(m) = \sum_{m=0}^{n+b} m p_{k,l}(m).\tag{10.82}$$

Consequently, the mean number of channels occupied by type k traffic is

$$\bar{L}_k = Y_k + L_k.\tag{10.83}$$

By Little's formula, we can easily calculate the mean soujourn time and mean virtual waiting time. The average number of type k jobs in the system is $\bar{L}_k/d_k$, and the average number of type k jobs in the virtual queue is L_k/d_k. The mean sojourn time and mean virtual waiting time for type k traffic are

$$\bar{W}_k = \frac{\bar{L}_k}{d_k \lambda_k (1 - C_k)},\tag{10.84}$$

$$W_k = \frac{L_k}{d_k \lambda_k (1 - C_k)}.\tag{10.85}$$

The values for the whole system is obtained by weighting with respect to the traffic streams, so that

$$\bar{W} = \sum_{k=1}^{K} \frac{\lambda_k(1 - C_k)}{\lambda(1 - C)} \bar{W}_k = \frac{1}{\lambda(1 - C)} \sum_{k=1}^{K} \frac{\bar{L}_k}{d_k},\tag{10.86}$$

$$W = \sum_{k=1}^{K} \frac{\lambda_k(1 - C_k)}{\lambda(1 - C)} W_k = \frac{1}{\lambda(1 - C)} \sum_{k=1}^{K} \frac{L_k}{d_k},\tag{10.87}$$

where $\lambda = \sum_{k=1}^{K} \lambda_k$ and $C = \sum_{k=1}^{K} C_k$.

Table 10.8   The generalized algorithm for a two chain example

State x	$q_1(x)$	$q_2(x)$	$q(x)$
0	0	0	1.
1	2	0	2
2	2	0.5	2.5
3	1.67	0.67	2.33
4	1.17	0.62	1.79
5	0.72	0.47	1.18
6	0.39	0.30	0.69
7	0.23	0.20	0.43
8	0.13	0.12	0.26
Sum	–	–	12.19

Table 10.9   The generalized algorithm for a two chain example

State x	$p(x)$
0	0.08
1	0.16
2	0.21
3	0.19
4	0.15
5	0.10
6	0.06
7	0.04
8	0.02
Sum	1.0

Example 10.5.1.   As an illustration of the algorithm, consider a $\sum_{k=1}^{2} M_k/G_k/6/8-$ GPS system. We have two traffic types. The first is arriving according to a Poisson process of intensity $\lambda_1 = 2$, mean service time of $\mu_1 = 1$, and bandwidth requirement $d_1 = 1$. The second, having the parameters $\lambda_2 = 1$, $\mu_2 = 1$, and $d_2 = 2$. The number of servers is $n = 6$ and the buffer $b = 8$ (including served jobs). We do not need to normalize in each step here, since the example is so small.

Using the algorithm described above, we proceed as follows (referring to Table 10.8). At the first row, we set $q_1(0) = q_2(0) = 0$ as per the definitions above. However, $q(0) = 1$ by default. In row one, source one has a non-zero entry (because $d_1 = 1$) given by (10.75); source two remains zero because $d_2 = 2$. Next we form the sum $q(1) = q_1(1) + q_2(1) = 2$. In the third row, the relative probability of the first source is calculated according to (10.75), and since $q(2 - d_2)$ now exists, the second source is calculated using the same formula. Proceeding in this manner, we can build up the entries in Table 10.8. The last row in the table gives the normalization constant, by which the actual probabilities can be found (the relative probabilities divided by this constant). For example, the global state probabilities can be calculated in this way, shown in Table 10.9. From these figures, we can compute important performance measures, such as blocking rates ($C_1 = 0.0212$ and $C_2 = 0.0563$), mean queue length in the system ($\bar{L} = 2.9788$) and mean sojourn time in the system ($\bar{W} = 0.8986$).

10.6 Summary

The three traffic distributions binomial, Poisson and negative binomial (referred to as the BPP family) can all be used to formulate systems that are reversible and have product form. The three traffic types correspond to different values of peakedness, $Z < 1$, $Z = 1$ and $Z > 1$, respectively. The binomial and negative binomial distributions can be regarded as extensions of Poisson traffic to allow for lower and higher variance than the mean. The binomial distribution may be used to model smooth traffic and the negative binomial distribution bursty traffic, such as data traffic.

The product form of systems offered Poisson traffic makes it possible to solve for state probabilities efficiently by an Erlang-type recursion (known as the Fortet-Grandjean or Kaufman-Roberts algorithm), or by convolution. Iversen (2013) describes the generalization of these methods to BPP traffic.

We also consider admission control by trunk reservation and suggest a simulation approach, which is relatively easy to implement. These are all models for loss systems.

A different class of models are systems with processor sharing. These represent a hybrid between loss and delay systems, and, even is the models do not have product form, it is possible to formulate iterative algorithms for such systems, see Iversen (2013). In this case, we restrict the discussion to Poisson traffic.

All these models are reversible, which means that they lend themselves to network analysis in a *as if independent* manner. We may use analytical methods or fixed-point methods to analyse networks consisting of such queues.

Multi-Service Network Analysis

11

Analyzing a network carrying different services with various characteristics such as mean arrival and service rates and bandwidth, together with priorities and adaptive routing, is not an easy task. We will however see that some approaches exist. The fundamental property is the reversibility of queueing systems, which holds for many important cases.

Reversibility means that the output process is Poisson distributed if the arrival process is Poisson. With this property, a network can be *decomposed* into its nodes, and the nodes studied separately. We will mainly use three approaches, an analytical approach to queueing systems, fixed-point methods and analysis based on effective bandwidths.

11.1 Fixed-point network analysis

As a result of reversibility, network components can in certain situations be regarded as independent (in the sense of a network product form solution), which is a fundamental assumption for construction of a tractable model for network analysis. This is the case for queueing network analysis, as well as for fixed-point methods, which are based on the assumption of independence of blocking rates in individual links.

In Chapter 10, we consider blocking in a single link with a given capacity C. We now turn to the calculation of end-to-end blocking probabilities in a network, where links are indexed by $l = 1, \ldots, m$, capacities by C_l and traffic classes by $k = 1, \ldots, K$. Associated with each origin-destination pair and traffic class is a set of routes. A class-k call requires $d_{l,k}$ capacity units on link l. The capacity available to class-k traffic depends on the capacity and load of the link.

The fixed-point model of a *single service* network with unit bandwidth requirement (introduced in Section 6.1) is a set of equations of the form

$$
\hat{v}_l = \sum_{r:l \in r} v_r \prod_{s \in r - \{l\}} (1 - \hat{B}_s),
$$

$$
\hat{B}_l = E_B(\hat{v}_l/\mu_l, C_l),
$$

(11.1)

where $\hat{v}_l$ and $\hat{B}_l$ denote approximations of the offered traffic rate and the link blocking rates obtained by iteration with repeated substitution. The traffic rate is the sum of thinned traffic from all routes r that use link l, and is also known as the *reduced load*. The function $E_B(\cdot)$ is the Erlang B-formula. The blocking rate on any route is given by

$$\hat{B}_r = 1 - \prod_{l \in r}(1 - \hat{B}_l).$$

Kelly (1986) showed that under fixed routing there exists a unique solution to the system 11.1. The solution may not be unique, however, under some dynamic routing schemes. As the modeled network grows, the solution to the fixed-point equations tend to become more accurate. Two limiting cases for which the fixed-point solution can be shown to converge to the exact blocking values:

(1) By keeping link capacities and traffic rates fixed and increasing the number of links and routes in the network;
(2) By keeping the network topology fixed and letting link capacity and traffic rates allow to grow in constant proportion.

An account of these limiting regimes can be found in Kelly (1991).

The fixed-point approximation is based on two assumptions on the network,

(1) The blocking in each link is assumed to be independent from blocking in other links;
(2) The offered traffic rate to each individual link is Poisson distributed, and the original offered traffic rate is thinned by blocking in other links on the route.

Here, *thinning* means that the effect of blocking upstreams or downstreams relative to a link decreases the offered traffic rate to the link. Clearly, these assumptions are idealizations. The link blocking rates cannot be independent, because a sharp increase in a certain traffic stream increases the blocking probabilities in all links along its path. Due to this dependence, the offered traffic subject to blocking is not actually Poisson either, but is actually smoother (see Section 5.5). Traffic thinning is the assumed random loss of any call, independently from link to link, retaining the traffic rate Poisson distributed but with a lower intensity.

To specify the multi-service network model, we need to introduce variables for origin-destination pairs as well as sets of feasible routes in the network-it is necessary to be able to identify origin-destination pairs, routes and links in the model.

Let C_l be the capacity of link measured in unit bandwidth per circuit. Also, let I be the set of all node pairs, each identified by (i, j). For each such node pair, there is an ordered set $\mathcal{R}_{ij}$ of routes, each indexed by r_{ij}. We drop the subscript of r whenever there is no risk for ambiguity. Thus, the triple $(i, j; r)$ uniquely defines a specific route.

The network is offered K different classes of traffic, indexed by k, and the triple $(i, j; k)$ defines an incoming call attempt of type k between nodes i and j. The bandwidth demand of a call k on link l is denoted $d_{l,k}$.

All call types arrive according to a Poisson process with an offered rate $\lambda_{ij,k}$. A call is set up on some route $r_{ij} \in \mathcal{R}_{ij}$ if there is available bandwidth to accommodate the call on each of the links of the route, and the resources are held for a duration of mean $\mu_{ij,k}^{-1}$. If none of the routes has required resources available, the call is blocked and lost. The end-to-end blocking probability of a call $(i, j; k)$ is denoted $B_{ij,k}$.

By a fixed-point method in a multi-service network, we mean the set of equations

$$\hat{v}_{l,k} = \sum_{r:l\in r} v_{r,k} \prod_{s\in r-\{l\}} (1 - \hat{B}_{s,k}),$$

$$\hat{B}_{l,k} = E(\hat{v}_{l,k}, \mu_{l,k}, d_{l,k}; C_{l,k}),$$

(11.2)

where $\hat{v}_{l,k}$ is the reduced load of traffic class k, $\hat{B}_{l,k}$ is the blocking on link l of class k traffic, and $E(\cdot)$ is a generalization of the Erlang B-formula to links carrying multiple services, for example the Fortet-Grandjean (Kaufman-Roberts) algorithm (10.49). We also allow a selected route to be a function of the blocking rates on the links that make up the route. The end-to-end blocking on route r of type k traffic is

$$\hat{B}_{r,k} = 1 - \prod_{l\in r}(1 - \hat{B}_{l,k}).$$

11.1.1 Admission control and adaptive routing

Trunk reservation as a means of traffic prioritization and admission control of circuit-switched services is discussed in Sections 6.3, 6.6 and 10.4. We associate a trunk reservation parameter $t_{l,k}$ with each link l and call type k. An incoming call is accepted subject to trunk reservation only if the unoccupied capacity $\bar{C}_l = C - \sum_k n_{l,k}d_{l,k}$ on any link l along a route satisfies

$$\bar{C}_l - t_{l,k} \geq d_{l,k}.$$

Some common routing strategies for circuit-switched traffic are described in Section 6.3. The focus in this section is on *least loaded routing*, where a primary (possibly direct) route is chosen, if it exists and has sufficient free resources. Otherwise, the route with the largest number of free end-to-end channels is selected. Under this scheme, the secondary route is chosen deterministically rather than randomly. For each route r, the link with the fewest number of free channels, called the *most congested link on route r* is identified by

$$\bar{l}_r = \min_{l\in(i,j;r)} \bar{C}_l,$$

and the route where this "bottleneck link" has the largest number of free channels compared to other routes is used, that is

$$\bar{r} = \max_{r\in\mathcal{R}_{ij}} \bar{l}_r.$$

This is referred to as a *min-max scheme* and the corresponding route the least loaded route. Thus, to each node pair (i, j) a set $\mathcal{R}_{ij}$ of alternative routes is associated, usually represented in a table. When there is an incoming call attempt $(i, j; k)$, the system tries to route the call onto the primary route, if it exists. If there is no such route, or if there are insufficient number of free channels to route the call on this route, the routing logic scans all alternative routes sequentially in the table for the min-max optimal route. If the selected route has sufficient number of free bandwidth, the call is set up, otherwise it is blocked and lost.

As shown in Section 6.6, dynamic routing strategies may lead to lower routing efficiency and network instability under heavy traffic. By using trunk reservation in conjunction with dynamic routing, such effects can be mitigated to a large degree.

11.1.2 The fixed-point model

Fixed-point methods can be formulated for networks with adaptive routing and trunk reservation. Two approaches are proposed in Greenberg and Srikant (1997) and Liu and Baras (1999). In the following, however, we use the approximate blocking algorithm for multiple services under trunk reservation presented in Section 10.4.

The routing schemes are supposed to be dependent on some function of blocking only. It is possible to incorporate more sophisticated routing schemes, defined for example on link state probabilities, but this would need the calculation of these state probabilities in each iteration, which may be rather expensive in terms of computational effort. We choose to consider routing strategies based on blocking probabilities, because we compute these anyway, and because most dynamic routing strategies tend to be "simple" in the sense that a node performing the routing decision usually does not have the full knowledge of the state of the network.

Consider a network where the link capacities C_l and a dynamic routing scheme are given. For the fixed-point model for a network with dynamic routing, we have the follwing parameters.

(1) The load arriving to a link j from chain k, $v_{j,k}$,
(2) The probability of admitting a job $A_{j,k}$ of chain k to link j, where the admittance is defined as unity minus blocking, $1 - B_{j,k}$,
(3) The probability $q_{r,k}$ that a call of class k is set up onto route r.

To evaluate the fixed-point equations, we first fix $A_{l,k}$ and the routes determined by $q_{r,k}$ in order to compute the offered traffic rates $v_{l,k}$. Initially, we may choose $B_{l,k} = 0$ so that $A_{l,k} = 1$ and any $q_{r,k}$, for example $q_{1,k} = 1$. Next, we calculate the blocking by using the Kaufman-Roberts algorithm, an algorithm for multi-service blocking under trunk reservation, or a model for a system with processor sharing. Then, based on $A_{l,k}$ (following from the blocking rates), new routes may be selected, which determine new values of $q_{r,k}$. This procedure is repeated until convergence is reached.

In addition to estimation of network blocking rates, fixed-point methods can be used to evaluate different trunk reservation settings and to assess the efficiency of various routing schemes.

Example 11.1.1.   We may use a fixed-point method to evaluate the following network model. The network topology is shown in Figure 11.1. A network is offered

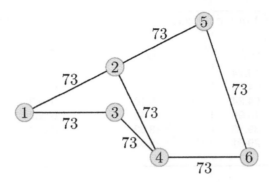

FIGURE 11.1

The topology for Example 11.1.1, with fixed-point analysis for two services.

two types of traffic, voice telephony and a video service. The minimum effective bandwidth service—telephony—determines the BBU of the node capacity. The node capacities of all nodes are equal and divided into 73 BBU, and telephony requires one and video five BBUs per connection. For simplicity, both are assumed to have holding time $\mu_k = 1$. The routing is fixed shortest route.

The telephony calls (service type 1) and the video calls (service type 2) are offered to each link in the network from external sources according to the load vectors

$$\lambda_1 = (15.0, 20.0, 27.5, 32.5, 22.5, 17.5, 10.0),$$

$$\lambda_2 = (1.5, 2.0, 2.8, 3.2, 2.2, 1.8, 1).$$

Due to routes using these links for transit traffic, the actual offer load to each link is

$$v_1 = (20.0, 25.0, 55.0, 50.0, 40.0, 30.0, 10.0),$$

$$v_2 = (2.0, 2.5, 5.5, 5.0, 4.0, 3.0, 1.0).$$

In order to calculate the blocking rates in the fixed-point equations, we can use the state-space recursion (10.49). In the fixed-point method, we need to formulate expressions for each service, that is all $v_{l,1}$ and $v_{l,2}$, where the first index represents the link, and the second the service. Thus, we have $K \cdot m$ equations, where K is the number of services and m the number of links.

Otherwise, the method is straightforward. We iterate the fixed-point equations until the fixed point is reached, or, at least so that we are sufficiently close to it. The result is shown in Table 11.1, where offered traffic is denoted v_k and blocking rates B_k. The total carried traffic is 226.3 and 21.2, respectively, totalling 247.5.

Example 11.1.2. It is also possible to use fixed-point method for network analysis with trunk reservation. Consider Example 11.1.1 with a trunk reservation parameter $t_2 = 2$, which means that no video service is allowed to connect if there are fewer

Table 11.1 Result of a multiservice fixed-point analysis of a network without trunk reservation

Link	ν_1	ν_2	B_1	B_2
1	19.74	1.87	0	0
2	24.93	2.46	0	0.001
3	54.24	5.10	0.083	0.385
4	48.90	4.49	0.052	0.266
5	39.13	3.63	0.014	0.084
6	29.31	2.67	0	0.006
7	10	1	0	0

Table 11.2 Result of a multiservice fixed-point analysis of a network with trunk reservation

Link	ν_1	ν_2	B_1	B_2
1	19.86	1.84	0	0
2	24.96	2.44	0	0.003
3	54.59	5.02	0.058	0.486
4	49.24	4.36	0.028	0.315
5	39.43	3.55	0.007	0.109
6	29.53	2.58	0	0.011
7	10	1	0	0

than $d_2 + t_2 = 5 + 2 = 7$ channels available. Using the method described in Section 10.3 in place of the Fortet-Grandjean (Kaufman-Roberts) algorithm, we get the result shown in Table 11.2. We notice that type 1 traffic has been promoted at the expense of traffic type 2 (which can be seen from traffic rates ν_k and blocking rates B_k). Also, the total traffic volume has increased to $227.6 + 20.8 = 248.4$. For the simulation of the trunk reserved system by simulation, 10,000 iterations were used.

11.2 Generalized queueing networks

The Jackson networks, introduced in Sections 7.2 and 8.2 were generalized in 1975 by Baskett et al. (1975), and are commonly referred to as BCMP networks. It was shown that queueing networks with more than one type of jobs have solutions in product form, provided that

(1) Each traffic type arrive according to a Poisson process, and each node is a reversible queueing system (that is, a Poisson arrival process gives a Poisson departure process). This comprises of the following queue types:

(a) Non-sharing $M/M/n$ queue, $n \geq 1$, with first come first serve (FCFS) discipline, where all jobs have the same exponential service time distribution. The service rate can, however, be state dependent;

(b) Queues with processor sharing (PS) or generalized processor sharing (GPS) disciplines;

(c) Infinite server systems;

(d) $M/M/n$ queue with last come first serve - pre-emptive resume (LCFS-PR). This case is not very useful in telecommunication contexts, and will not be considered further.

(2) The jobs are classified into K chains, where each chain is characterized by its own transition probabilities p_{ijk} from node i to node j, or possibly leaving the network with probability $p_{i0,k}$, and mean service rate μ_k (apart from the non-sharing $M/M/n$ queueing system (type (a)), where the service rate must be identical for all chains in a queue.)

We can allow the arrival intensity from outside a network, $\lambda_{ij} = \sum_{k=1}^{K} \lambda_{ijk}$ to depend on the current number of jobs in the network. Furthermore, the service rates μ_{ijk} may depend on the number of jobs in a system. By using these properties, we can model queueing networks which are either closed, open or mixed. Thus, the BCMP network class includes extensions to the one-dimensional open networks discussed in Chapter 7 and the one-dimensional closed networks of Chapter 8.

Firstly, we have a look at analysis of open multi-service queueing networks, which form an extension to Chapter 7. The strategy is much the same as for single-service networks. The main difference is the choice of queueing model and related adaptations.

Initially, we consider each chain k separately and solve the related traffic equation (7.2). Solving these traffic equations gives the arrival rates of each traffic type k to each queue. The next step depends on the choice of queueing model. We consider the two cases (a) and (b) above.

In case (a), a non-sharing $M/M/n$ queue, $n \geq 1$, all chains have to have the same service rate $\mu_k = \mu$. In the sequel, we drop the subscript referring to the origin-destination pair. Since a sum of Poisson processes is another Poisson process, we can write $\lambda = \sum_{k=1}^{K} \lambda_k$ for the total arrival process. Thus, each queue can be analyzed as in the one-dimensional case, using (5.17) in the $M/M/1$ case, and (5.19) in the $M/M/n$ case. These queues are independent, and network performance measures can be obtained like in Chapter 7.

Now, we focus on closed multi-dimensional networks, which are much harder to analyze than open networks. For this purpose, it is possible to use extensions to the methods presented in Chapter 8. Multi-dimensional convolution is probably the most flexible method and, in addition, relatively easy to use. It is also possible to use multi-dimensional mean value analysis (MVA) and decomposition. Approximation of a closed network by an open network should be relatively easy with the approach described in Section 8.5.

11.2.1 Convolution algorithm

The convolution algorithm for a multi-chain network is essentially the same as in the single chain case. First, each chain is considered separately, and the corresponding traffic equation (8.1) is solved for relative traffic load.

At an arbitrary reference node we assume the arrival rate is equal to one. For each chain we may choose a different node as reference node. For chain k in queue i the relative arrival intensity $v_{i,k}$ is obtained from

$$v_{i,k} = \sum_{j=1}^{n} r_{ij,k} v_{j,k}, \quad k = 1, 2, \ldots, K, \tag{11.3}$$

where n is the number of queues, K is the number of chains and $r_{ij,k}$ is the probability that a job belonging to chain k is transferred from node i to node j, where (i, j) is an existing link.

Any queue l can be chosen as reference. The relative load at queue l from jobs belonging to chain k is

$$\rho_{l,k} = v_{l,k}/\mu_{l,k}, \tag{11.4}$$

where $\mu_{l,k}$ is the mean service rate at queue l for jobs of chain k.

Based on the relative loads, we can obtain the relative state probabilities for each queue. Finally, by convolution, we compute the normalization constant and the normalized state probabilities for any of the queues in the network (by changing the order of convolution), and its related performance measures.

Example 11.2.1. Consider a small network with an $M/M/1 - PS$ queue in series with an infinite server, whose output is fed back to the $M/M/1 - PS$ queue. In the network, there is only one routing possibility; from the queue to the IS and back to the queue. There are two chains, with two jobs from chain one with $v_1 = 2$, $\mu_1 = 1$, and three jobs from chain two with $v_2 = 1$ and $\mu_2 = 1$. The states are denoted (n_1, n_2) and range from $(0, 0)$ to $(2, 3)$. For each state, the relative joint state probabilities are computed according to the model chosen. In the $M/M/1$ case, we use (10.69), which reduces to (10.67) in the case of two chains. The corresponding relative state probabilities for the infinite server is given by (8.3). Denoting the relative state probabilities $q_1(n_1, n_2)$ and $q_2(n_1, n_2)$ for the two queues, we can readily compute the convolution $q_{12}(n_1, n_2) = (q_1 * q_2)(n_1, n_2)$. The normalization constant is given by $q_{12}(2, 3)$, and the actual state probabilities are $p_1(n_1, n_2) = q_1(n_1, n_2)q_2(2 - n_1, 3 - n_2)/q_{12}(2, 3)$.

From the figures we can for example conclude that the percentual idle time of the queue is 0.31%. Performance measures for the queue can be calculated as shown in Section 10.5. The length of queue one is as usual the expectation $\sum_{\mathbf{n}=(0,0)}^{(2,3)} (n_1 + n_2) p_1(n_1, n_2) = 4.0031$, where $\mathbf{n}$ is the state vector of the network.

Note that in case we have a multichannel queue, Equation (10.70) has to be used instead of (10.69).

11.2.2 Mean value analysis

Another way to analyze multi-class closed queueing networks with non-sharing FIFO, processor sharing or delay (IS) queues is by *multi-dimensional mean value analysis*. The principles of the single class mean value analysis discussed in Section 8.4 are relatively easily adapted to multiple dimensions. We consider a closed network with J queues and K job classes.

The main difference is that the state in the multi-class case is a vector of numbers of jobs per class, $\mathbf{N} = (N_1, N_2, \ldots, N_K)$, and the values of such vectors should be ordered into an increasing sequence where an entry differ from a previous entry by having one more job. The recursion must start at $(0, 0, \ldots, 0)$ and the states be evaluated in the order shown in Figure 11.2.

Initially, we determine the relative load $v_{j,k}$ of chain k at queue j by solving the traffic equation (11.3). For chain k and queue j, let $W_{j,k}(\mathbf{n})$ be the delay, $L_{j,k}(\mathbf{n})$ the queue length, $\lambda_{j,k}(\mathbf{n})$ the throughput for some state vector of jobs $\mathbf{n}$, and $\mu_{j,k}$ the queue service rate. Here, n_k is a component in $\mathbf{n}$ such that $0 \leq n_k \leq N_k$. Then for an $M/M/1$ queue, we have

$$W_{j,k}(\mathbf{n}) = (1 + L_{j,k}(\mathbf{n} - \mathbf{e}_k))/\mu_{j,k},$$

$$\lambda_{j,k}(\mathbf{n}) = n_j v_{j,k} / \sum_{j=1}^{J} v_{j,k} W_{j,k}(\mathbf{n}),$$

$$L_{j,k}(\mathbf{n}) = \lambda_{j,k}(\mathbf{n}) W_{j,k}(\mathbf{n}), \tag{11.5}$$

$$L_j(\mathbf{n}) = \sum_{k=1}^{K} L_{j,k}(\mathbf{n}),$$

where $\mathbf{e}_k$ is a vector with a one in position k and zeros elsewhere. Thus $\mathbf{n} - \mathbf{e}_k$ is short notation for the vector $(n_1, n_2, \ldots, n_{k-1}, n_{k-1}, n_{k+1}, \ldots, n_K)$, the state vector with one job type k less.

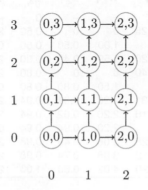

FIGURE 11.2

Diagram over states for a two-dimensional MVA up to state $(2, 3)$.

The sequence of calculation with $K = 2$ jobs is initiated with $L_{j,k}(0,0) = 0$ for each queue $j = 1, 2, \ldots, J$. Next, the delays $W_{j,1}(1,0) = (1 + 0)/\mu_{j,1}$ and $W_{j,2}(0,1) = (1 + 0)/\mu_{j,2}$ are determined, together with the throughputs according to (11.5). When all delays and throughputs for each queue are known, the total queue lengths $L_{j,k}(1,0)$, $L_{j,k}(0,1)$ and $L_j(1,0) = \sum_k L_{j,k}(1,0)$ and $L_j(0,1) = \sum_k L_{j,k}(0,1)$ can be calculated. Next, for state $(1,1)$ the new delays follow from $W_{j,1}(1,1) = (1 + L_j(0,1))/\mu_{j,1}$ and $W_{j,2}(1,1) = (1 + L_j(1,0))/\mu_{j,2}$, respectively. Continuing in this way, as illustrated in Figure 11.2, the performance metrics for state (N_1, N_2) is obtained.

The extension of the mean value analysis algorithm to multiple server queues is straightforward, as is the formulation of approximate MVA schemes. Such extensions for the single chain case are discussed in Section 8.4.

The time and space requirements of the multiple-class MVA are considerably larger than those of the single-class algorithm. These are roughly proportional to

$$KJ \prod_{k=1}^{K} (N_k + 1)$$

arithmetic operations, and

$$J \prod_{k=1}^{K} (N_k + 1)$$

memory positions.

Example 11.2.2. Consider the same problem as in Example 11.2.1. The MVA calculations according to (11.5) are shown in Table 11.3. The length of queue one is $L_1(2,3) = 4.0031$ as expected.

Table 11.3   Mean value analysis calculations for a two-chain network example

n_1	n_2	L_1	$W_{1,1}$	$\lambda_{1,1}$	$W_{2,1}$	$\lambda_{2,1}$	L_2	$W_{1,2}$	$\lambda_{1,2}$	$W_{2,2}$	$\lambda_{2,2}$
0	0	0	–	–	–	–	0	–	–	–	–
1	0	0.50	2.00	0.25	0	0	0.50	2.00	0.25	0	0
0	1	0.50	0	0	1.00	0.50	0.50	0	0	1.00	0.50
1	1	1.20	3.00	0.20	1.50	0.40	0.80	2.00	0.20	1.00	0.40
2	0	1.20	3.00	0.40	0	0	0.80	2.00	0.40	0	0
0	2	1.20	0	0	1.50	0.80	0.80	0	0	1.00	0.80
2	1	2.06	4.40	0.31	2.20	0.31	0.94	2.00	0.31	1.00	0.31
1	2	2.06	4.40	0.16	2.20	0.62	0.94	2.00	0.16	1.00	0.62
0	3	2.06	0	0	2.20	0.94	0.94	0	0	1.00	0.94
2	2	3.02	6.12	0.25	3.06	0.49	0.98	2.00	0.25	1.00	0.49
1	3	3.02	6.12	0.12	3.06	0.74	0.98	2.00	0.12	1.00	0.74
2	3	4.00	8.03	0.20	4.02	0.60	1.00	2.00	0.20	1.00	0.60

11.3 Flow analysis by effective bandwidth

Multi-service networks with admission control and alternative routing are in general difficult to analyze and design. A particular difficulty in general networks are complex dependencies between traffic streams and queue performance throughout the network. In Jackson and BCMP networks, these dependencies are circumvented by formulating conditions in which the queues in the network act as if independent of each other.

A similar result can be formulated in terms of effective bandwidths. It turns out that if the capacity is above a certain threshold, the effective bandwidth of the departure process from the queue is equal to the effective bandwidth of arrival process. In this case, the queue essentially becomes "invisible" to, or decoupled from, the rest of the network. The threshold is referred to as the decoupling bandwidth. One of the advantages of this approach is the versatility of effective bandwidth as traffic descriptor for modeled or measured traffic. The decoupling bandwidth also provides a guideline of how large capacity the queues must have in order to achieve decoupling and support the required bandwidth of a given traffic stream, end-to-end.

Under these conditions, the network can be analyzed similarly to a circuit-switched network; for any queue with capacity C, offered K traffic streams where n_k and α_k are the number of sources and effective bandwidth of stream k, respectively, we have the capacity constraint

$$\sum_{k \in K} n_k \alpha_k \leq C. \tag{11.6}$$

The constraint guarantees a limited buffer overflow probability. The relation between effective bandwidth and buffer overflow in a queue are discussed in Chapter 9. The fundamental result is restated below.

Theorem 11.3.1. *Consider a collection $k \in K$ of independent source types, with n_k sources of each type and bounded, stationary and ergodic discrete-time arrival processes $\{A_n^{(k)}\}$, each satisfying a pathwise large deviation principle with convex rate function $I(\cdot)$ and having the logarithmic moment generating function*

$$\Lambda_k(\delta) = \lim_{n_k \to \infty} \frac{1}{n_k} \ln(\mathbf{E}(\exp(\delta S_{n_k}^{A^{(k)}}))) < \infty \text{ for } \delta < \infty. \tag{11.7}$$

Suppose the traffic streams share a deterministic buffer with some work conserving service policy at rate C. Then, for the effective bandwidths of arrival processes, we have

$$\sum_{k \in K} n_k \alpha_k(\delta) \leq C \equiv \lim_{b \to \infty} \frac{1}{b} \ln(\mathbf{P}(W \geq b)) \leq -\delta, \tag{11.8}$$

where $\alpha_k(\delta) = \Lambda_k(\delta)/\delta$, b is the buffer size of the queue and W denotes the stationary workload process in the queue.

A process is called *ergodic* if its statistical properties can be determined from a single, sufficiently long sample of the process. A buffer for which (11.8) holds is said to satisfy a δ-constraint, which represents the (approximate) performance guarantee of a queue with respect to buffer overflows.

11.3.1 Effective and decoupling bandwidths

For the results in this section, we will assume that the necessary technical conditions are satisfied for all traffic streams.

Assumption 11.3.2. All traffic processes in the network satisfy a pathwise large deviation principle (9.22).

By Assumption 11.3.2, all rate functions and derived functions exist and the rate functions are convex. For many traffic models, we can calculate the logarithmic moment generation function $\Lambda(\delta)$ (9.24) and the rate function $I(x) = \Lambda^*(x)$ (9.23) for arrival and departure processes. These functions form the basis of network analysis using effective bandwidths.

The objective is to study impact on queues by means of the effective bandwidths of arriving traffic. We observe heterogeneous traffic streams arriving to a discrete-time queue with deterministic service rate where the different traffic types share a buffer. The main idea is that if the service rate is above a certain threshold with respect to the arrival processes, the effective bandwidth of the departure process is equal to that of the arrival process. When this is the case, the queue is decoupled from or made "invisible" to the rest of the network. The main results are stated as propositions as technical details and proofs are largely omitted. Proofs of the results in this section can be found in de Veciana et al. (1993).

Proposition 11.3.3. *Let $\{A_n\}$ be a bounded, stationary and ergodic discrete-time arrival process of some aggregate traffic stream with effective bandwidth $\alpha(\delta) = \Lambda(\delta)/\delta$ and rate function $\Lambda^*(\cdot)$ entering a discrete-time queue with deterministic service rate C, and satisfying a pathwise LDP. The effective bandwidth $\alpha_D(\delta)$ of the departure process $\{D_n\}$ satisfying an LDP with convex good rate function is given by*

$$\alpha_D(\delta) = \begin{cases} \alpha(\delta) & \text{if } \alpha^*(\delta) \le C \\ C - \frac{1}{\delta}\Lambda^*(C) & \text{otherwise} \end{cases}, \tag{11.9}$$

where $\alpha^*(\delta)$ is defined by

$$\Lambda(\delta) = \sup_{\alpha}\{\alpha\delta - \Lambda^*(\alpha)\}, \tag{11.10}$$

or

$$\alpha^*(\delta) = \frac{\mathrm{d}}{\mathrm{d}\delta}\Lambda(\delta). \tag{11.11}$$

Definition 11.3.1. We refer to $\alpha(\delta)$ as the *decoupling bandwidth* of a traffic stream. For the single buffer case the decopuling constraint $\alpha^*(\delta) \le C$ guarantees that the effective bandwidth of the departure process is the same as that of the arrival process for the given δ-constraint.

Fact 11.3.4. If the stationary arrival process has a bounded sustainable peak arrival rate such that $\alpha(\infty) \triangleq \lim_{\delta \to \infty} \alpha(\delta) < \infty$, then the decoupling bandwidth satisfies the following inequality

$$\alpha(\infty) \geq \alpha^*(\delta) \geq \alpha(\delta). \tag{11.12}$$

We notice that the decoupling bandwidth $\alpha*(\delta)$ usually is larger than the effective bandwidth $\alpha(\delta)$ of the source according to (11.12). We illustrate Proposition 11.3.3 by three examples.

Example 11.3.1. For a Gaussian traffic model, let μ be the mean and σ^2 the large sample variance of n sources of the arrival process A_j, where

$$\sigma^2 = \lim_{n \to \infty} \frac{1}{n} \operatorname{Var}\left(\sum_{j=1}^{n} A_j\right) < \infty.$$

Suppose that the traffic is offered a single server queue with capacity C and deterministic service time, and we require a δ-constraint to be satisfied. The logarithmic moment generation function of the Gaussian distribution is $\Lambda(\delta) = \mu\delta + \delta^2\sigma^2/2$. The effective bandwidth of the arrival process is $\alpha(\delta) = \Lambda(\delta)/\delta = \mu + \delta\sigma^2/2$.

The rate function $\Lambda^*(\alpha)$ is found from the Fenchel-Legendre transform of $\Lambda(\cdot)$,

$$\Lambda^*(\alpha) = \max_{\delta}\{\delta\alpha - \Lambda(\delta)\},$$

with maximum at $\alpha - \mu - \delta\sigma^2 = 0$ or $\delta = (\alpha - \mu)/\sigma^2$, which gives $\Lambda^*(\alpha) = (\alpha - \mu)^2/2\sigma^2$. The decoupling bandwidth is $\alpha^*(\delta) = d\Lambda(\delta)/d\delta = \mu + \delta\sigma^2$. Thus, we have

$$\alpha_D(\delta, C) = \begin{cases} \mu + \frac{\delta\sigma^2}{2} & \text{if } \mu + \delta\sigma^2 \leq C, \\ C - \frac{(C-\mu)^2}{2\sigma^2\delta} & \text{otherwise.} \end{cases} \tag{11.13}$$

The effective bandwidths for the model parameters $\mu = 1$, $\sigma^2 = 1$ and $C = 2$ are shown in Figure 11.3.

Example 11.3.2. Let some type of traffic be described by a discrete-time on-off model. The traffic is offered a single server queue with deterministic service rate and capacity C. We require a δ-constraint to be fulfilled. The state transition rates from off state to on is $(1 - p_{00})$, and form on to off, $(1 - p_{11})$. When in off state, no traffic is generated, whereas in on state, traffic is generated at the constant peak rate $h > C$.

The logarithmic moment generation function is

$$\Lambda(\delta) = \ln\left(\frac{p_{00} + p_{11}\exp(\delta h) + \sqrt{(p_{00} + p_{11}\exp(\delta h))^2 + 4(1 - p_{00} - p_{11})\exp(\delta h)}}{2}\right), \tag{11.14}$$

and the decoupling bandwidth

$$\alpha^*(\delta) = h\exp(\delta h - \Lambda(\delta))\left(\frac{p_{11}\exp(\Lambda(\delta)) + (1 - p_{00} - p_{11})}{2\exp(\Lambda(\delta)) - (p_{00} + p_{11}\exp(\delta h))}\right). \tag{11.15}$$

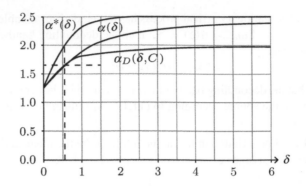

FIGURE 11.3

Decoupling bandwidth and effective bandwidth of the departure process for a Gaussian source.

We also wish to know the value of the rate function at C, $\Lambda^*(C) = \max_\delta\{\delta C - \Lambda(\delta)\}$. Computation of the Fenchel-Legendre transform is not necessary since we only need a function value at a single point. The derivative of the argument of the transform gives the condition for maximum; it is achieved at the δ-value such that $C - \alpha^*(\delta) = 0$. Denoting the numerical root by $\hat{\delta}$, the point $(\hat{\delta}, C - \Lambda^*(C)/\hat{\delta})$ is approximately $(0.555, 1.653)$, and $\Lambda^*(C) = 0.193$. Thus, we have

$$\alpha_D(\delta, C) = \begin{cases} \frac{\Lambda(\delta)}{\delta} & \text{if } \alpha^*(\delta) \le C, \\ C - \frac{\Lambda^*(C)}{\delta} & \text{otherwise,} \end{cases} \tag{11.16}$$

where the numerical value of $\Lambda^*(C)$ for $\hat{\delta}$ is used. The effective bandwidths for the model parameters $p_{00} = 0.5$, $p_{11} = 0.5$, $h = 2.5$ and $C = 2$ are shown in Figure 11.4.

Example 11.3.3. For a Poisson traffic arrival process of intensity λ, we have

$$\Lambda(\delta) = \lambda(e^\delta - 1),$$
$$\alpha(\delta) = \lambda(e^\delta - 1)/\delta,$$
$$\alpha * (\delta) = \lambda e^\delta,$$
$$\Lambda^*(\alpha) = \alpha(\ln \alpha - \ln \lambda) + \lambda - \alpha.$$

We have

$$\alpha_D(\delta, C) = \begin{cases} \frac{\lambda(e^\delta - 1)}{\delta} & \text{if } \lambda(e^\delta) \le C, \\ C - \frac{C(\ln C - \ln \lambda) + \lambda - C)}{\delta} & \text{otherwise,} \end{cases}$$

The effective bandwidths for the model parameters $\lambda = 1$ and $C = 2$ are shown in Figure 11.5.

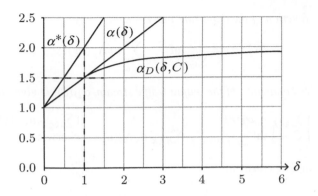

FIGURE 11.4

Decoupling bandwidth and effective bandwidth of the departure process for an on-off source.

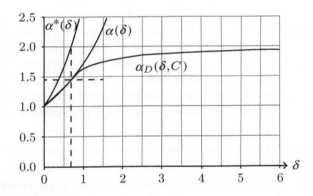

FIGURE 11.5

Decoupling bandwidth and effective bandwidth of the departure process for a Poisson source.

Our approach will be to partially identify the rate function $\Lambda_D^*(\cdot)$ from which, by way of the previous lemma, we can determine $\Lambda_D(\delta)$. This in turn permits us to study the output's effective bandwidth $\alpha_D(\delta) = \Lambda_D(\delta)/\delta$.

Proposition 11.3.3 can be extended to the case where the arrival process is an aggregate of independent traffic streams as follows. For a proof, see de Veciana et al. (1993).

Proposition 11.3.5. *Consider a collection of independent sources, with n_k sources of each type $k \in K$, with discrete-time bounded, stationary and ergodic arrival processes $\{A_{n_k}^{(k)}\}$ offered a discrete-time queue with deterministic service rate C and aggregate departure process $\{D_n\}$ each satisfying a pathwise LDP. Suppose the streams share a deterministic buffer subject to a work conserving service policy with rate C.*

Then, the rate function of the departure process is given by

$$\Lambda_D^*(\alpha) = \inf_{\sum_{k \in K} n_k \alpha_k = \alpha} \sum_{k \in K} n_k \Lambda_k^*(\alpha_k), \qquad (11.17)$$

and the effective bandwidth of the output traffic stream, $\alpha_D(\delta)$, is given by

$$\alpha_D(\delta) = \begin{cases} \sum_{k \in K} n_k \alpha_k(\delta) & \text{if } \sum_{k \in K} n_k \alpha_k^*(\delta) \leq C \\ C - \frac{1}{\delta} \inf_{\sum_{k \in K} n_k \alpha_k = C} \sum_{k \in K} n_k \Lambda_k^*(\alpha_k) & \text{otherwise} \end{cases}$$

$$(11.18)$$

Thus, the decoupling constraint

$$\sum_{k \in K} n_k \alpha_k^*(\delta) \leq C, \qquad (11.19)$$

is a sufficient and necessary condition for the effective bandwidth of the output to equal that of the input stream. Otherwise, the effective bandwidth is reduced and increases hyperbolically in δ to the service rate C.

In Propositions 11.3.3 and 11.3.5 we determine the effective bandwidth for the aggregate departure process. However, it is often desirable to know the effective bandwidth of the departure process from a queue of individual traffic streams, since these may take different routes in the network after leaving the queue. We may split the departure processes into two streams, one representing a particular traffic stream we wish to analyze, and the second representing the aggregate of the remaining traffic streams.

Proposition 11.3.6. *Consider two independent discrete-time arrival processes $\{A_{n_1}^{(1)}, A_{n_2}^{(2)}\}$ satisfying the conditions of Proposition 11.3.5. Suppose the processes share a deterministic buffer in a queue with deterministic service rate C via a work conserving service policy and the system satisfies the effective bandwidth constraint*

$$\alpha_1(\delta) + \alpha_2(\delta) < C, \qquad (11.20)$$

for some δ-constraint. If the departures corresponding to the first stream $\{D_{n_1}^{(1)}\}$ satisfy an LDP with convex good rate function, it is given by

$$\Lambda_{D^{(1)}}^*(\alpha) = \Lambda_1^*(\alpha) \qquad (11.21)$$

and a sufficient condition for the effective bandwidth of the departures of the first stream to equal that of the input is

$$\alpha_1^*(\delta) + \alpha_2(0) = \alpha_1^*(\delta) + \mathbf{E}(A_1^{(2)}) \leq C, \qquad (11.22)$$

An analogous result holds for the second departure stream.

Equation (11.22) can be interpreted as follows. If the arrival process of traffic stream two offers a load close to its mean rate $\alpha_2(0) = \mathbf{E}(A_1^{(2)})$ and has service priority,

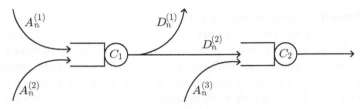

FIGURE 11.6

Analysis of a tandem queue fed by three different traffic streams.

the effective capacity available for stream one is $C - \mathbf{E}(A_1^{(2)})$. The effective bandwidth of the departure process of stream one equals the arrival process if $\alpha_1^*(\delta) \leq C - \mathbf{E}(A_1^{(2)})$, which is equivalent to saying that the buffer does not overflow, conditioned on the δ-constraint.

The decoupling constraint (11.22) is sufficient but not necessary to guarantee that the effective bandwidths of the departure process and the arrival process are equal. In case the system gives priority to the first traffic stream, the result of Proposition 11.3.3 is valid with a decoupling constraint of $\alpha_1^*(\delta) \leq C$, which clearly is weaker than condition (11.22).

Example 11.3.4. Consider the tandem queue shown in Figure 11.6, fed by three traffic streams, $A_{n_1}^{(1)}$, $A_{n_2}^{(2)}$ and $A_{n_3}^{(3)}$. Subject to some δ-constraint, let the effective and decoupling bandwidths be $\alpha_k(\delta)$ and $\alpha_k^*(\delta)$, respectively, for $k = 1, 2, 3$. The δ-constraint

$$\lim_{b \to \infty} \frac{1}{b} \ln \mathbf{P}(W_i > b) \leq -\delta, \quad i = 1, 2,$$

where W_i is the workload process at queue i. By Proposition 11.3.1, the constraint is satisfied at the first queue if

$$\alpha_1(\delta) + \alpha_2(\delta) \leq C_1.$$

If the service rate of the first queue is large enough so that

$$\alpha_1(0) + \alpha_2^*(\delta) \leq C_1,$$

then by Proposition 11.3.6, the effective bandwidth of the departure process $D_{n_2}^{(2)}$ is equal to that of the arrival process $A_{n_2}^{(2)}$, that is, $\alpha_2(\delta)$. In the second queue, the δ-constraint is satisfied if

$$\alpha_2(\delta) + \alpha_3(\delta) \leq C_2.$$

Note that we only need to consider decoupling on aggregate traffic streams on routes passing multiple queues, in this case traffic stream two. The flow analysis should therefore be conducted per route.

In a general network, the decoupling constraints are more complex. We consider *feed-forward* networks, where no traffic stream revisits an already visited queue along a route. Formally, a feed-forward network is such that in any cut in the network, dividing the queues into two disjoint sets, all routes flow in the same direction.

Denote by $\mathcal{R}$ the set of routes in the network, and $\mathcal{R}_i$ the set of routes passing through queue i. In order to guarantee decoupling at queue i, the traffic streams must be decoupled with respect to all previous queues passed on reach route $r \in \mathcal{R}_i$, as well as in queue i. For each node i, we have the *nodal decoupling constraint*

$$\alpha_r^*(\delta) + \sum_{k \in R(i) - \{r\}} \alpha_k(0) \leq C_i, \quad r \in \mathcal{R}_i, \tag{11.23}$$

where $\alpha_r(\delta)$ and $\alpha^*(\delta)$ are the effective and decoupling bandwidths for traffic on route r. If no two routes share a queue downstream in the network, the condition (11.23) suffices to guarantee decoupling. In practice, however, alternative routing lead to routes sharing a downstream queue. In this case, a stronger decoupling condition is necessary in addition to requiring decoupling with respect to previous queues on the routes. Suppose two traffic streams passing through a queue i on routes $r_1, r_2 \in \mathcal{R}_i$ share a downstream queue. Then we must have

$$\alpha_1^*(\delta) + \alpha_2^*(\delta) + \sum_{k \in \mathcal{R}_i - \{r_1, r_2\}} \alpha_j(0) \leq C_i. \tag{11.24}$$

Thus, a network is analyzed with respect to traffic routes. Decoupling of downstream disjoint routes passing through a queue i is guaranteed by condition (11.23), whereas alternative routes to the same destination (passing through i) is guaranteed with the stronger condition (11.24).

Using the decoupling bandwidths, we have an approximate method of resource allocation. By choosing node capacities so that (11.22) is satisfied throughout the network, decoupling is guaranteed, the δ-constraint is fulfilled and we may use the effective bandwidths of all traffic streams in flow analysis end-to-end.

11.3.2 Design methodology for multi-service networks

Whenever we can determine the performance metrics of a network, it is possible—at least in principle—to formulate a design strategy. At the very least network design can be done by a trial and error strategy, but any randomized method should be able to cope with most of the design steps involved.

We assume that the node locations are defined, the traffic sources are known as well as the traffic distribution between the locations and a set of performance conditions to be met. The relation between cost and capacity of the resources are also assumed to be known.

It is straightforward to formulate some network design strategies; trial and error, randomization and local search. We may use Figure 1.2 to formulate the design process, starting with a feasible topology.

In a trial and error strategy, we generate network topologies arbitrarily and perform flow assigment and capacity allocation, before evaluating performance and cost.

In each trial, we record the configuration with the lowest cost which satisfies the performance constraints. The network cost is also recorded for comparison. This procedure does not have any well-defined stopping criterion.

Closely related to trial and error are randomization strategies. Such a strategy requires some means of generating topologies at random. The difference compared to trial and error is that the network topologies generated then are guaranteed to be random, which makes it possible to estimate a probability of having reached an optimal configuration. In each simulation, a random topology is generated and processed as in the trial and error strategy. By dividing the number of times an optimal configuration appears by the total number of runs, the probability of optimality can be estimated. The probability of an existing better configuration is decreasing with the number of runs, so an upper limit on this probability can be used as a stopping criterion.

Another strategy closely related to trial and error is local search. We begin with an initial topology, possibly generated by some approximation method. If we have reasons to believe that the global optimum should lie in the neighborhood of the initial topology, we may after evaluating it, perturbate the candidate topology slightly, for example by changing the end points of some links.

When planning for a design methodology, there is a large number of considerations that should be made. We summarize the most critical decisions:

Network Quality Metrics

When searching for a network candidate which is optimal in some sense, these usually evaluate not only to different costs, but also to different performance or quality measures. Such measures may be end-to-end delay, packet loss, some reliability measure such as the number of spanning trees of a topology (discussed in Chapter 12), or some function of such measures.

Letting G be a network configuration, $q(G)$ a quality measure and $c(G)$ the network cost, we may formulate a network problem formally as

$$\min_{G} c(G) \quad \text{such that } q(G) \geq q_0, \tag{11.25}$$

where q_0 is some given limit on network quality. The constraint can also be a vector of values. Alternatively, we may evaluate

$$\min_{G} c(G)/q(G), \tag{11.26}$$

to obtain the best ratio of cost to quality measure.

Traffic and Queueing Models

The choice of queueing model is largely dependent on the transport technology used in the network, the traffic model used and to what degree we wish to make approximations. The traffic may be described by a set of models in a multi-service context, such as a collection of BPP models. We may also describe the aggregate traffic as a single traffic stream, typically described by the effective bandwidth of the aggregated traffic.

All traffic sources of interest are modeled, either by using any of the well known traffic models or by estimation. Next, the admission region is defined based on these models. The reason is to establish a relationship between the effective bandwidths between traffic sources and basic bandwidth units (BBU) or channels, which are used in queueing analysis.

Another consideration is whether to adopt a multi-channel or a single-server system. The latter is easier to handle computationally and may be used as an approximation of the former in certain load regions (such as in heavy traffic). The choice of queueing model also influences the set of methods available for performance evaluation.

We may wish to view a network as a circuit-switched network, if the carried services are predominantly of circuit-switched type, such as voice telephony. The traffic model may be Poisson arrivals of calls with a suitable bandwidth requirement. Alternatively, we can use queues with deterministic service rate and the effective bandwidths of some suitable traffic model. Yet another possibility is assuming queues with (generalized) processor sharing. These three examples suggest three different solution approaches: a fixed-point method with multi-dimensional Erlang recursion, analysis by decoupling bandwidths, and convolution or mean value analysis, respectively.

Performance Evaluation

The adopted traffic and queueing models largely determine the set of methods available for network analysis, as whether the network should be considered open, closed or mixed. Also the size of the network, that is, the number of traffic types, jobs per type, nodes, channels or routes, also indicate whether an exact method is viable or an approximate method is necessary. Some of the methods available to some subsets of the models are fixed-point methods, state-space recursions, convolution, mean value analysis, decomposition, network approximations and simulation. Sometimes a combination of methods can be used.

Topological Design

There are two types of topological design with different objectives; capacitated network design (Chapter 4) and survivable network design (Chapter 12). When the objectives for a survivable network are met, the requirements for an optimal capacitated network are often met too, being a subgraph of the survivable network.

A common approach to incorporate both survivable and capacity related constraints is to search for an optimal capacitated topology within some class of networks satisfying some reliability criterion, such as the class of topologies having minimum degree two or greater.

It is useful to determine an initial, tentative topology. Analysis on such a topology helps in understanding the problem and may serve as a first approximation. The tentative topology can later be modified slightly (or optimized) if necessary.

Topological design is $\mathcal{NP}$-hard, and it is therefore advisable to use several methods on the same problem to have an indication of the optimality of a topology.

In particular Monte Carlo methods are useful in that a probability of the existence of a better solution than the one obtained can be calculated. This probability can be made arbitrarily small by running the Monte Carlo method sufficiently many times.

Flow Assignment

The optimal routing problem is closely related to the capacity allocation in the network. The former is concerned with the routing of traffic so that available resources are used optimally by, for example, minimizing delay or the number of hops on a route. The latter, on the other hand, require some nominal values of aggregated traffic, which is a result of routing.

The simplest routing principle is usually some variant of shortest path routing. These are often taken as nominal (primary) routes. In case alternative routes are allowed, these might or might not be accounted for in the network design. The role of alternative routes are typically different in circuit-switched networks and packet networks. In the former, only one or more primary routes are active at a time, and the other routes provide alternatives in case the primary route fails. In the latter case, traffic may be split onto several paths during normal operation, which provides diversity.

It is common to represent the flows by their mean value. Superposition of flows are then handled like for constant flows.

Capacity Allocation

The capacity allocation to links and nodes depend on some flow requirement on these. There are a number of approaches to solve this sub-problem. Capacity allocation can be determined analytically for an open Jackson network (Chapter 7), which may serve as an approximation to more complex networks. The Moe greedy heuristic works for different network types. In fact, it is applicable whenever a performance metric that depends on link capacity can be determined (such as delay or loss). It also usually give good approximations.

Another approximative capacity allocation method is by using decoupling bandwidths. These bounds tend to be unnecessarily conservative, however. Finally, for discrete capacities, an integer program can be formulated and solved with dynamic programming or branch-and-bound, for example.

Cost Evaluation

The selection criterion for a network design is usually the cost, given that given design constraints are met. The resource cost function therefore plays a very important role. Usually, transmission resources are only available in certain discrete capacities with a decreasing density, that is, cost per capacity unit. Cost functions with a decreasing density are called (downwards) concave, and this property represents the "economy of scale" of the resources. The cost may also consist of different components, such as an "installation cost" and a "utilization cost".

In many situations, it is more convenient to work with a continuous cost function, defined for any capacity in some fesibility interval. We can then fit a cost model, such as a power law or logarithmic model, to available data. As a first approximation, a linear cost function is often used.

The network cost is commonly represented by the cost of the links. This is usually the most costly resource of a network. On the other hand, the link costs indicate the capacity requirements on the nodes as well, and it should be straightforward to include node costs based on available capacity figures, if necessary. It should be noted, that when the number of nodes remains fixed, whereas the number of links is variable, the cost difference between two network configurations is usually greater with respect to the link costs than with respect to the node costs.

In some situations, it may be of interest to include estimates of the operational costs of the resources as well (such costs are not considered in this text).

11.4 Summary

In order to analyze multi-service networks, three approaches are discussed; fixed-point methods, queueing network analysis and a network calculus based on effective bandwidth. All three methods are based on independence of the resources in the network—analytically or approximately.

Fixed-point methods are very flexible for the incorporation of technical details of a system, and are probably the method of choice for circuit-switched loss networks. The principle assumes approximate independence between queues which may or may not be a good approximation. In large networks with routing diversity, these methods tend to give more accurate results than for networks which are dominated by a few large traffic streams. Fixed-point methods can be used to incorporate admission control by trunk reservation and adaptive routing, see Greenberg and Srikant (1997) and Liu and Baras (1999).

The results for queueing networks, first introduced by J. R. Jackson, are extended to multiple job classes, called BCMP networks (Baskett et al., 1975). Such a network can be open, closed or mixed. As with single-service networks, open multi-service networks are much easier to analyze than closed networks. In an open network, each queue can be analyzed in isolation from the rest of the network, and the results superposed to find network performance metrics. A closed network is analyzed by extension of the methods discussed in Chapter 8 to multi-class networks: convolution, mean value analysis, and network approximation.

When the offered traffic is heterogeneous and no tractable queueing model can be found, effective bandwidths are suitable as traffic descriptors. A network calculus based on effective bandwidths is discussed, where queues in a network are decoupled, given that the server capacity exceeds a certain bound, the so-called decoupling bandwidth. In effect, the queues can then be analyzed separately from the rest of the network (Wischik, 1999; de Veciana et al., 1993). The decoupling bandwidths must be determined on a per route basis, and can be used for approximate resource allocation.

Survivable Networks

The design of survivable networks, where the term 'survivable' refers to the ability of a network to operate when some of its components (links and/or nodes) are subject to certain failure probabilities, is an important network engineering problem. In the basic network reliability problem it is usually assumed that there are no routing or capacity constraints. Thus, traffic loss due to over-saturation of a link is not considered to be a network failure situation.

We can distinguish between deterministic and probabilistic measures. A deterministic measure does not take the failure probability into account, but is only a measure of the number of potential routes; it is strictly topological. Reliability measures are typically restricted to either nodes or links, that is, it is assumed that either nodes or links—but not both—are subject to failure.

The most widely used probabilistic reliability measure is the probability polynomial. It is usually defined for a graph $G = (V, E)$ where each edge is uniformly assigned the probability of operation $(1 - p)$ (when p refers to the probability of failure), whereas nodes are assumed to be fully operational with probability one. Graphs with a failure probability assigned to each edge are sometimes called random graphs or probability graphs.

Just as with capacitated network design, survivable network design encompasses a number of subtleties and $\mathcal{NP}$-hard problems.

12.1 Connectivity and cuts

Firstly, we need to define what exactly is meant by reliability, as well as some way to measure it. Measures that are studied in network reliability are $\{s, t\}$- or two-terminal reliability, k-terminal reliability, and all-terminal reliability. Using the undirected graph $G = (V, E)$ as a topological model, these measures correspond to the restriction of the connectivity analysis to subsets of the vertex set V of cardinality $2, 2 < k < |V|$, and $|V|$, respectively.

In other words, we select the appropriate measure depending on whether we want to study the reliability between two specific vertices, the mutual reliability in a k-subset of the vertices, or the mutual reliability between all vertices. We will mostly use all-terminal reliability measures in this text, since these are the most natural measures in a core network design context. The first thing we need is a way to find out whether two vertices in a graph are connected or not.

The connectivity of two vertices in a graph, introduced in Definition 2.3.1, is discussed in Section 2.3. Thus, two, vertices i and j in a graph are connected if there is a path from i to j. Depth-first search or breadth-first search can be used to determine whether two vertices are connected or not. Connectivity (between two vertices) is central to reliability, and this concept can be generalized to more than two vertices. However, rather than using the term 'connectivity' to mean a Boolean variable (being connected or not), in this chapter it will be used to connote some measure of network resilience, or 'degree of connectivity'.

In a graph, two basic reliability measures are the rather intuitive concepts of connectivity and cuts.

Definition 12.1.1. A vertex-cut for two vertices i and j is a set of vertices such that when removed, i is disconnected from j. The two-terminal vertex-connectivity $\kappa(i, j)$ is the size of the smallest vertex-cut disconnecting i and j. The (all-terminal) vertex-connectivity $\kappa(G)$ of a graph G is the smallest vertex-cut that disconnects G. $\kappa(G)$ equals the minimum of $\kappa(i, j)$ over all pairs of distinct vertices i and j.

The 'removal' of a vertex should here be understood as a vertex that 'cannot be traversed.' Note that a complete graph with n vertices has no vertex-cuts at all, but by convention it is set to $n - 1$. Similarly, an edge-cut is defined.

Definition 12.1.2. An edge-cut for two vertices i and j is a set of edges such that when removed, i is disconnected from j. The two-terminal edge-connectivity $\lambda(i, j)$ is the size of the smallest edge-cut disconnecting i and j. The (all-terminal) edge connectivity $\lambda(G)$ of a graph G is the smallest edge-cut of G that disconnects G. $\lambda(G)$ equals the minimum of $\lambda(i, j)$ over all pairs of distinct vertices i and j.

The following relations hold for the above measures in a graph G with n vertices and m edges:

$$\kappa(G) \leq \lambda(G) \leq \delta(G) \leq \frac{1}{n} \sum_{i=1}^{n} d_i = 2m/n, \tag{12.1}$$

where $\delta(G)$ is the minimum degree of the graph and d_i the node degree of node i. Note that the minimum degree of the graph is easily verified, but it is only an upper bound to the more precise and useful connectivity measures $\kappa(G)$ and $\lambda(G)$. A stronger reliability measure can be defined by requiring that the paths of connection are disjoint.

Definition 12.1.3. Given a connected undirected graph $G = (V, E)$ and two distinct vertices $i, j \in V$, a set of paths from i to j is said to be vertex-disjoint if no vertex (other than i and j) is on more than one of the paths. Similarly, the set is called edge-disjoint if none of the paths share an edge with any other path.

Denote the greatest number of vertex-disjoint paths between two vertices i and j by $\kappa'(i, j)$, and the greatest number of edge-disjoint paths between i and j by $\lambda'(i, j)$. Then the following relations hold:

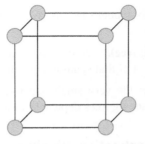

FIGURE 12.1

A 3-vertex-connected graph (the cube graph).

Theorem 12.1.1 (Menger's theorem for vertex-connectivity).   *Let G be a finite undirected graph and i and j two nonadjacent vertices. Then the size of the minimum vertex-cut for i and j (the minimum number of vertices whose removal disconnects i and j) is equal to the maximum number of pairwise vertex-independent paths from i to j, that is, $\kappa(i, j) = \kappa'(i, j)$.*

Theorem 12.1.2 (Menger's theorem for edge-connectivity).   *Let G be a finite undirected graph and i and j two distinct vertices. Then the size of the minimum edge-cut for i and j (the minimum number of edges whose removal disconnects i and j) is equal to the maximum number of pairwise edge-independent paths from i to j, that is, $\lambda(i, j) = \lambda'(i, j)$.*

Menger's theorems assert that the two-terminal vertex-connectivity $\kappa(i, j)$ equals $\kappa'(i, j)$ and the two-terminal edge-connectivity $\lambda(i, j)$ equals $\lambda'(i, j)$ for every pair of vertices i and j. This fact is actually a special case of the max-flow min-cut theorem, which can be used to find $\kappa(i, j)$ and $\lambda(i, j)$. The vertex-connectivity and edge-connectivity of G can then be computed as the minimum values of $\kappa(i, j)$ and $\lambda(i, j)$, respectively. These theorems thus allow a re-definition of connectivity as simply the sizes of the smallest vertex-cuts $\kappa(G)$ and edge-cuts $\lambda(G)$, respectively.

A graph is called k-vertex-connected if its vertex-connectivity is k or greater. Similarly, it is called k-edge-connected if its edge-connectivity is k or greater.

Example 12.1.1.   Consider the cube graph in Figure 12.1. It has a vertex-cut of 3 and is therefore, according to Theorem 12.1.1, 3-vertex-connected. It is also, according to Theorem 12.1.2, 3-edge-connected (both $\kappa(G)$ and $\lambda(G)$ equal 3).

12.2 Spanning trees

An alternative measure of the connectivity of a graph is the number $\tau(G)$ of spanning trees the graph G has. Graphs with the maximum number of spanning trees are the most reliable when the failure probability of the edges is high. This statement will be made more precise in the subsequent sections. We recall the definition of spanning tree:

Definition 12.2.1 (Spanning tree).   A spanning tree of a connected graph G is a connected, acyclic subgraph of G that spans every vertex.

Example 12.2.1.   Consider the very small network consisting of four nodes in Figure 12.2. It has the 16 spanning trees depicted in Figure 12.3.

12.2.1 The deletion-contraction principle

Let $G = (V, E)$ be a connected graph and $e = \{u, v\} \in E$. The deletion $G - e$ is the graph $(V, E - \{e\})$. The contraction G/e is obtained by merging the two vertices u and v together. For contraction to make sense, we require that the edge e is not a loop.

Proposition 12.2.1.   *Let $e \in E$ and $\mathcal{T}(G)$ be the set of spanning trees of G. Then, we have the bijections:*

$$\{T \in \mathcal{T}(G) | e \notin T\} \leftrightarrow \mathcal{T}(G - e),$$

and

$$\{T \in \mathcal{T}(G) | e \in T\} \leftrightarrow \mathcal{T}(G/e).$$

That is, the spanning trees of G that do not contain e are in bijection with the spanning trees of the deletion $G-e$, and the spanning trees of G that do contain e are in bijection with the spanning trees of the contraction G/e.

The first bijection is quite obvious: the set of spanning trees of G not containing e equals the set of spanning trees of G with e removed. In the second bijection, the edge e is made 'permanent' by merging the two nodes. Thus, we have

$$\tau(G) = \tau(G - e) + \tau(G/e),$$

for any graph G and edge e. This gives a recursive algorithm to compute $\tau(G)$ for any graph. Unfortunately, it is computationally inefficient to do so—the complexity of the algorithm is $O(2^n)$. It is, however, a theoretically important tool.

Edge contraction can also be used to find minimum cuts in the following randomized algorithm, presented in Motwani and Raghavan (1995). An edge $e \in E$ in the

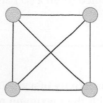

FIGURE 12.2

The complete four-node graph.

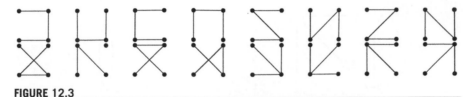

FIGURE 12.3

The 16 spanning trees of the complete four-node graph.

graph $G = (V, E)$ is chosen randomly and is contracted, that is, the vertices at each end are merged. Any edge between the two merged vertices are removed, so that there are no loops. All other edges are retained. Each contraction reduces the number of vertices of G by one.

The edge contraction does not change the size of the cut between two separate vertices i and j, since no egde between them has been contracted. Had it been, i and j would have been merged. As a result, a cut between i and j in a graph after some contraction step is also a cut in the original graph. The contraction is repeated until only two vertices remain. The remaining number of edges between these two vertices is the size of the cut.

The algorithm finds the size of *any* cut. By randomization, we can search for a minimum cut and determine the probability that one has been found. If the algorithm is run $N^2/2$ times, the probability of not finding a minimum cut is

$$\left(1 - \frac{2}{N^2}\right)^{N^2/2},$$

which can be made arbitrarily small by using sufficiently large N.

Example 12.2.2.   To illustrate how the deletion-contraction principle works, consider again the four-node network in Figure 12.2. The top row of Figure 12.4 shows the first step of the deletion-contraction. The first component is the graph with the diagonal from the upper left-hand corner to the bottom right-hand corner removed. The second component is the contraction: the two corresponding vertices are merged and the edges (apart from the deleted one) are preserved.

In the second contraction-deletion step, shown in Figure 12.5, these two components are decomposed further. The second diagonal is now removed from both previous components, which gives the four components shown. It is now easy to see that each component has exactly four spanning trees, so the total number of spanning trees for the graph in Figure 12.2 is 16.

12.2.2 Kirchhoff's matrix-tree theorem

The number of spanning trees can be computed efficiently for any given graph G by Kirchhoff's theorem, which relates the number of spanning trees of G to the eigenvalues of the Laplacian matrix of G, where the Laplacian matrix equals the difference between the degree matrix of G and its adjacency matrix (Colbourn (1993)).

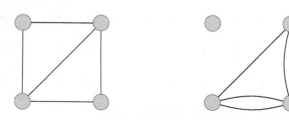

FIGURE 12.4

The first deletion-contraction step.

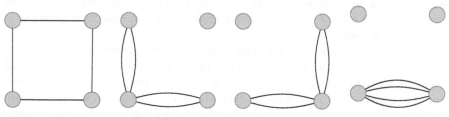

FIGURE 12.5

The second deletion-contraction step.

Definition 12.2.2. Let $G(V, E)$ be a graph with $|V|$ labeled vertices and D its degree matrix, that is, the $|V| \times |V|$ matrix such that

$$D_{ii} = \deg(v_i) \quad \text{for all } v_i \in V,$$

and $A = (a_{ij})$ its adjacency matrix, that is:

$$a_{ij} = \begin{cases} 0 & \text{if } i = j \\ 1 & \text{if } i \neq j \text{ and } v_i \text{ is adjacent to } v_j \\ 0 & \text{otherwise} \end{cases}.$$

Then the Laplacian matrix of G is $L(G) = D - A$. Note that multiple edges are allowed in the adjacency matrix.

Example 12.2.3. Let G be the small network as depicted in Figure 12.2. The Laplacian matrix $L(G)$ is then the difference between the degree matrix and the adjacency matrix, that is:

$$L(G) = \begin{pmatrix} 3 & 0 & 0 & 0 \\ 0 & 3 & 0 & 0 \\ 0 & 0 & 3 & 0 \\ 0 & 0 & 0 & 3 \end{pmatrix} - \begin{pmatrix} 0 & 1 & 1 & 1 \\ 1 & 0 & 1 & 1 \\ 1 & 1 & 0 & 1 \\ 1 & 1 & 1 & 0 \end{pmatrix} = \begin{pmatrix} 3 & -1 & -1 & -1 \\ -1 & 3 & -1 & -1 \\ -1 & -1 & 3 & -1 \\ -1 & -1 & -1 & 3 \end{pmatrix}.$$

Note that each row and column of $L(G)$ sums to zero, so it is certainly true that $\det(L(G)) = 0$. An amazing result is therefore the following statement. For a proof, see for example Nederlof (2008).

Theorem 12.2.2 (The Kirchhoff matrix-tree theorem). *Let $G(V, E)$ be a connected graph with $n = |V|$ labeled vertices, and let $\lambda_1, \lambda_2, \ldots, \lambda_{n-1}$ be the non-zero eigenvalues of the Laplacian matrix Q of G. Then the number of spanning trees $\tau(G)$ of G is*

$$\tau(G) = \frac{1}{n}(\lambda_1, \lambda_2, \ldots, \lambda_{n-1}).$$

Equivalently, the number of spanning trees is equal to the absolute value of any cofactor (signed minor) of the Laplacian matrix of G. This is obtained by choosing an entry a_{ij} in the matrix and crossing out the entries that lie in row i and column j and taking the determinant of the reduced matrix. Thus, in practice, it is easy to construct the Laplacian matrix for the graph, cross out for example the first row and the first column, and take the determinant of the resulting matrix to obtain the number of spanning trees $\tau(G)$.

Example 12.2.4. Once again we let G be the graph in Figure 12.2. We have already determined the Laplacian matrix in Example 12.2.3. By cancelling the first row and the first column the reduced matrix is obtained as

$$M_{11} = \begin{pmatrix} 3 & -1 & -1 \\ -1 & 3 & -1 \\ -1 & -1 & 3 \end{pmatrix},$$

and finally, taking the determinant of M_{11}, $\det(M_{11}) = 16$, gives the number of spanning trees in the graph.

Example 12.2.5. Let G be the seven-node network. Its Laplacian matrix $L(G)$ is

$$L(G) = \begin{pmatrix} 3 & -1 & -1 & -1 & 0 & 0 & 0 \\ -1 & 4 & -1 & 0 & -1 & 0 & -1 \\ -1 & -1 & 3 & 0 & 0 & 0 & -1 \\ -1 & 0 & 0 & 3 & -1 & -1 & 0 \\ 0 & -1 & 0 & -1 & 3 & -1 & 0 \\ 0 & 0 & 0 & -1 & -1 & 3 & -1 \\ 0 & -1 & -1 & 0 & 0 & -1 & 3 \end{pmatrix}.$$

and the number of spanning trees is 972 by the Kirchhoff matrix-tree theorem.

12.2.3 Graph strength

The strength of a graph relates the concepts of spanning trees and edge-connectivity of a graph. It can also be used as a reliability criterion. It is particularly useful for finding the connectivity 'bottlenecks' in a graph. The strength $\sigma(G)$ of an undirected

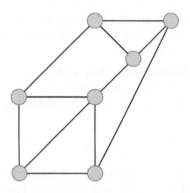

FIGURE 12.6

A network with seven nodes and 972 spanning trees.

simple graph $G = (V, E)$ can be seen as a number corresponding to partitioning of the graph into subgraphs by disconnecting it where it is weakest. The strength is the minimum ratio of deleted edges to the number of subgraphs created.

Definition 12.2.3. Let Π be the set of all partitions of V, and $\partial \pi$ be the set of edges crossing over the sets of the partition $\pi \in \Pi$, then

$$\sigma(G) = \min_{\pi \in \Pi} \frac{|\partial \pi|}{|\pi| - 1}.$$

There are several known algorithms for computing the strength of a graph. One of them, an approximate, but intuitively illuminating algorithm is presented by Galtier et al. (2008). It is based on an alternative definition, formulated in terms of spanning trees.

Definition 12.2.4. Let T be the set of spanning trees of G, Π the set of all partitions of V, and $\partial \pi$ the set of edges crossing over the sets of the partition $\pi \in \Pi$, then

$$\sigma(G) = \max \left(\sum_{T \in T} \lambda_T : \forall T \in T, \ \lambda_T \geq 0, \ \forall e \in E, \ \sum_{T \ni e} \lambda_T \leq w(e) \right),$$

where λ_T are real numbers and $w(e)$ some edge weight.

The dual problem to the one formulated in Definition (12.2.4) is

$$\sigma(G) = \min \left(\sum_{e \in E} w(e) y_e : \forall e \in E, \ y_e \geq 0, \ \forall T \in T, \ \sum_{e \in T} y_e \geq 1 \right). \quad (12.2)$$

Theorem 12.2.3. *Given a connected graph G and a positive real $\epsilon \leq 1/2$, there exists an algorithm of computational time $O\left(m \log(n)^2 \log\left(\frac{m}{n}\right) / \epsilon^2 \right)$ that returns a* *set of trees $T_1, \ldots .T_p$ of G, associated to real positive numbers $\lambda_1, \ldots, \lambda_p$ with*

$$\sum_{i \in \{1,\ldots,p\}:T_i \ni e} \lambda_i \leq 1, \quad \text{for all } e \in E,$$

and

$$\sum_{i \in \{1,\ldots,p\}} \lambda_i \geq \frac{1}{1+\epsilon}\sigma(G).$$

The dual expression in Equation (12.2) can be used directly to formulate an algorithm for computing the strength.

Algorithm 12.2.4 (Approximate Graph Strength). Input an undirected graph $G = (V, E)$

STEP 0: Initiate by assigning to each edge $e \in E$ a small weight $w(e) = \delta = O(n^{-3/\epsilon})$;

STEP 1: Compute a minimum spanning tree T with respect to w;

STEP 2: For each $e \in T$, multiply $w(e)$ by $(1 + \epsilon)$;

STEP 3: If $w(T) < 1$ go to STEP 1;

STEP 4: Calculate $\sigma(G) = \sum_{e \in E} w(e)$.

Output graph strength $\sigma(G)$

The algorithm determines, in each iteration, the minimum spanning tree of the graph, and whenever an edge e belongs to the minimum spanning tree, its weight is increased by a small amount. The edges in the 'weakest cut' in the graph will consequently receive the largest weights, because they are used more often than other edges.

An important property of the strength is that each subgraph (which is not a singleton) generated by partitioning the graph according to Definition 12.2.3 has better strength than the original graph. Let $P = \{S_1, \ldots, S_p\}$ be a partition of G that achieves the strength of G, that is

$$\sigma(G) = \frac{w(\delta\{S_1, \ldots, S_p\})}{p - 1},$$

then, for all $i \in \{1, \ldots, p\}$, we denote by $G(S_i)$ the restriction of G to S_i, and we have

$$\sigma(G(S_i)) \geq \sigma(G(S)).$$

If c is the minimum cut of G, the strength lies between $c/2$ and c. The upper bound follows immediately from the definition of the minimum cut: we have two disjoint sets if the edges in the minimum cut are removed, and the Definition 12.2.3 requires $\sigma(G)$ to be smaller than or equal to this number.

The interpretation of the strength is readily provided by the Tutte-Nash-Williams theorem.

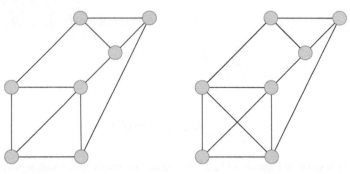

FIGURE 12.7

The optimal 3-connected solution of Example 12.4.1 (left) and a network with strength 2 (right).

Theorem 12.2.5 (Tutte-Nash-Williams). *A graph G contains k edge-disjoint spanning trees if and only if the strength of G is larger than or equal to k, that is, $\sigma(G) \geq k$.*

The theorem suggests that the strength of a graph may be an appropriate reliability measure in certain situations. This may be the case when failures are dependent along a route, or when a fast switchover to an independent circuit is desired for some other reason. It may also be desired to have a number disjoint trees if one wishes to separate different types of traffic onto edges dedicated for particular traffic types (for example, payload and signaling).

Example 12.2.6. Let G_1 be the first graph in Figure 12.7 and G_2 the second graph. Exact calculation of the strength of G_1 using the Definition 12.2.3 gives the fractions:

$$\frac{3}{1}, \frac{5}{2}, \frac{6}{3}, \frac{8}{4}, \frac{10}{5}, \frac{11}{6},$$

so the strength $\sigma(G_1) = \frac{11}{6} = 1.83$. We can therefore conclude, referring to Theorem 12.2.5, that G_1 does not have two edge-disjoint spanning trees.

Similarly, for G_2 we have:

$$\frac{3}{1}, \frac{6}{2}, \frac{8}{3}, \frac{9}{4}, \frac{11}{5}, \frac{12}{6},$$

so the strength $\sigma(G_1) = \frac{12}{6} = 2$, and so two edge-disjoint spanning trees exist in G_2. Note that it is not obvious how to find the two disjoint spanning trees. One possible pair of disjoint spanning trees is depicted in Figure 12.8. Computing the graph using Algorithm 12.2.4 yields the approximate values $\hat{\sigma}(G_1) = 1.77640$ and $\hat{\sigma}(G_2) = 1.93614$. We therefore conclude that $\hat{\sigma}(G_1) < \hat{\sigma}(G_2)$.

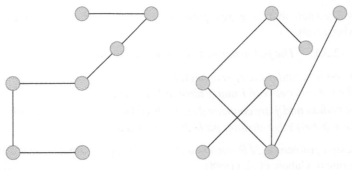

FIGURE 12.8

An edge-disjoint spanning tree (left). The complementary edge-disjoint spanning tree (right).

12.3 A primal-dual algorithm

We define the survivable network design problem formally.

Definition 12.3.1 (The Minimum Cost Survivable Network Design Problem).
Given a cost matrix (c_{ij}) *and required connectivity matrix* $R = (r_{ij})$, *the Minimum Cost Survivable Network (MCSN) problem is that of finding a graph* $G = (V, E)$ *with minimum total cost*

$$\sum_{\{i,j\} \in E} c_{ij}$$

and vertex-connectivity s_{ij} *between distinct vertices* i, j *satisfying*

$$s_{ij} \geq r_{ij} \quad \forall i, j, \quad i \neq j.$$

It is possible to formulate a primal-dual algorithm for the survivability network design problem. Suppose we have a set of vertices V, an undirected complete graph $G = (V, \bar{E})$, edge costs $c_e \geq 0$ and a connectivity requirement given by a matrix $R = (r_{ij})$ between each node pair i, $j \in V$. The matrix R is assumed to be symmetric, so that $r_{ij} = r_{ji}$ for all i, j. The problem can be formulated so that we want to select a subset of edges $E \subset \bar{E}$ so that the nodes i and j are at least r_{ij}-connected.

We associate a number r_i to each vertex $i \in V$, where $r_i = 2$ for nodes that need high protection (such as sites belonging to the backbone network), $r_i = 1$ for nodes belonging to, for example, an access network site, and $r_i = 0$ for optional nodes. Then,

$$r_{ij} = \min\{r_i, r_j\}, \quad i, j \in V.$$

The matrix R is assumed to be symmetric, so that $r_{ij} = r_{ji}$ for all i, j. The goal of the design is to select a subset of edges $E \subset \bar{E}$ so that the the nodes i and j are at least r_{ij}-connected, at minimum network cost.

First, we can note that there are equivalent definitions of r_{ij}-connectivity. We have the following result.

Theorem 12.3.1. *The following statements are equivalent:*

- *(1) The nodes i and j are r_{ij}-connected.*
- *(2) All edge cuts between i and j have at least r_{ij} edges.*
- *(3) The nodes i and j are connected in $(V, \bar{E} \setminus E)$ for any edge set E with $|E| < r_{ij}$.*
- *(4) There are r_{ij} edge-disjoint paths between i and j.*

The design problem is $\mathcal{NP}$-complete, but can be approximated by a primal-dual algorithm due to Gabow et al. (1998).

We construct a so-called *proper function* $f : 2^E \to \mathbb{N}$ such that

(1) $f(\varnothing) = 0$

(2) $f(S) = f(V - S)$, for all $S \subseteq V$ (Symmetry)

(3) $f(A \cup B) \le \max\{f(A), f(B)\}$ (Maximality), for A, B disjoint

We may illustrate the symmetry property of the function as the number of edges between the subgraph formed by the set S and the subgraph formed by the set $V - S$. These two numbers must be equal since the edges must have one end point in each set. The maximality principle says that the maximum of the function $f(\cdot)$ of two disjoint sets, for example, such as formed by dividing $G = (V, E)$ into two subgraphs, is never less than for the union of the sets or the whole graph G in this case. We note that the strength of Section 12.2 is larger in the components of the graph than in the original graph. This follows the maximality principle. It can be interpreted as there is always a 'weakest cut' in a graph G which, if connecting A and B, the function value of the union $f(A \cup B)$ is lower than or equal to the value for the components A and B.

The problem can be formulated as an integer program by introducing the decision variables $x_{ij} \in \{0, 1\}$ representing the inclusion or exclusion of link (i, j),

$$\min \sum_{(i,j) \in E} c_{ij} x_{ij}$$
$$x(\delta(S)) \ge f(S), \quad S \subseteq V, \tag{12.3}$$
$$x_{ij} \in \{0, 1\}, \quad (i, j) \in E.$$

The quantity $\delta(S)$ is called the coboundary of the set S, equaling the set of edges with exactly one endpoint in S.

For the survivable network design problem, the proper function is given by

$$f(S) = \max_{i \in S, j \notin S} r_{ij}. \tag{12.4}$$

The algorithm is building the network in n phases and each phase is ending with a 'clean-up' phase. The ending condition is when the number of phases are sufficient to construct a graph fulfilling all the connectivity requirements r_{ij}. The final number of phases of the algorithm is

$$f_{\max} = \max_S f(S). \tag{12.5}$$

In each phase, the set E of edges included in the graph $G = (V, E)$ is incremented to fulfill the connectivity requirements.

We let E_{n-1} denote the set of edges selected in the first $n - 1$ phases and let $\delta_A(S)$ denote the set $A \cap \delta(S)$ for any $A \subseteq E$. In phase n we may choose edges from the set $E_n \subseteq \bar{E} - E_{n-1}$ of edges not yet selected to be included in the design. This set is such that whenever $f(S) \geq n$ and $|\delta_{E_{n-1}}(S)| = n - 1$, then $|\delta_E(S) \geq 1|$. This means that when there are unfulfilled connectivity requirements ($f(S) \geq n$) and we have the connectivity $|\delta_{E_{n-1}}(S)| = n - 1$ satisfied in the phase just completed, then $|\delta_{E_n}(S)| \geq 1$, or that the remaining set of edges can provide at least one additional edge increasing the current connectivity r_{ij}.

In each phase, we update the set of currently selected edges $E_n = E_{n-1} \cup E_h$, where E_h is the additional edges found in phase n, called the augmentation phase h. We have the ending condition $|\delta_{E_n}(S)| \geq \min\{f(S), n\}$, which is the minimum of the maximal connectivity requirements for all subsets S and phase number n, respectively. Consequently, the final set of edges $E_{f_{\max}}$ is a feasible solution to the integer problem (12.3).

We now define the augmentation phase h as

$$\min \sum_{(i,j) \in E_h}$$
$$x(\delta(S)) \geq h(S), \quad S \subseteq V \tag{12.6}$$
$$x_{ij} = \{0, 1\}, \quad (i, j) \in E_h,$$

where $E_h = \bar{E} - E_{n-1}$ and we define a function $h(S)$ so that $h(S) = 1$ whenever $f(S) \geq n$ and $|\delta_{E_{n-1}}(S)| = n - 1$, and $h(S) = 0$ otherwise. This function defines the unit increase in connectivity, whenever the ending condition has not yet been met in phase $n - 1$. The function $h(\cdot)$ is called the uncrossable function. Note that the condition $f(S) \geq n$ corresponds to the condition that there are still unfulfilled connectivity requirements.

The algorithm consists of two steps, augmentation in n phases, and a cleanup phase to ensure the optimality bound on the cost.

The algorithm can be viewed a generalization of Dijkstra's algorithm for shortest paths. The approximation yields a performance guarantee of $2f_{\max} - 1$ for $f_{\max} \geq 2$ and 2 for $f_{\max=1}$.

Algorithm 12.3.2 (Main Algorithm). Input an undirected graph $G = (V, E)$, edge costs $c_{ij} \geq 0$, a proper function $f(\cdot)$ and its maximum value $f_{\max} = \max_S f(S)$

STEP 0: Set $E_0 = \varnothing$;
STEP 1: **for** $n = 1$ **to** $f_{\max}$ (the number of phases)
$\qquad h_n(S) = 1$ if $f(S) \geq n$ and

$|\delta_{E_{n-1}}(S)| = n - 1$ and $h_n(S) = 0$ otherwise;

$E_n = \bar{E} - E_{n-1}$;

Determine E_h by calling the Uncrossable Algorithm with the arguments V, E_n, $\mathbf{c}$, h_n.

Update $E_n = E_{n-1} \cup E_h$.

Output $E_{f_{\max}}$, the number of edges.

We assume that the edge costs equal edge distances, so that $c_{ij} = d_{ij}$. Any cost can be used in the algorithm, however.

Algorithm 12.3.3 (Uncrossable Algorithm). Input an undirected graph $G = (V, E_h)$, edge costs $c_{ij} \geq 0$, and an uncrossable function $h(\cdot)$

STEP 0: Set $E_0 = \varnothing$, $i = 0$, $d(v) = 0$ for all $v \in V$. Denote by $\mathcal{C}$ the set of active sets C;

STEP 1: **while** $\mathcal{C} > 0$

$\quad i = i + 1$;

$\quad$ **for** all $v \in C$, increase $d(v)$ uniformaly by Δ
$\quad$ until some edge $(u, v) \in E_h$ satisfies $d(u) + d(v) = c_{uv}$ for $e_h = (u, v) \in \delta(C)$ of some $C \in \mathcal{C}$;
$\quad$ Update $E_p = E_{p-1} \cup \{(u, v)\}$.

STEP 2: Let all sets C that were active during the last phase be unmarked. Mark the set V;

STEP 3: **for** $j = i$ **down to** 1

$\quad$ If the edge (u, v) is special $C(u, v)$ is unmarked and $\delta_{E_h}(C(u, v)) \subseteq \delta_{E_h}(A(u, v))$, set then $E_p = E_p - (u, v)$
$\quad$ Mark (u, v).

Output E_p.

The set S is called violated if $|\delta_E(S)| < h(S)$. A set called active if it is a minimal violated set. In each iteration, an edge $(u, v) \in E_h$ is selected from the coboundary of some currently active set C and is added to E. The edge may be in the coboundary of either one or two active sets.

We have some initial collection $\mathcal{C}$ of active sets, and in each iteration we select some edge in the coboundary of at least one, and at most two, active sets from $\mathcal{C}$. Once an edge in the coboundary of an active set C is selected, then C is no longer violated and hence no longer active. Thus, there will be at most $|\mathcal{C}|$ iterations of the edge addition step.

The cleanup phase is removing 'uneconomical' edges that are not required to fulfill the connectivity requirement. We therefore investigate the edges in the set E_p in reverse order. Edges added in a later phase are more expensive than edges added early on. We therefore scan the edges of E in the reverse of their selection step. Edges that can be removed without violating the connectivity requirement are called special. Note that the case when $f_{\max} = 1$ reduces to finding connected components and $f_{\max} = 2$ reduces to finding a 2-connected components for the graph.

Example 12.3.1. We assume that we have the network in Figure 12.9. We assume that we have a symmetric requirement matrix

$$r_{ij} = \begin{pmatrix} - & 1 & 2 & 1 & 2 & 1 \\ 1 & - & 1 & 1 & 1 & 1 \\ 2 & 1 & - & 1 & 2 & 1 \\ 1 & 1 & 1 & - & 1 & 1 \\ 2 & 1 & 2 & 1 & - & 1 \\ 1 & 1 & 1 & 1 & 1 & - \end{pmatrix},$$

from we easily construct the function $f(S) = \max_{i \in S, j \notin S} r_{ij} = (2, 1, 2, 1, 2, 1)$. We also have a matrix with edge costs

$$c_{ij} = \begin{pmatrix} - & 23 & 20 & 32 & 45 & 51 \\ 23 & - & 10 & 23 & 23 & 37 \\ 20 & 10 & - & 15 & 29 & 32 \\ 32 & 23 & 15 & - & 32 & 20 \\ 45 & 23 & 29 & 32 & - & 32 \\ 51 & 37 & 32 & 20 & 32 & - \end{pmatrix},$$

and set $E_0 = \oslash$ and the set of active sets $\mathcal{C} = \{\{1\}, \{2\}, \{3\}, \{4\}, \{5\}, \{6\}\}$ that is, all individual vertices are forming the components of the active set, since no connection does yet consist and all requirements are unfulfilled. As the algorithm progresses, the minimum cost $\Delta = \min_{i \neq j} c_{ij}/2$ that has to be paid in order to add an edge is calculated, and after that, the costs are updated. We now have the variables needed to call Algorithm 12.3.2. $h_1(S) = 1$, since both $|\delta_{E_0}(S)| = 0$ and $f(S) \geq 1$. The minimum cost that has to be paid to add the first edge to the graph is $\Delta = 5$, and we keep track of the cost in the vector $\mathbf{p}(S)$ which adds the edge $E_1 = \{2, 3\}$ to the design. This vector is in each step updated as $\mathbf{p}^{(n)}(S)$. Next, the costs are updated as $c_{ij}^{(n)} = c_{ij}^{(n-1)} - p_i^{(n)}(S) - p_j^{(n)}(S)$. The active sets are now $\mathcal{C} = \{\{1\}, \{2, 3\}, \{4\}, \{5\}, \{6\}\}$. After five iterations, the first step is complete with $E_5 = \{\{2, 3\}, \{3, 4\}, \{4, 6\}, \{1, 3\}, \{2, 5\}\}$. The cleaning up step does not affect the edges.

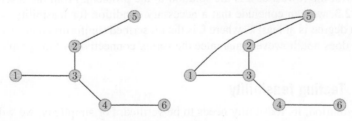

FIGURE 12.9
The Goemans-Williamson algorithm in Example 12.3.1.

The next phase starts with the active set of edges $C = \{\{1\}, \{5\}\}$. The algorithms is completed with $E = \{\{2, 3\}, \{3, 4\}, \{4, 6\}, \{1, 3\}, \{2, 5\}, \{1, 5\}\}$. Since all requirements are satisfied, the algorithm terminates. The results of the design algorithm is shown in Figure 12.9.

Note that the edge $(1, 5)$ is a consequence of requiring 2-connectivity for vertices 1,3 and 5. This requirement is satisfied for vertex 3 after the first phase, so only the vertices 1 and 5 become active in the second phase.

12.4 Local search

In this section, a survivable network design problem is discussed where the design criterion is expressed as a minimum requirement on the vertex-connectivity $\kappa(G)$ in the graph G. Associated with the problem is a cost matrix defining the cost of all potential edges in the graph. The optimal solution is the graph which fulfills the reliability requirement and has the lowest cost.

A requirement connectivity matrix $R = (r_{ij})$ is a matrix defining the two-terminal connectivity requirement for each pair i, j of vertices. Similarly, the 'actual' connectivity matrix $S = (s_{ij})$ is defined for each candidate solution.

The MCSN problem is $\mathcal{NP}$-complete, so the best we can hope for is to find an approximately optimal solution. A simple and yet often fruitful approach to find a good candidate is to search for a solution in a neighborhood of an initial good guess. Such a local search procedure can be constructed given that:

(a) A reasonably good initial solution can be found;
(b) A transformation for generating similar—but slightly different—solutions by making changes to the initial solution can be defined. This transformation defines the search neighborhood.

In order to solve the problem, we therefore need a way to generate candidate solutions, and a way to test these candidates for feasibility, that is, verifying that the design criterion $s_{ij} \geq r_{ij}$ for all $i, j \in V$, $i \neq j$ are met. Clearly, the feasible solution with the lowest cost is the solution to the problem. From the discussion in Section 12.3, we can conclude that a necessary condition for feasibility is that the minimum degree is at least k, where k is the prescribed (uniform) connectivity. This condition does not, however, guarantee the vertex-connectivity of the graph.

12.4.1 Testing feasibility

For each solution, its feasibility needs to be verified. For simplicity, we will let the connectivity requirement be $r_{ij} = k$ for all i, j. This uniform connectivity requirement leads to a more efficient algorithm than for the general case, although the principles of the method remain the same. The algorithm for verifying feasibility is based on the max-flow min-cut algorithm.

If the edges in a network are assigned unit capacities, the max-flow algorithm will give zero or unit augmenting capacities and thus provides a way to verify feasibility of the candidate solution. First, we introduce some results on the complexity of the max-flow min-cut algorithm. The following theorems (for proofs, see Papadimitriou and Steiglitz, 1982) are valid for directed graphs $G = (V, E)$ with unit edge capacities.

Theorem 12.4.1.   *For directed graphs $G = (V, E)$ with unit edge capacities, the max-flow algorithm requires at most $O(|V|^{2/3} \cdot |E|)$ time.*

If, in a directed graph $G = (V, E)$ with unit edge capacity, each vertex has either indegree 1 or 0, or outdegree 1 or 0, the graph is called *simple*. An upper bound on the complexity of calculating the two-terminal connectivity is stated in the following theorem (Papadimitriou and Steiglitz, 1982).

Theorem 12.4.2.   *For simple networks, the max-flow algorithm requires at most $O(|V|^{1/2} \cdot |E|)$ time, where $|E|$ is the number of edges.*

Now we arrive at the following theorem. The proof is very instructive and can be used to construct an algorithm for verification of solution feasibility.

Theorem 12.4.3.   *Let $G = (V, E)$ be an undirected graph and i, j two distinct vertices in V. Then it is possible to find the vertex-connectivity $s_{ij} \geq r_{ij}$ in $O(|V|^{2.5})$ time.*

Proof.   Create a directed graph $G' = (V', E')$ from $G = (V, E)$ by replacing every vertex $v \in V$ by two vertices $v_1, v_2 \in V'$ (see Figure 12.10). Consider the flow network obtained by assigning a unit capacity to every arc of G'. Then the maximum flow between vertices i and j in this network is the vertex-connectivity r_{ij}, because the unit capacity edge from v_1 to v_2 means that the vertex v in the original graph G can lie on, at most, one path between i and j. The flow network corresponding to G' is simple, so by Theorem 12.4.2 we can calculate the maximum flow in $O(|V|^{2.5})$ time.

Now we can formulate the following procedure for verification of the connectivity between two vertices.

Proposition 12.4.4.   *Let $G = (V, E)$ be a network where the capacity of every edge is 1. Then, for each pair of vertices i, $j \in V$, the value of a maximum flow in G equals the number of edge-disjoint directed paths in G.*

FIGURE 12.10

Convert a node v in an undirected graph (left) to two nodes v_1 and v_2 in a directed graph (right).

12.4.2 Generating an Initial Solution

An initial solution can be constructed using an algorithm based on a greedy heuristic described in Papadimitriou and Steiglitz (1982). As noted earlier, the degree of vertex i must be at least $\max_j r_{ij}$. Define the deficiency of a vertex i in an undirected graph G as

$$\text{deficiency}(i) = \max_j r_{ij} - \deg(i). \tag{12.7}$$

Next, add edges to the graph G until all the deficiencies are non-positive. We then need to test the resulting graph for feasibility using the max-flow algorithm with unit edge capacities.

Algorithm 12.4.5 (Generation of Initial Solution). Given a set of vertices V and a cost matrix C defined on the possible edges.

STEP 0: Initiate by ordering the vertices (randomly) and create an array of size $|V|$ containing the deficiency of each vertex;

STEP 1: Starting from the left, add an edge between a vertex with the largest deficiency and one with the next largest deficiency. Of all the vertices with next largest deficiency, we choose the one that results in the smallest increase in cost; all other ties are resolved by choosing the earliest vertex in the array. Multiple edges are not allowed.

STEP 2: If all deficiencies are equal to or less than zero, stop: an initial solution has been found; else go to STEP 1.

Output initial solution $G_0 = (V, E)$.

The procedure is used and illustrated in detail in Examples 12.4.1 and 12.4.2 below.

12.4.3 The search neighborhood

We can create a neighborhood of a given initial feasible solution by using the X-change transformation. We will assume that the costs are proportional to distances, as no capacity is considered in this problem. Without loss of generality, we let $c_{ij} = d_{ij}$ for the edges (i, j).

Definition 12.4.1 (The X-Change Neighborhood). Let the set of graphs feasible in an instance of the MCSN problem be denoted by F. That is, F consists of all graphs with a given number of vertices and vertex-connectivity satisfying

$$s_{ij} \geq r_{ij} \quad \text{for all} \quad i, j, i \neq j$$

Consider a graph $G = (V, E) \in F$ in which the edges (i, l) and (j, k) are present and the edges (i, k) and (j, l) are absent. Define a new graph $G' = (V, E')$ by removing edges (i, l) and (j, k) and adding edges (i, k) and (j, l). That is

$$E' = E \cup \{(i, k)(j, l)\} - \{(i, l)(j, k)\}.$$

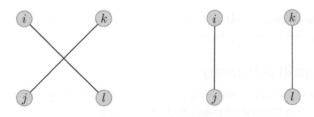

FIGURE 12.11

Connections before the X-change (left), and connections after the X-change (right).

Then, if $G' \in F$, we say it belongs to an X-change neighborhood of G, and the set of all X-changes of G defines the X-change neighborhood. If the new cost is less than the old solution, that is

$$d_{ik} + d_{jl} < d_{il} + d_{jk}, \qquad (12.8)$$

then the X-change is called favorable. The transformation is illustrated in Figure 12.11.

The reason for selecting the X-change neighborhood here from many possible search neighborhoods is that the transformation preserves the degrees of the vertices. This property makes the feasibility test of the new graph more efficient. In fact, if we had to check the entire graph for feasibility after each favorable X-change candidate was discovered, the local search algorithm would be very slow, but it turns out that a complete check is not necessary. In the case that $r_{ij} = k$, it is only necessary to check that two vertex-connectivities are preserved to establish the feasibility of the new graph.

Proposition 12.4.6. *If an X-change on a feasible network destroys feasibility by reducing s_{ab} below r_{ab}, where $a, b \in V$, then*

$$s_{ik} < r_{ab} \text{ or } s_{jl} < r_{ab},$$

where the X-change removes edges $\{i, k\}$ and $\{j, l\}$.

This follows immediately from the fact that adding new edges cannot decrease the connectivity, and thus we only need to check that the connectivity between the nodes i and k, and between j and l has been preserved. Since the connectivity between any of i, j, k, and l and any of the rest of the nodes is satisfied before the X-change, the only way the connectivity between any other pair of nodes (a and b, say) can be affected is by changing paths, for example from $a \to i \to l \to b$ to $a \to i \to k \to b$.

Since the X-change has the property that preserves the number of edges and the degree of every vertex, it is desirable to have an assortment of starting solutions in order to widen the search space, possibly with different numbers of edges and different vertex degrees. This may sometimes be obtained by randomly reordering the vertices before applying the greedy heuristic to create an initial solution.

These starting solutions tend to have low cost and a small number of edges, see Papadimitriou and Steiglitz (1982).

12.4.4 Algorithm summary

We summarize the discussion above as an algorithm, assuming that the cost matrix C and the uniform connectivity requirement $r_{ij} = k$ are given.

Algorithm 12.4.7 (Main Algorithm).   Given a set of vertices V, a cost matrix $C = (c_{ij})$ for possible edges and the uniform connectivity requirement $r_{ij} = k$.

STEP 1: Generate an initial solution using Algorithm 12.4.5.

STEP 2: Test the initial solution for feasibility using the max-flow min-cut algorithm and Proposition 12.4.4. If the solution is infeasible, permute the nodes and go to STEP 1.

STEP 3: Start local search; Compute the cost c_{init} of the initial solution. Find vertices for an X-change and transform the graph.

STEP 4: Test the new graph for feasibility, using Proposition 12.4.6. If the new solution is infeasible, X-change back to the original solution and go to STEP 3.

STEP 5: Compute the cost of the new solution c_{new}. If $c_{new} < c_{init}$, accept the new solution, let $c_{init} = c_{new}$ and go to STEP 3.

Output best topology $G = (V, E)$ and its associated cost.

Note that we need to specify some sort of terminating condition for the algorithm, for example the maximum number of iterations.

Example 12.4.1.   We will solve the MCSN problem for the seven nodes depicted to the left in Figure 12.12. This example is taken from Papadimitriou and Steiglitz (1982), Steiglitz et al. (1969).

Let the edge costs be the integer part of the Euclidean distances between the corresponding vertices. Let, for example, the distance between vertex A and vertex B be 20 units and suppose the vertices are enumerated as in the right-hand side of Figure 12.12. The cost matrix is then

$$C = \begin{pmatrix} 0 & 20 & 20 & 28 & 31 & 44 & 28 \\ 20 & 0 & 28 & 20 & 14 & 28 & 20 \\ 20 & 28 & 0 & 44 & 42 & 56 & 20 \\ 28 & 20 & 44 & 0 & 14 & 20 & 40 \\ 31 & 14 & 42 & 14 & 0 & 14 & 31 \\ 44 & 28 & 56 & 20 & 14 & 0 & 44 \\ 28 & 20 & 20 & 40 & 31 & 44 & 0 \end{pmatrix}. \tag{12.9}$$

We use Algorithm 12.4.5 to generate an initial solution. The procedure is summarized in Table 12.1.

1. Initially, all vertices have the same deficiency and we start from vertex 0, that is, the first vertex to the left. Any of the other vertices are candidates, but the

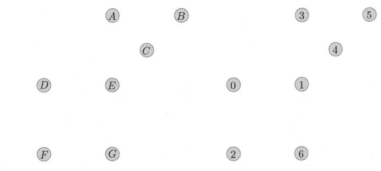

FIGURE 12.12

Seven given vertices that are to be connected so that the resulting network is 3-vertex-connected at minimum possible cost (left). A random permutation of the vertices yields initial vertex ordering (right).

Table 12.1   Initial solution generated by the heuristic algorithm

3	3	3	3	3	3	3
2	2	3	3	3	3	3
2	2	2	3	3	3	2
2	2	2	2	2	3	2
2	2	2	2	1	2	2
1	2	1	2	1	2	2
1	1	1	1	1	2	2
1	1	1	1	1	1	1
0	1	1	0	1	1	1
0	0	1	0	0	1	1
0	0	0	0	0	0	1
0	-1	0	0	0	0	0

two closest are vertices 1 and 2. Since 1 precedes 2, we choose vertex 1, add the edge {0, 1}, reduce the deficiency of vertices 0 and 1 by 1, and go onto the next row.

2. The leftmost vertex with the highest deficiency is now vertex 2. Among the vertices with the same deficiency, vertex 6 is closest, so we add the edge {2, 6}, reduce the deficiency of vertices 2 and 6, and proceed to the next row.

3. Starting from vertex 3, which is now the leftmost vertex with the highest deficiency, we see that vertex 4, of the two remaining vertices with deficiency 3, is closest. We add {3, 4}, reduce the deficiencies, and proceed.

4. The only vertex with deficiency 3 is now vertex 5. Since no other vertex has deficiency 3, we have to select a candidate from vertices with deficiency 2.

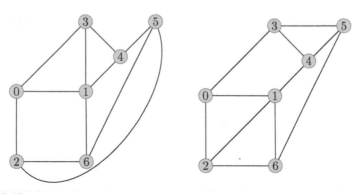

FIGURE 12.13

An initial feasible solution (left). The optimal 3-connected solution (right).

Of all such candidates, vertex 4 is the closest. We add {5, 4}, reduce the deficiency of vertices 4 and 5 by 1, and proceed.

5. Again vertex 0 is the leftmost vertex with the highest deficiency. Since there is already an edge {0, 1}, vertex 1 is forbidden. The closest is then vertex 2, so we add {0, 2} and reduce deficiencies.

6. Starting from vertex 1, we see that vertices 3 and 6 are closest, and since 3 precedes 6, we add {1, 3}.

7. Now only two vertices have deficiency 2: 5 and 6. Since there is no edge between them we add {5, 6}.

8. Starting from vertex 0 again, the closest of possible vertices is now 3, and so we add {0, 3}.

9. The closest possible candidate to vertex 1 is now vertex 4: we add {1, 4}.

10. Starting from vertex 2 and since there is already an edge {2, 6}, we have no choice but to connect 2 to 5.

11. Only vertex 6 remains with positive deficiency. The closest possible vertex to connect to is now vertex 1. All deficiencies are now non-positive, so we stop.

The resulting graph is shown in Figure 12.13. The initial solution is clearly feasible. We therefore accept it as starting solution for our search. Its cost, which follows from the cost matrix, is 270.

For the local search, we pick two edges without common vertices and where the vertices belonging to the different edges are not neighbors, and perform the X-change transformation. Let us say we have picked the edges {1, 3} and {2, 5}. Deleting these edges and creating the edges {1, 2} and {3, 5} produce a new graph. Since the node pairs {1, 3} and {2, 5} still have connectivity 3 after the X-change (verified using Proposition 12.4.6 and the max-flow algorithm), the new solution is feasible. Furthermore, it has a cost of 242, so we accept it as a better solution than the initial. (This is, as a matter of fact, the optimal solution, see Papadimitriou and Steiglitz, 1982.)

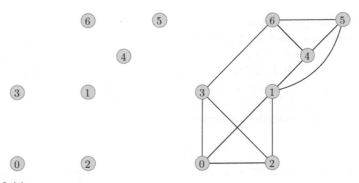

FIGURE 12.14

Another node enumeration (left). An infeasible initial solution (right).

Table 12.2   Initial solution generated by the heuristic algorithm

3	3	3	3	3	3	3
2	3	2	3	3	3	3
2	2	2	3	2	3	3
2	2	2	2	2	3	2
2	2	2	2	1	2	2
1	2	2	1	1	2	2
1	1	1	1	1	2	2
1	1	1	1	1	1	1
0	0	1	1	1	1	1
0	0	0	0	1	1	1
0	0	0	0	0	1	0
0	-1	0	0	0	0	0

Example 12.4.2.   Suppose that we start with the vertices enumerated as on the left side of Figure 12.14. It turns out that this enumeration does not yield a feasible initial configuration. The steps of the algorithm are listed in Table 12.2.

12.5 **The reliability polynomial**

The reliability polynomial is a probabilistic measure of network reliability. It is defined for a connected graph $G = (V, E)$ in which each edge is associated with the probability p of operating (failing with probability $q = 1 - p$) and all edges operate or fail independently of each other. This is a simple model which has been used extensively in network design and analysis.

The reliability polynomial ties together topological aspects (the coefficients) with the operational probability p. Often, the probability p is not known, but the polynomial can nevertheless be used to compare two network topologies and give insight into their reliability properties. It should be kept in mind that the model is an idealization: faults are rarely independent or fault rates uniform across a network.

Consider a graph $G(V, E)$ in which each edge $e \in E$ is operational with the same probability p, independently of other edges. We introduce a structure function

$$\phi : 2^E \mapsto \{0, 1\}, \tag{12.10}$$

which is a mapping from the state-space—the set of all possible states (which is of size $2^{|E|}$)—to the binary set $\{0, 1\}$ representing 'failed' and 'operational' states of the network, respectively.

Let the set of operational edges $S \subseteq E$ be the state of the network. That is, the network is in state S when all edges of S are operational and all edges $E \notin S$ (not in S) are failing. The state S is then operational when $\phi(S) = 1$, and failing when $\phi(S) = 0$.

Consider a state S in which $|S| = i$ edges are operational out of a total $|E| = m$ edges. Then the probability of the network being in a state S is $\mathbf{P}(S) = p^i (1-p)^{m-i}$. The reliability polynomial can formally be defined as

$$R(G; p) = \sum_{S \subseteq E} \mathbf{P}(S)\phi(S), \tag{12.11}$$

since the 2^m states are disjoint events covering all possibilities. $R(G; p)$ is therefore a polynomial of degree at most m in one variable, the edge operational probability p. Thus, if every edge operates independently, the reliability polynomial $R(G; p)$ evaluated at p gives the probability that the graph is in an operational state.

There exist several forms of the polynomial. The two most useful in this context are the forms where the coefficients are expressed in the number of operational states and the number of cutsets, respectively. Let N_i be the number of operational states with i edges in operation. Let $F_i = N_{m-i}$. Then the all-terminal reliability polynomial is

$$R(G; p) = \sum_{i=0}^{m} F_i (1-p)^i p^{m-i},$$

which is called the F-form of the reliability polynomial. Similarly, if C_i is the number of sets consisting of i edges whose removal renders the network failed, then

$$R(G; p) = 1 - \sum_{i=0}^{m} C_i (1-p)^i p^{m-i}.$$

It follows that $R(G; 0) = 0$ and $R(G; 1) = 1$, provided that $\phi(\oslash) = 0$ and $\phi(E) = 1$. Note that

$$F_i + C_i = \binom{m}{i}. \tag{12.12}$$

12.5.1 **The deletion-contraction principle**

Another strategy is to find transformations that have a predictable effect on each coefficient. A basic device for studying transformations is the factoring theorem, also called pivotal decomposition. Consider an edge e of G. Deleting e gives a new graph $G - e$. Contracting e by identifying its end points, removing e, but saving all multiple edges and other loops that arise, gives a multigraph $G \cdot e$. It is easy to see that

$$R(G; p) = p \cdot R(G \cdot e; p) + (1 - p) \cdot R(G - e; p).$$

This implies recursive formulae for the coefficients, for example

$$F_i(G) = F_i(G \cdot e) + F_{i-1}(G - e).$$

12.5.2 **Bounds and approximations**

Suppose that we were able to find lower and upper bounds, $N_i^{(L)} \leq N_i \leq N_i^{(U)}$, on each coefficient. Then

$$\sum_{i=0}^{m} N_i^{(L)} p^i (1 - p)^{m-i} \leq R(G; p) \leq \sum_{i=0}^{m} N_i^{(U)} p^i (1 - p)^{m-i}.$$

The exact calculation of all of the coefficients is $\mathcal{NP}$-hard. However, some of them can easily be calculated.

If fewer than $n - 1$ edges are in operation, the graph is disconnected. Therefore, $F_i = 0$ for $i > m - n + 1$. If the smallest cutset has size c (the edge-connectivity is c), there is no way to remove fewer than c edges and disconnect the graph. Thus $F_i = \binom{m}{i}$ for $i < c$. The coefficient F_{m-n+1} is exactly the number of spanning trees of the graph, which can readily be calculated using Kirchhoff's theorem.

According to a lemma by Sperner (Colbourn, 1993), the following relation holds:

$$(m - i)N_i \leq (i + 1)N_{i+1}.$$

Therefore, given F_i, a lower bound on F_{i-1} and an upper bound on F_{i+1} can be derived. Together with exact coefficients, we can compute lower and upper bounds on every coefficient in the F-form.

12.5.3 **A randomized algorithm**

Another approach for computation of the coefficients of the reliability polynomial approximately has been proposed by Beichl et al. (2009). Given a graph G, we construct a tree where each node is a connected subgraph in G. The root consists of G itself and each child of a node in a given level are the possible connected subgraphs in G with one edge removed as compared with the previous level.

Each level i is composed of exactly $i!$ copies of each of the connected subgraphs having exactly $|E| - i$ edges. Each level of the tree has exactly $i!C_{|E|-i}$ nodes. The number of nodes in the tree can be estimated by the following randomized algorithm proposed by Beichl et al. (2009). The counting of the number of subtrees is done by a procedure proposed by Knuth (1975).

Algorithm 12.5.1 (Estimation of the Coefficients of the Reliability Polynomial). Given a graph $G = (V, E)$.

STEP 0: Set $a_0 = 1$ and let C be an empty vector of coefficients;
STEP 1: for $k = 1$ **to** $|E| - |V| + 1$
  Let D_k be the set of all edges which if removed do not disconnect G;
  Set $a_k = |D_k|$;
  Uniformly select and edge e from the set D_k;
  Let the new graph G be $G - \{e\}$, that is, G with the edge e removed;
STEP 2: for $k = 0$ **to** $|E| - |V| + 1$
  Set $C_{|E|-k} = \prod_{0 \leq i \leq k} a_i/k!$;
STEP 3: for $k = 0$ **to** $|V| - 2$
  Set $C_k = 0$;

Output the set of coefficients C.

Algorithm 12.5.1 can be run N number of times to improve the coefficient estimates. Then, in step 1, we let a_k be an average of the magintudes of $|D_k|$. Step 3 follow from the definition of the reliability polynomial; these coefficients are always zero.

Example 12.5.1. Consider the graph G_1 in Figure 12.17. This graph is studied in Rodionova et al. (2004) and further discussed in Section 12.6. The graph has the reliability polynomial coefficients $(1, 11, 55, 163, 310, 370, 224)$. Using the algorithm with 400 iterations gives the coefficients $(1, 11, 55, 163, 309, 370, 224)$, which is a very good approximation.

Example 12.5.2. Considering the two topologies from Chapter 4, shown in Figures 12.15 and 12.16. In the spanner approximation we have the approximate polynomial

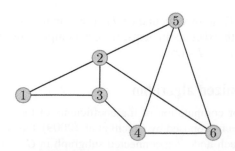

FIGURE 12.15

The 1.2-spanner (Mansour-Peleg) design from Chapter 4.

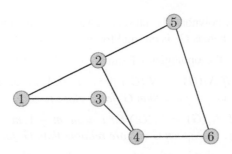

FIGURE 12.16
The Gomory-Hu design from Chapter 4.

coefficients $(1, 9, 35, 70, 62)$ and in the Gomory-Hu design $(1, 7, 15)$. Since the coefficients for the latter are lower than those for the former, the spanner approximation is more reliable. Although this may be concluded by inspection, by this calculation the reliability can be quantified.

12.6 **Optimal topologies and circulants**

As mentioned earlier, the reliability polynomial is useful for comparing network topologies with the same number of vertices and edges. The following two topologies are given in Rodionova et al. (2004).

Example 12.6.1. Consider the two graphs G_1 and G_2 in Figure 12.17. The coefficients of corresponding reliability polynomials $R(G_1; p)$ and $R(G_2; p)$ are

$$(1, 11, 55, 163, 310, 370, 224),$$

$$(1, 11, 55, 163, 309, 368, 225).$$

Using the relation (12.12), we can immediately see that G_1 has fewer edge-cuts of size 4 and 5 than G_2 (20 and 92, compared to 21 and 94), whereas G_2 has a larger number of spanning trees than G_1 (225 compared to 224). By taking the difference

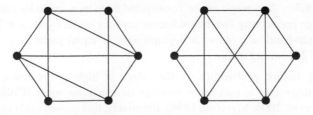

FIGURE 12.17
Two graphs, G_1 (left) and G_2 (right), optimal for different regions of p.

between the reliability polynomials and solving for p, we conclude that G_1 is optimal for $p > 1 - \sqrt{2}/2$, whereas G_2 is optimal for $p < 1 - \sqrt{2}/2$.

The comparison of the topologies in Example 12.6.1 illustrates the following facts:

Theorem 12.6.1.   *If $N_i(G) = N_i(G')$ for $i = 0, 1, 2, \ldots, k$, and $N_{k+1}(G) > N_{k+1}(G')$ then G is more reliable than G' for p close to zero.*

Theorem 12.6.2.   *If $N_i(G) = N_i(G')$ for $i = m, m - 1, m - 2, \ldots, m - k$, and $N_{m-k-1}(G) > N_{m-k-1}(G')$ then G is more reliable than G' for p close to one.*

A network topology that is optimal for all p—when it exists—is called uniformly optimally reliable. Such topologies must have a dominant reliability polynomial (for all p). A common situation, however, as in Example 12.6.1, is that the reliability polynomials 'cross,' so that two different topologies are optimal in different regimes of p.

For the ensuing discussion, we introduce the class of regular graphs, circulants, and θ-graphs. Circulants are regular and highly symmetric graphs with important properties, which have rendered them particular attention in survivable network design. Of course, circulants can rarely be deployed in practice, since geographical, technical, and other restrictions prevent such symmetric networks to be formed.

Definition 12.6.1 (Regular graph).   A regular graph is a graph without loops and multiple edges where every vertex has the same node degree. If the node degree is k, the graph is called k regular (or regular of degree k).

A 0-regular graph consists of isolated vertices, a 1-regular graph consists of isolated edges, and a 2-regular graph consists of isolated cycles. A 3-regular graph is also known as a cubic graph.

Definition 12.6.2 (Circulant).   A circulant is a graph $C_n^{a_1, 12, \ldots, a_h}$ of order n, where $0 < a_1 < a_2 < \ldots < a_h < (n + 1)/2$, which has $i \pm a_1, i \pm a_2, \ldots, i \pm a_h (\mathrm{mod}\, n)$ adjacent to each node i.

Example 12.6.2.   Figure 12.18 shows four different circulants: the 6-cycle graph C_6^1, the $C_6^{1,2}$ circulant, the Möbius ladder $C_6^{1,3}$ (also denoted M_6), and the complete graph $C_6^{1,2,3}$ (also denoted K_6).

Definition 12.6.3 (θ-graph).   A θ-graph (theta graph) is a graph such that there is a pair of distinct vertices joined by three pairwise internally vertex-disjoint paths.

Example 12.6.3.   A θ-graph can be constructed from an n-cycle by adding a chord, that is, an edge connecting two non-adjacent vertices in the cycle, see Figure 12.19 (left). The right-hand side shows a θ-graph with as equal paths as possible. Both graphs have 72 spanning trees.

When the failure probability of the edges is high (p is low), an optimal network topology has a maximum number of spanning trees (Colbourn, 1993, Weichenberg et al. 2004, Myrvold, 1996). Intuitively, this corresponds to maximizing the number of graph connections (see Table 12.3).

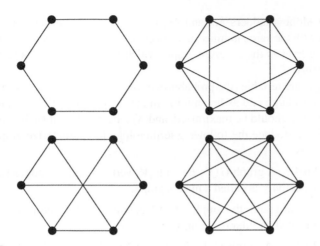

FIGURE 12.18

The circulants C_6^1 (top left), $C_6^{1,2}$ (top right), $C_6^{1,3}$ (also known as the Möbius ladder M_6) (bottom left), and $C_6^{1,2,3}$ (the complete graph K_6) (bottom right).

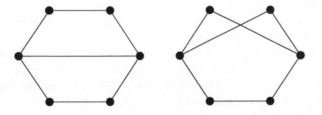

FIGURE 12.19

Two θ-graphs: 6-cycle with a chord (left), θ-graph with as equal path-lengths as possible (right).

Table 12.3 Uniformly optimally reliable topologies for edge numbers up to $n + 1$

$|E|$	Optimal Topology
$n - 1$	any spanning tree
n	an n-cycle
$n + 1$	a θ-graph with path lengths as even as possible

For some cases, the graphs with maximum number of trees are known (see Table 12.3). It has also been proved that these graphs are uniformly optimally reliable Myrvold (1996).

Also, the Möbius ladders M_6 and M_8 have the most spanning trees among all cubic graphs with the same number of vertices, whereas the 10-vertex graph with the most spanning trees is the Petersen graph, which is not a Möbius ladder (Jakobson and Rivin, 1999).

When edges are reliable (the probability of operation p is high), the most reliable networks have as large and as few minimum cuts as possible, that is λ, the size of a minimum cut of G should be maximized, and N_λ, the number of such cuts, should be minimized. We introduce the following terminology to characterize graphs in terms of minimum cuts:

Definition 12.6.4.   A graph $G(V, E)$ with $|V|$ vertices and $|E|$ edges having maximum edge-connectivity is called a max-λ graph.

For an r-regular graph, this implies that λ equals r, and N_λ equals n. As a matter of fact, all connected circulants are max-λ.

Definition 12.6.5.   A graph $G(V, E)$ with $|V|$ vertices and $|E|$ edges is called a super-λ graph if it is max-λ and every edge disconnecting set of order λ isolates a point of degree λ.

A super-λ graph therefore has as large and few minimum edge-cuts as possible. Thus, when edges have low failure probability, the optimal network topology should be super-λ, if such a topology exists. Intuitively, super-λ graphs minimize the probability of low-order cut events, which are the most likely disconnection scenario under the assumption of low failure probability.

The family of circulants has been proposed as topologies for reliable networks due to their properties and relative flexibility—it can be defined for most combinations of number of nodes and degree (Weichenberg et al. 2004). In fact, the only circulants that are not super-λ are the cycles and $C_{2m} < 2, 4, \ldots, m - 1$, with $m \geq 3$ and m an odd integer.

Example 12.6.4.   The 6-cycle in Figure 12.18 is not super-λ, because removal of any two edges disconnects the graph.

We summarize this section as a set of rules of thumb. In general terms, when we seek a network topology that is optimally reliable, we are searching for topologies which are (having):

(1) More spanning trees;
(2a) Larger minimum network cut (maximizes the size of the minimum network cuts);
(2b) Fewer minimum network cuts (minimizes the number of minimum network cuts);

It has also been suggested (although not proved for general graphs) that optimal topologies should have or be (Colbourn, 1993):

(3a) Maximum girth (the length of the shortest cycle);
(3b) Fewest cycles of length equal to the girth;
(4) As regular as possible.

Finally, an important issue from practical point of view is the possible restriction of network topologies to planar graphs, that is, a graphs that can be drawn in the plane without any edges intersecting (apart from at the vertices). The restriction to planar graphs naturally limits the space of possible solutions. Many circulants, for example, are non-planar.

12.7 Summary

A network topology should be designed with respect to reliability as well as capacity. Often, design with respect to the reliability requirements generates topologies which are readibly suitable as a capacity design as well. The survivable network design can be seen as a extension of the discussion in Chapter 4.

There are several reliability metrics available, and these may lead to different solution methods and configurations. Sometimes a rather simple measure, such as 2-connectivity, is sufficient to base a topology design on. However, there are much more refined measures, such as the number of spanning trees and the reliability polynomial. Technological aspects, such as the possibility to reroute and reconfigure a network subject to link failure, should also be taken into consideration when choosing reliability criterion.

The chapter presents algorithms and methods for various performance criteria some of which are exact, but for most problems approximations or randomized algorithms have to be used. Some types of graphs with particularly interesting reliability properties are also discussed.

The literature on network reliability and the survivable network design problem is vast. Good surveys can be found in Papadimitriou and Steiglitz (1982) and Myrvold, 1996. Graph strength is discussed in Galtier et al. (2008). Material on the reliability polynomial is presented in Colbourn (1993) and Beichl et al. (2009). Some graph types with special properites are analyzed in Weichenberg et al. 2004 and Rodionova et al. (2004).

Bibliography

Bard, Y., 1979. Some extensions to multiclass queueing network analysis. In: Proceedings of the Third International Symposium on Modelling and Performance Evaluation of Computer Systems: Performance of Computer Systems. North-Holland Publishing Co., pp. 51–62.

Baskett, F., Chandy, K.M., Muntz, R.R., Palacios, F.G., 1975. Open, closed and mixed networks of queues with different classes of customers. Journal of the ACM 22 (2), 248–260.

Baumann, R., Heimlicher, S., Strasser, M., Weibel, A., 2006. A Survey on Routing Metrics, TIK Report 262, ETH Zürich.

Bean, N.G., Gibbens, R.J., Zachary, S., 1995. Asymptotic analysis of single resource loss systems in heavy traffic, with applications to integrated networks. Advances in Applied Probability, 273–292.

Beasley, D., Bull D.R., Martin, R.R., 1993, An Overview of Genetic Algorithms: Part 1, Fundamentals (accessed from Citeseerx 7.11.2012).

Beichl, I., Cloteaux, B., Sullivan, F., 2009. An approximation algorithm for the coefficients of the reliability polynomial. Congressus Numerantium 197, 143–151.

Bertsekas, D.P., Gallager, R.G., 1987. Data Networks. Prentice-Hall, USA.

Biggs, N.L., 1985. Discrete Mathematics. Oxford Science Publications, UK.

Botvich, D.D., Duffield, N.G., 1995. Large Deviations, the Shape of the Loss Curve, and Economies of Scale in Large Multiplexers (accessed from Citeseerx 10.10.2012).

Buffet, E., Duffield, N.G., Exponential Upper Bounds via Martingales for Multiplexers with Markovian Arrivals. Report DIAS-APG-92-16 (accessed on 12.11.2012).

Buzen, J.P., 1973. Computational Algorithms for Closed Queueing Networks with Exponential Servers. Communications of the ACM 16 (9), 527–531.

Chandy, K.M., Herzog, U., Woo, L., 1975. Approximate analysis of general queueing networks. IBM Journal of Research and Development 19 (1), 36–42.

Choudhury, G.L., Leung, K.K., Whitt, W., 1994. Resource-Sharing Models with State-Dependent Arrivals of Batches. AT&T Bell Laboratories.

Christiano, P., Kelner, J.A., Madry, A., Spielman, D., Teng, S.-H., 2010. Electrical flows, laplacian systems, and faster approximation of maximum flow in undirected graphs. In: Proceedings of the 43rd ACM Symposium on Theory of Computing, STOC 2011, San Jose, CA, USA, 6–8 June 2011, pp. 273–282.

Cohen, J.W., 1969. The Single Server Queue. North Holland, Amsterdam.

Colbourn, Charles J., 1993. Some Open Problems on Reliability Polynomials. DIMACS Tecnical Report, No. 93–28, Canada.

Corne, D.W., Oates, M.J., Smith, G.D. (Eds.), 2000. Telecommunications Optimization: Heuristic and Adaptive Techniques. John Wiley & Sons Ltd., UK.

Cremonesi, P., Schweitzer, P.J., Serazzi, G., 2002. A unifying framework for the approximate solution of closed multiclass queuing networks. IEEE Transactions on Computers 51 (12), 1423–1434.

Crosby, S., Leslie, I., Lewis, J.T., O'Connell, N., Russell, R., Toomey, F., 1995. Bypassing modelling: an investigation of entropy as a traffic descriptor in the Fairisle ATM network. In: Proceedings of the 12th UK Teletraffic Symposium.

Crosby, S., Leslie, I., Huggard, M., Lewis, J.T., Toomey, F., Walsh, C., 1996. Bypassing modelling: further investigations of entropy as a traffic descriptor in the Fairisle ATM network. In: Proceedings WATM'95 First Workshop on ATM Traffic Management, Ecole Nationale Superieure des Telecommunications.

Dahlqvist, G., Björk, A., 1974. Numerical Methods. Prentice-Hall, Engelwood Cliffs.

de Veciana, G., Courcoubetis, C., Walrand, J., 1993. Decoupling Bandwidths for Networks: A Decomposition Approach to Resource Management, Technical Report UCB/ERL M93/50, EECS Department, University of California, Berkeley.

Dijkstra, E.W., 1959. A note on two problems in connexion with graphs. Numerische Mathematik 1, 269–271.

Dinitz, Y., 1970. Algorithm for solution of a problem of maximum flow in a network with power estimation. Doklady Akademii nauk SSSR 11, 1277–1280.

Duffield, N.G., 1993. Exponential Bounds for Queues with Markovian Arrivals. Report DIAS-APG-93-01 (accessed on 12.11.2012).

Duffield, N.G., Huggard, M., Russell, R., Toomey, F., Walsh, C., 1995. Fast bounds for ATM Quality of Service parameters. In: Proceedings of the 12th IEE UK Teletraffic Symposium, Old Windsor, 15–17 March 1995, Paper 9, pp 1–9 (accessed on 12.11.2012).

Edmonds, J., Karp, R.M., 1972. Theoretical improvements in algorithmic efficiency for network flow problems. Journal of the ACM 19 (2), 248–264.

Ford, L.R., Fulkerson, D.R., 1962. Flows in Networks. Princeton University Press, New Jersey.

Fortet, R., Grandjean, C., 1964. Congestion in a loss system when some calls want several devices simultaneously. Proceedings of the Electrical Communications 39, 513–526.

Fulkerson, D.R., 1961. An out-of-kilter method for minimum cost flow problems. Journal SIAM 9 (1), 18–27.

Gabow, H.N., Goemans, M.X., Williamson, D.P., 1998. An efficient approximation algorithm for the survivable network design problem. Mathematical Programming 82, 13–40.

Galtier, J., 2008. New algorithms to compute the strength of a graph. INRIA Rapport de recherche, 6592 ISSN 0249-6399.

Gerla, M., Kleinrock, L., 1977. On the topological design of distributed computer networks. IEEE Transactions on Communications 25 (1), 48–60. http://dx.doi.org/10.1109/TCOM.1977.1093709.

Gibbens, R.J., Kelly, F.P., 1995. Network programming methods for loss networks. IEEE Journal on Selected Areas in Communications 13 (7), 1189–1198 (Invited Paper).

Gibbens, R.J., Hunt, P.J., Kelly, F.P., 1990. Bistability in Communication Networks. Disorder in Physical Systems. Oxford University Press. pp. 113–128.

Gibbens, R.J., Sargood, S.K., Van Eijl, C., Kelly, F.P., Azmoodeh, H., Macfadyen, R.N., Macfadyen, N.W., 2000. Fixed-point methods for the end-to-end performance analysis of IP networks. In: 13th ITC Specialist Seminar on Internet Trac Measurement and Modelling.

Gilks, W.R., Richardson, S., Spiegelhalter, D.J. (Eds.), 1996. Markov Chain Monte Carlo in Practice. Chapman & Hall/CRC, USA.

Goldberg, A.V., Tarjan, R.E., 1986. A new approach to the maximum flow problem, annual ACM symposium on theory of computing. In: Proceedings of the Eighteenth Annual ACM Symposium on Theory of Computing, pp. 136–146.

Goldberg, A.V., Tardos, 'E., Tarjan, R.E., 1990. Network flow algorithms, vol. 9. Springer-Verlag, USA. Reprint from Algorithms and Combinatorics.

Goodrich, M.T., Tamassia, R., 2002. Algorithm Design. John Wiley & Sons, Delhi.

Goyal, N., Rademacher, L., Vempala, S., 2008. Expanders via Random Spanning Trees (accessed from Citeseerx 7.11.2012).

Greenberg, A.G., Srikant, R., 1997. Computational techniques for accurate performance evaluation of multirate, multihop communication networks. IEEE/ACM Transactions on Networking 5 (2), 266–277.

Grimmett, G.R., Stirzaker, D.R., 1992. Probability and Random Processes, second ed. Oxford University Press.

Gross, D., Harris, C.M., 1998. Fundamentals of Queueuing Theory, third ed. John Wiley & Sons Inc., USA.

Hajek, B., 2006. Notes for ECE 467 Communication Network Analyis (accessed on 10.10.2012).

Hodousek, O., 2003. Evaluation of the Erlang-B formula. In: RTT 2003 - Proceedings. FEI, Slovak University of Technology, Bratislava, pp. 80–83.

Ingolfsson, A., Akhmetshina, E., Budge, S., Li, Y., Wu, X., 2002. A survey and experimental comparison of service level approximation methods for non-stationary $M/M/s$ queueing systems. INFORMS Journal on Computing Spring 19 (2), 201–214.

ITU-T, 1993. Recommendation E.600: Terms and Definitions of Traffic Engineering.

Iversen, V.B., 2005. Modelling restricted accessibility for wireless multi-service systems. Springer Lecture Notes on Computer Science, vol. 3883, pp.93–102.

Iversen, V.B., 2007. Reversible fair scheduling: the teletraffic revisited. In: Proceedings from 20th International Teletraffic Congress ITC20, Ottawa, Canada, June 17–21, 2007. Springer Lecture Notes in Computer Science, vol. LNCS 4516, pp. 1135–1148.

Iversen, V.B., 2013. Handbook: Teletraffic Engineering. ITU-D. Techical University of Denmark.

Jakobson, Dmitry, Rivin, Igor, 1999. On Some Extremal Problems in Graph Theory. Available from: <arXiv:math.CO/9907050>.

Kelly, F.P., 1986. Blocking probabilities in large circuit-switched networks. Advances in Applied Probability 18 (2), 473–505.

Kelly, F.P., 1991. Loss networks. Annals of Applied Probability 1 (3), 319–378.

Kelly, F.P., 1996. Notes on effective bandwidths. In: Kelly, F.P., Zachary, S., Ziedins, I.B. (Eds.), Stochastic Networks: Theory and Applications. Oxford University Press, pp. 141–168.

Kleinrock, L., 1970. Analytic and simulation methods in computer network design. In: AFIPS Conference Proceedings, SJCC 1970, AFIPS, pp. 569–579.

Kleinrock, L., 1976. Queueing Systems: Volume II Computer Applications. Wiley Interscience, New York.

Knuth, D.E., 1975. Estimating the efficiency of backtrack programs. Mathematics of Computation 29, 121–136.

Kuphaldt, T.R., 2006. Lessons in Electric Circuits, vol. I – DC, fifth ed. <www.ibiblio.org/obp/electricCircuits> (accessed in November 2012).

Ladd, S.R., 2000. Optimization by Evolution. Coyote Gulsh Productions, USA.

Lam, S.S., Lien, Y.L., 1982. Optimal routing in networks with flow-controlled virtual channels. ACM SIGMETRICS, 38–46.

Leighton, Rao, S., 1999. Multicommodity Max-Flow Min-Cut Theorems and Their Use in Designing Approximation Algorithms (accessed from Citeseerx in November 2012).

Leighton, T., Makedon, F., Plotkin, S., Stein, C., Tardos, É., Tragoudas, S., 1993. Fast Approximation Algorithms for Multicommodity Flow Problems (accessed from Citeseerx in November 2012).

Leong, T., Shor, P.W., Stein, C., 1993. Implementation of a multicommodity flow algorithm. In: Johnson, D.S., McGeoch, C.C. (Eds.), Network Flows and Matching: First DIMACS

Implementation Challenge. DIMACS Series in Discrete Mathematics and Theoretical Computer Science, vol. 12. American Mathematical Society, Providence, RI, pp. 387–407.

Lewis, J., Russell, R., 1996. An Introduction to Large Deviations for Teletraffic Engineers. <ftp://www.stp.dias.ie/DAPG/LDtut96.ps>.

Liu, M., Baras, J.S., 1999. Fixed Point Approximation for Multirate Multihop Loss Networks with Adaptive Routing. Technical Research Report CSHCN T.R. 99–21.

Luenberger, D.G., 1989. Linear and Nonlinear Programming, second ed. Addison-Wesley Publishing Company, Inc., USA.

Mansour, Y., Peleg, D., 1998. An Approximation Algorithm for Minimum-Cost Network Design. Institute of Science, Rehovot, Israel.

Mathis, M., Semke, J., Mahdavi, J., Ott, T., 1997. The macroscopic behavior of the TCP congestion avoidance algorithm. Computer Communication Review 27 (3).

Motwani, R., Raghavan, P., 1995. Randomized Algorithms. Cambridge University Press, UK.

Myrvold, W., 1996. Reliable Network Synthesis: Some Recent Developments. University of Victoria, Canada.

Nederlof, Jesper, 2008. Inclusion Exclusion for Hard Problems. Utrecht University, M.Sc. thesis.

O'Connell, N., 1999. Large Deviations with Applications to Telecommunications. BRIMS, Hewlett-Packard Labs, Bristol, UK.

Olga, K.R., Alexey, S.R., Hyunseung, C., 2004. Network probabilistic connectivity: optimal structures. ICCSA, vol. 4, pp. 431–440.

Papadimitriou, C.H., Steiglitz, K., 1982. Combinatorial Optimization: Algorithms and Complexity. Prentice-Hall, USA.

Pirinen, A., 2001. Congestion and Dimension Characterizations by Economical Factors. øAbo Akademi University, Finland.

Polya, G., 1988. How to Solve It. Princeton University Press, USA.

Press, W.H., Teukolsky, S.A., Vetterling, W.T., Flannery, B.P., 2007Numerical Recipes. The Art of Scientific Computing, third ed. Cambridge University Press, New York.

Reiser, M., Lavenberg, S.S., 1980. Mean-value analysis of closed multichain queueing networks. Journal of the ACM 27 (2).

Sedgewick, R., Flajolet, P., 1996. An Introduction to the Analysis of Algorithms. Addison-Wesley Publishing, USA.

Shaffer, C.A., 1997. A Practical Introduction to Data Structures and Algorithm Analysis. Prentice-Hall, USA.

Skiena, S.S., 1998. The Algorithm Design Manual. Springer-Verlag, New York.

Sniedovich, M., 2006. Dijkstra's algorithm revisited: the dynamic programming connexion. Journal of Control and Cybernetics 35 (3), 599–620.

Steiglitz, K., Weiner, P., Kleitman, D.J., 1969. The design of minimum cost survivable networks. IEEE Transactions on Circuit Theory 16, 455–460.

Stewart, W.J., 1994. Introduction to the Numerical Solution of Markov Chains. Princeton University Press, USA.

Thompson, J.R., 2000. Simulation – A Modeler's Approach. John Wiley & Sons, USA.

Walukiewicz, S., 1991. Integer Programming. PWN-Polish Scientific Publishers, Warszawa.

Weichenberg, G.E., Chan, V.W.S., M, Médard, 2004. High-reliability architectures for networks under stress. In: IEEE INFOCOM.

Whitley, D., 1993. A Genetic Algorithm Tutorial. Technical Report CS-93-103. Colorado State University, USA.

Whitt, W., 1984. Open and closed models for networks of queues. AT&T Bell Laboratories Technical Journal 63 (9), 1911–1979.

Wischik, D., 1999. The output of a switch, or effective bandwidths for networks. Queueing Systems 32, 383–396.

Zhao, Y.Q., Liu, D., 2004. The censored markov chain and the best augmentation. Journal of Applied Probability 33, 623–629.

Index

Printed and bound by CPI Group (UK) Ltd, Croydon, CR0 4YY

03/10/2024

01040326-0016